U0172424

Plant Illustrations of Nanji Islands National Marine Nature Reserve

南麂列岛国家级海洋自然保护区植物图鉴

主编 陈林 陈先夏

南麂列岛国家级海洋自然保护区生物多样性系列图鉴

Nanji Islands National Marine Nature Reserve Biodiversity Illustration Monograph Series

南麂列岛国家级海洋自然保护区管理局
南京林业大学生物与环境学院

華中科技大學出版社
http://www.hustp.com
中国·武汉

图书在版编目（C I P）数据

南麂列岛国家级海洋自然保护区植物图鉴 / 陈林, 陈先夏主编. — 武汉 : 华中科技大学出版社, 2022.10

ISBN 978-7-5680-8273-0

Ⅰ. ①南… Ⅱ. ①陈… ②陈… Ⅲ. ①海洋—自然保护区—植物—平阳县—图集 Ⅳ. ①Q948.525.54-64

中国版本图书馆CIP数据核字(2022)第078203号

南麂列岛国家级海洋自然保护区植物图鉴　　陈 林 陈先夏 主编

Nanji Liedao Guojiaji Haiyang Ziran Baohuqu Zhiwu Tujian

出版发行：华中科技大学出版社（中国·武汉）　　电话：（027）81321913

地　　址：武汉市东湖新技术开发区华工科技园　　邮编：430223

出 版 人：阮海洪

策划编辑：段园园 陈晓彤　　责任监印：朱 玢

责任编辑：陈 骏　　装帧设计：段自强

印　刷：深圳市彩鑫金域印刷有限公司

开　本：787 mm × 1092mm　1/16

印　张：31.5

字　数：300千字

版　次：2022年10月第1版 第1次印刷

定　价：468.00元

投稿热线：13710226636（微信同号）

本书若有印装质量问题，请向出版社营销中心调换

全国免费服务热线：400-6679-118 竭诚为您服务

《南麂列岛国家级海洋自然保护区生物多样性系列图鉴》

Nanji Islands National Marine Nature Reserve Biodiversity Illustration Monograph Series

《南麂列岛国家级海洋自然保护区植物图鉴》

编委会

专家顾问：方炎明　王贤荣　蔡厚才
主　　编：陈　林　陈先夏
副 主 编：伊贤贵　李　蒙　张　敏　段一凡　朱　弘　陈　舜　陈万东　谢尚微
编　　委：（以姓氏笔画为序）
马雪红　叶　琦　田昌芬　朱淑霞　伍尔魏　刘　欣　严继萍　李龙娜
李雪霞　吴水涵　吴其超　吴　桐　吴　曦　邱　靖　汪知蓓　宋炎峰
陆瑜心　陈　玉　周华近　倪孝品　曹小朦　彭智奇　曾贵侯　潘靖宜
摄　　影：（以姓氏笔画为序）
陈　林　伊贤贵　朱　弘　彭智奇　朱淑霞　吴其超　张开文　李　蒙
蒋　蕾　邱　靖　何春梅　童毅华　王　璐　倪孝品　伍尔魏　曾贵侯
参编单位：南京林业大学生物与环境学院
南麂列岛国家海洋自然保护区管理局
亚热带森林生物多样性保护国家林业和草原局重点实验室
南京林业大学树木标本馆（NF）

Plant Illustrations of Nanji Islands National Marine Nature Reserve

Editorial Committee

Advisors:

Fang Yanming　Wang Xianrong　Cai Houcai

Chief Editors:

Chen Lin　Chen Xianxia

Associate Editors:

Yi Xiangui　Li Meng　Zhang Min　Duan Yifan　Zhu Hong　Chen Shun　Chen Wandong　Xie Shangwei

Editors: (alphabetically sorted)

Cao Xiaomeng　Chen Yu　Li Longna　Li Xuexia　Liu Xin　Lu Yuxin　Ma Xuehong　Ni Xiaopin　Pan Jingyi

Peng Zhiqi　Qiu Jing　Song Yanfeng　Tian Changfen　Wang Zhibei　Wu Erwei　Wu Qichao　Wu Shuihan　Wu Tong　Wu Xi　Yan Jiping　Ye Qi　Zeng Guihou　Zhou Huajin　Zhu Shuxia

Photograph: (alphabetically sorted)

Chen Lin　He Chunmei　Jiang Lei　Li Meng　Ni Xiaopin　Peng Zhiqi　Qiu Jing　Tong Yihua　Wang Lu　Wu Erwei　Wu Qichao　Yi Xiangui　Zeng Guihou　Zhang Kaiwen　Zhu Hong　Zhu Shuxia

Support departments:

College of Biology and the Environment, Nanjing Forestry University

Nanji Islands National Marine Nature Reserve Administration

Key Laboratory of Subtropical Forest Biodiversity Conservation, National Forestry and Grassland Administration

Herbarium of Nanjing Forestry University (NF)

总 序

南麂列岛国家级海洋自然保护区是1990年国务院批准建立的我国首批五个国家级海洋类型自然保护区之一，1998年成为我国最早纳入联合国教科文组织（UNESCO）世界生物圈保护区网络的海洋类型自然保护区，2002年被列为联合国开发计划署、全球环球基金、中国政府（UNDP/GEF/SOA）中国南部沿海生物多样性管理项目（SCCBD）四个示范区之一，2005年被《中国国家地理》杂志等全国35家知名媒体评为“中国最美十大海岛”，2014年被列为东亚海环境管理伙伴关系计划（PEMSEA）中国第四期项目十三个示范区之一，2015年被国家海洋局评为全国十大美丽海岛之一。

南麂列岛位于浙江省东南海域，隶属于温州市平阳县，地理坐标为27° 24′ 30″ N ~ 27° 30′ 00″ N，120° 56′ 30″ E ~ 121° 08′ 30″ E，与大陆最近距离为28海里，北离温州市区50海里，南至台湾基隆港140海里，离钓鱼岛约162海里。其地理位置和生态环境独特，处于中亚热带海域，台湾暖流和江浙沿岸流的交汇处，流系复杂、锋面发达，独特多样的生态环境为海洋生物的繁衍和生长提供了十分理想的自然条件，孕育出了高度的生物多样性，这在国内独一无二，在世界上也非常罕见，具有重要的国际保护意义、科学研究和生态价值。国内外学者对南麂列岛的生物多样性一直十分关注，从建区前至今，已进行过百余次科学调查考察活动，同时保护区也在持续开展科研监测，现已鉴定出各类生物2700余种，其中海洋生物2155种，约占我国已发现海洋生物总数的十分之一，是东海海洋生物多样性的典型代表。特别是贝藻类生物资源十分丰富，是我国主要的海洋贝藻类天然博物馆和“南种北移、北种南移”的引种驯化基地，从而使南麂列岛获得了“贝藻王国”的美誉，引起了国内外海洋生物学界的广泛关注和高度重视。

生物多样性使地球充满生机，也是人类生存和发展的基础。保护生物多样性有助于维护地球家园，促进人类可持续发展，已成为全人类的共识。2005年，时任浙江省委书记习近平视察南麂时指出：“南麂是一个宝岛，南麂自然保护区是浙江省唯一的国家级海洋类自然保护区，这里拥有得天独厚的自然景观和丰富多样的海洋生物资源，具有重要的科学和生态价值，一定要高度重视这里的生态环境，把生物多样性保护好。”海洋生物多样性保护一直是南麂列岛保护区的核心工作，在各级政府的高度重视下，不断地完善保护措施和管理手段，有效地保护了南麂列岛的生态环境和生物多样性，为维持海洋生态系统平衡和促进海洋生物繁衍生长提供了重要保障。据调查显示，南麂列岛核心区贝藻类生物量已数倍于区外，海洋生态环境稳定，保护成效显著，形成生物多样性保护推动绿色发展和人与自然和谐共生的良好局面。

经过长期积累、深入思考和科学总结，保护区管理局组织人员精心编制了《南麂列岛国家级海洋自然保护区生物多样性系列图鉴》丛书，以南麂列岛海洋生物多样性为重心，涵盖海、陆、空三个领域，分列贝类、大型底栖海藻、鱼类、虾蟹类、微小型藻类、植物与鸟类等十余册图鉴。丛书第一次较为系统地总结了南麂列岛保护区生物及其栖息生境的情况，图文并茂，理论体系科学，覆盖面广，兼顾了科研和科普作用，是对南麂列岛保护区推进生物多样性保护、维持海洋生态系统平衡和促进海洋生物繁衍生长等工作成果的全面展示，是公众认识南麂生物多样性与助力生态文明建设的重要途径。

把握时代发展趋势，顺应时代发展潮流。南麂列岛保护区将坚持走好“碧海银滩就是金山银山”海洋生态发展路子，锚定“蓝色增长”发展要素，以“向海图强、人海和谐、依海富民、开放共赢”为工作要求，全力推进“生态、科技、美丽、富民”四岛联建，着力打造世界人与生物圈的“耀眼明珠”。南麂海洋的保护和发展将努力为推进我国海洋事业的发展、实现生态文明理念、构建地球生命共同体和建设清洁美丽的世界探索“南麂道路”，打造“南麂样板”。

南麂列岛国家级海洋自然保护区管理局局长

陈光夏

2022年5月28日于平阳

Series Foreword

Nanji Islands National Marine Nature Reserve is one of the first five national marine nature reserves approved by the State Council in 1990. In 1998, it became the first marine nature reserve of China included in the UNESCO World Network of Biosphere Reserves (WNBR). In 2002, it was listed as one of the four demonstration zones of the South China Coastal Biodiversity Management Project (SCCBD) by the United Nations Development Programme (UNDP), the Global Environment Facility (GEF), and the State Oceanic Administration (SOA). In 2005, it was praised as one of the "Top Ten Beautiful Islands in China" by 35 well-known media such as Chinese National Geography. It was also named as one of the 13 demonstration zones of the Phase Four Projects of the Partnerships in Environmental Management for the Seas of East Asia (PEMSEA) in China in 2014, and also rated as one of the "Top Ten Beautiful Islands in China" again by the State Oceanic Administration in 2015.

Nanji Islands is located in the southeast sea area of Zhejiang Province, which is affiliated to Pingyang County, Wenzhou City, with the geographic coordinates of 27°24′30″N ~ 27°30′00″N, 120°56′30″E ~ 121°08′30″E. The closest distance to the mainland is 28 nautical miles, 50 nautical miles from Wenzhou City in the north, 140 nautical miles from Keelung Port, Taiwan Province in the south, and about 162 nautical miles from the Diaoyu Islands. There is a unique geographical location and ecological environment for Nanji Islands. It is located at the confluence of Taiwan Warm Current and Jiangsu-Zhejiang Coastal Current in the mid-subtropical sea areas, which provides an ideal natural habitat for marine organisms due to the special, complex and diverse ecological environment. The high biodiversity of Nanji Islands is unique in China and rare in the world, which gives a great significance for global conservation, scientific research, and ecological value. Domestic and foreign scholars keep on paying close attention to the biodiversity of the Nanji Islands during the past decades. There were more than 100 scientific investigations and continuous scientific monitoring since the establishment of the reserve. It is a typical representative of marine biodiversity in the East China Sea, with more than 2700 species of organisms identified up to now, which includes 2155 species of marine organisms that accounted for about 1/10 of the total marine organisms discovered in China. Additionally, Nanji Islands is recognized as the main natural museum and the gene bank of marine shellfish and algae, as well as the breeding base for the species migration, introduction and cultivation between northern China and south China. Thus, such abundant marine resources make it enjoy a reputation of the "Kingdom of Shellfish and Algae", which has aroused widespread concern and great attention of marine biology at home and abroad.

Biodiversity makes the earth full of vitality, and it is equally the basis of human survival and development. It has become the consensus of all mankind that biodiversity conservation is conducive to safeguarding the earth and promoting sustainable development of human beings. In 2005, when Xi Jinping, the Provincial Party Secretary of Zhejiang at that time inspected the Nanji Islands, he pointed out that: "Nanji is a treasure island. Nanji Islands National Marine Nature Reserve is the only national marine nature protected areas in Zhejiang Province, which has unique natural landscape and rich marine bio-resources, with important scientific and ecological values. We must attach great importance to the ecological environment here and protect biodiversity well." Marine biodiversity conservation has always been the vital task of this region. With the strong support from all relevant parties at home and abroad, especially with the high attention of governments at all levels, the ecological environment and biodiversity of Nanji Islands have been effectively protected through years of continuous improvement of conservation and management measures, providing an important guarantee for maintaining the balance of the marine ecosystem and enhancing the reproduction and growth of marine organisms. Survey indicated the biomass of shellfish and algae in the core area of Nanji Islands is several times higher than that outside the area. There are stable marine ecological environments, remarkable conservation outcomes, which forming a good situation for biodiversity conservation promoting the green development and the harmonious coexistence between human and nature.

After a long-term accumulation, in-depth thinking and scientific summary, the reserve administration has meticulously compiled the “Nanji Islands National Marine Nature Reserve Biodiversity Illustration Monograph Series”. The series focuses on the marine biodiversity, covering organisms of marine, terrestrial, and airborne contains more than ten illustrations of shellfish, macrobenthic algae, fish, shrimp and crabs, microalgae, plants and birds. It is among the first time to systematically summarize the status of organisms and their habitats in Nanji Islands. The series is taking into account the role of scientific research and science popularization, with well illustration, scientific theoretical system, and extensive coverage. It is a comprehensive display of the achievements of Nanji Islands National Marine Nature Reserve in promoting biodiversity conservation, maintaining the balance of marine ecosystem and enhancing the reproduction and growth of marine organisms, which is an important way for the public to understand Nanji biodiversity and help the construction of ecological civilization.

In order to conform to the development trend of the times, Nanji Islands National Marine Nature Reserve will adhere to the marine ecological development path of "Blue waters and silver beaches are invaluable assets", anchor the development elements of "blue growth". With the requirements of "strengthening with the sea, harmonious between people and the sea, enriching people by the sea, opening for mutual benefit", we will fully promote the joint construction of the four islands of "ecology, science and technology, beauty and prosperity", and strive to create the "Human and Biosphere Dazzling Pearl" of the world. Achievements on the marine conservation and development of Nanji Islands will strive to explore the "Nanji Road" and create a "Nanji Mode" for promoting the China's marine development, improving the concept of ecological civilization, building a shared future for all life on the earth and a clean and beautiful world.

陈宪夏

Chen Xianxia

Director of

Nanji Islands National Marine Nature Reserve Administration

Pingyang, May 28, 2022

序

岛屿是生物地理界的“特区”，也是生物学研究的“天然实验室”。作为科学家的向往地，岛屿与一些著名的生物学家和生物学理论关联。加拉帕戈斯群岛号称《物种起源》的孕育地，达尔文（Charles R. Darwin，1809–1882）于1835年考察加拉帕戈斯群岛之后提出了著名的生物进化论。1854–1862年，进化论者华莱士（Alfred R Wallace，1823–1913）潜心研究马来群岛的动物区系，发现了区分亚洲与大洋洲动物的分界线，即著名的“华莱士线”。麦克阿瑟（Robert H. MacArthur，1930–1972）和威尔逊（Edward O. Wilson，1929–2021）专注岛屿物种丰度 - 面积关系，提出了影响深远的岛屿生物地理学理论。岛屿生物区系值得持续深入探究。

南麂岛，不仅是科学家的向往地，也是远行者的心仪地。南麂岛鸟瞰时形似昂首向东方飞奔的鹿类动物——麂子，又位于浙江省南部海域，故得此名。南麂岛的蓝天碧海、阳光沙滩、奇礁异石、悬崖峭壁、静谧港湾，使之成为《中国国家地理》杂志评出的中国最美十大海岛之一。了解岛屿生物也是社会大众的期待。

以南麂岛为主岛的南麂列岛，地处亚热带海域，气候温和湿润，四季分明，属中亚热带海洋性季风气候区。作为国家级海洋自然保护区，以及联合国教科文组织世界生物圈保护区，南麂列岛的岛屿和海洋生态系统十分独特，海洋贝类、藻类、鱼类等生物资源丰富，陆源种子植物和脊椎动物众多。海洋和岛屿生物区系共同构成了南麂列岛生物多样性的基本骨架，厘清南麂列岛动植物区系是采取生物多样性保护行动的前提。

绿水青山就是金山银山。南麂列岛良好的生态环境既是当地的自然禀赋，也是区域性经济发展的基础。为建设生态文明，岛屿建设者和守望者不断探求人与自然和谐共生、共建地球生命共同体的真谛，积极营造尊重自然、顺应自然和保护自然的良好氛围，在研究与保护岛屿生物多样性方面迈出了坚实的步伐、取得了显著的成效。

《南麂列岛国家级海洋自然保护区植物图鉴》正是在这样的大背景下问世的。该书得到南麂列岛国家海洋自然保护区管理局的大力支持，由南京林业大学树木学研究团队根据科学考察数据汇编而成，收录了25种蕨类植物、6种裸子植物和426种被子植物。《南麂列岛国家级海洋自然保护区植物图鉴》记录的既是岛屿植物区系的全貌，也是岛屿自然历史的一个“剖面”。该书对自然保护专业人员、高等院校师生、社会公众都具有重要参考价值。

中国林学会树木学分会主任委员

南京林业大学教授

方炎明

2022年3月24日于南京

Foreword

Islands are "Special Zone" of a biogeographic realm and "Natural Laboratory" for biological research. As a dream destination for scientists, the island is associated with some well-known biologists and biological theories. Charles R. Darwin (1809-1882) put forward the famous theory of biological evolution after visiting the Galapagos Islands in 1835, which is known as the birthplace of "The Origin of Species". From 1854 to 1862, another evolutionist, Alfred R. Wallace (1823-1913), devoted himself to the study of the fauna of the Malay Archipelago, and discovered the boundary between Asia and Oceania, the famous "Wallace Line". Robert H. MacArthur (1930-1972) and Edward O. Wilson (1929-2021) focused on the relationship between island species abundance and area, and put forward a far-reaching theory of island biogeography. Island biota deserves continued in-depth exploration.

The Nanji Islands is not only the yearning place for scientists, but also the favorite place for travelers. It is named by the vivid outline as a running deer toward east with a bird's-eye view, and the location of southern oceans of Zhejiang Province. Nanji Islands is one of the "Top Ten Most Beautiful Islands in China", ranked by Chinese National Geography, due to the uniqueness in blue sky, sandy beaches, rocks, cliffs and calm harbors. Thus, learning about the island creatures is also the expectation of the public.

Nanji Islands, with the Nanji Island as the main island, is located in the subtropical sea, with a mild and humid climate and distinct seasons, belonging to the mid-subtropical marine monsoon climate. As a National Marine Nature Reserve and UNESCO World Biosphere Reserve, the ecosystem of island and marine is unique, with abundant marine shellfish, algae, fish and other biological resources, as well as numerous terrestrial seed plants and vertebrates. The marine and island biota here together constitute the basic skeleton of the biodiversity in Nanji Islands, and clarifying the flora and fauna is a prerequisite for the conservation actions.

Lucid waters and lush mountains are invaluable assets. The good ecological environment is not only the natural endowment of Nanji Islands, but also the foundation of regional economic development. For the construction of ecological civilization, island builders and watchers constantly seek the true meaning of harmonious coexistence between human and nature, building a shared future for all life on earth, and actively create a good atmosphere of "Respecting nature, conforming to nature, and protecting nature". There are solid progress and notable success in the research and protection of island biodiversity.

This book was published in this context. Supported by the Nanji Islands National Marine Nature Reserve Administration, the book was compiled by the dendrology research team of Nanjing Forestry University based on the scientific investigation data, including 25 species of ferns, 6 species of gymnosperms, and 426 species of angiosperms. This book is not only a comprehensive record of Nanji Islands flora, but also a "profile" of the island's natural history, with an important reference value for the nature conservation professionals, teachers and students of colleges and universities, and the social publics.

方炎明

Fang Yanming

Chairman of Dendrology Society in the Chinese Society of Forestry

Professor of Nanjing Forestry University

Nanjing, March 24, 2022

前　言

南麂列岛国家级海洋自然保护区是1990年9月经国务院批准设立的我国首批5个国家级海洋类型自然保护区之一；1998年12月加入联合国教科文组织世界生物圈保护区网络；2005年被《中国国家地理》杂志等全国35家媒体评为“中国最美十大海岛”之一。南麂列岛位于浙江省东南部，西距浙江省温州市鳌江镇28海里，北距温州市50海里，南距台湾基隆港140海里，由85个岛屿组成，其中面积大于500平方米的岛屿52个。保护区总面积201.06平方千米，包括海域面积189.93平方千米，陆域面积11.13平方千米，其中主岛南麂岛面积7.64平方千米。保护区以海洋贝藻类、海洋性鸟类、野生水仙花及其生态环境为主要保护对象，享有“贝藻王国”的美誉。同时，南麂列岛也是国家AAAA级旅游景区，海岛风光旖旎，景色宜人，山秀、石奇、滩美、草绿、海蓝、空远，有海滨浴场、猴子拜观音、美龄居等自然与人文景观多处，被誉为“碧海仙山”。

南麂列岛处于台湾暖流和江浙沿岸流的交汇处，气候温和湿润，季节影响显著、四季分明，属于典型的中亚热带海洋性季风气候区，生物种类多样，生物区系复杂。南麂列岛已有500余年的开发历史，原生植被受到了较大的破坏，现有植被较为单调，主要以人工林以及自然演替的次生植被为主，优势类群群落结构简单，灌木层、草本层物种较为单一，具备次生演替的初期特征。人工林及人工次生林单优势特征明显，主要以台湾相思 *Acacia confusa*、木麻黄 *Casuarina equisetifolia* 和黑松 *Pinus thunbergii* 等为建群种，郁闭度高，多样性低；自然植被主要为灌丛和草甸，以野梧桐 *Mallotus japonicus*、矮小天仙果 *Ficus erecta*、滨柃 *Eurya emarginata* 等灌木和卤地菊 *Melanthera prostrata*、山菅 *Dianella ensifolia*、五节芒 *Miscanthus floridulus*、结缕草 *Zoysia japonica* 等草本为主；零星分布的亚热带常绿阔叶次生林林相较为残破，林窗较多，自然林分竞争激烈，呈多优势发展趋势，其中南麂主岛和大檑山屿有较完整的亚热带常绿阔叶次生林、人工次生林、灌草丛和草甸，大檑山屿、小檑山屿和竹屿等岛上的野生水仙 *Narcissus tazetta* var. *chinensis* 更是南麂列岛陆生特色植物的典型代表，在岛屿生态系统中占据着极其特殊的地位。

南麂列岛的陆生植物区系具有明显的大陆岛屿植物区系特点，属内种系贫乏，以矮小的海岛灌丛和草本植物为主，具有明显的热带起源和南北过渡特征，与华东中亚热带、华南南亚热带及日本滨海植物区系有着密切联系。笔管榕 *Ficus subpisocarpa*、鹅掌柴 *Schefflera heptaphylla*、山蒟 *Piper hancei*、了哥王 *Wikstroemia indica*、肉叶耳草 *Hedyotis strigulosa*、蔓九节 *Psychotria serpens* 等亚热带成分和滨柃 *Eurya emarginata*、柃木 *E. japonica*、滨海前胡 *Peucedanum japonicum*、滨海珍珠菜 *Lysimachia mauritiana*、滨海白绒草 *Leucas chinensis*、肾叶打碗花 *Calystegia soldanella*、单叶蔓荆 *Vitex rotundifolia*、普陀狗娃花 *Aster arenarius*、芙蓉菊 *Crossostephium chinense* 和矮生薹草 *Carex pumila* 等典型的滨海岛屿区系特有成分在本岛普遍分布。

作为典型的海洋岛屿生态系统保护区，科研机构在1989年、1992年、2003年和2012年分别对南麂列岛进行了四次生物资源与生态环境调查研究，主要对象是潮间带贝藻类、底栖生物资源和浅海生态环境；而对岛屿陆生植物多样性与植被的调查比较零星分散，缺乏全面系统的植物本底调查。为全面了解南麂列岛陆生植物多样性与植被状况，更加科学有效地保护南麂列岛的生物多样性与岛屿自然生态系统，进一步有针对性地开展物种保护、资源可持续利用、生态景观规划及长期动态监测。2017年至2019年间，在南麂列岛国家海洋自然保护区管理局“南麂列岛森林植物物种资源调查”项目的专款资助下，南京林业大学研究团队完成了南麂列岛的森林植物区系与植被综合调查研究，获得了大量的一手基础资料。我们在此基础上，结合《浙江植物志》《温州植物志》等历史资料，以图版与文字的形式收录了南麂列岛国家海洋自然保护区主要的维管植物103科320属457种（含少量亚种、变种、变型和品种），其中蕨类植物11科18属25种、裸子植物4科5属6种、被子植物88科297属426种。科的排列上，蕨类植物按 PPG I（2016）系统，裸子植物按 Christenhusz（2011）系统，被子植物按 APG IV（2016）系统排列，属种按拉丁名字母的顺序排列。

本书收录了南麂列岛大部分野生、外来逸生、入侵等植物以及少量栽培植物（含品种），具有一定的广度及深度，可以作为农、林、环保等部门的参考资料，同时也可成为广大植物爱好者、园艺工作者以及植物相关科研与教学人员观察自学教本。本书是南麂列岛国家海洋自然保护区管理局与南京林业大学合作的科研成果之一，编写过程中获得了相关单位、部门、领导、同事和朋友的大力支持和帮助，在此表示衷心感谢！由于时间和水平所限，疏漏之处在所难免，敬请各位读者不吝指正。

编者

2022 年 9 月

Preface

Nanji Islands National Marine Nature Reserve is one of the first five national marine nature reserves approved by the State Council in September 1990. It was joined the UNESCO World Network of Biosphere Reserves (WNBR) in December 1998, and rated as one of the "Top Ten Most Beautiful Islands in China" by 35 media such as Chinese National Geography in 2005. Nanji Islands is located in the southeast of Zhejiang Province, 28 nautical miles from Aojiang Town, the closest point to the mainland, 50 nautical miles from Wenzhou City in the north, and 140 nautical miles from Keelung Port, Taiwan Province in the south. It is composed of 85 islands, including 52 islands with an area of more than 500 m^2. The total area of the reserve is 201.06 km^2, including sea area of 189.93 km^2 and land area of 11.13 km^2, of which Nanji Island, the main island, covers an area of 7.64 km^2. The reserve takes marine shellfish algae, marine birds, wild daffodils and their ecological environment as the main protection objects, and enjoys the reputation of "Shellfish Kingdom". Simultaneously, as a National AAAA level Tourist Attraction, Nanji Islands is reputed as "Blue Sea & Celestial Mountain" for the beautiful and pleasant island sceneries and natural and cultural landscapes, such as "Monkey Kneeling Avalokitesvara", "Mei Ling House", etc.

Nanji Islands is located at the confluence of Taiwan Warm Current and Jiangsu-Zhejiang Coastal Current, with mild and humid climate, significant seasonal influence and distinct seasons. It belongs to a typical mid-subtropical maritime monsoon climate zone, with high biodiversity and complex biota. Nanji Islands has been developed over 500 years, and the primary vegetation has been greatly destroyed. The present vegetation is relatively simple, mainly with artificial forest and natural secondary vegetation, representing initial characteristics of secondary succession by simple community structure, fewer shrubs and herbs. There is an obvious characteristic of single dominance in artificial and artificial secondary forests with high canopy density and low diversity, which dominated by *Acacia confusa*, *Casuarina equisetifolia* and *Pinus thunbergii* as constructive species. The natural vegetation is mainly of thickets and meadows, with shrubs of *Mallotus japonicus*, *Ficus erecta*, *Eurya emarginata*, and herbs of *Melanthera prostrata*, *Dianella ensifolia*, *Miscanthus floridulus* and *Zoysia japonica*. Meanwhile, the sporadic subtropical evergreen broad-leaved secondary forests are relatively dilapidated, with more forest gaps and fierce competition among natural stands, presenting a multi-dominance development trend. Among them, there are relatively complete subtropical evergreen broad-leaved secondary forests, artificial secondary forests, scrub and meadows on Nanji Island and Daleishan Island. The typical plant communities of *Narcissus tazetta* var. *chinensis* on Daleishan Island, Xiaoleishan Island and Bamboo Island have an extremely special position in island ecosystem.

The terrestrial flora of Nanji Islands has obvious characteristics of continental island flora, dominated by dwarf shrubs and herbs. It has obvious characteristics of tropical origin and north-south transition, which closely related to the flora of East China, South China and coastal Japan. The subtropical components such as *Ficus subpisocarpa*, *Schefflera heptaphylla*, *Piper hancei*, *Wikstroemia indica*, *Hedyotis strigulosa*, *Psychotria serpens* and typical endemic components of coastal island flora such as *Eurya emarginata*, E. *japonica*, *Peucedanum japonicum*, *Lysimachia mauritiana*, *Leucas chinensis*, *Calystegia soldanella*, *Vitex rotundifolia*, *Aster arenarius*, *Crossostephium chinense* and Carex pumila are widely distributed on this island.

As a typical marine island ecosystem natural reserve, Nanji Islands conducted four investigations on biological resources and ecological environments in 1989, 1992, 2003 and 2012, respectively. The main objects were intertidal shellfish algae, benthic biological resources and shallow sea ecological environment. However, the surveys on diversity and vegetation of terrestrial plants were rather scattered and lacked a comprehensive and systematic research. In order to comprehensively understand the diversity of terrestrial plants and vegetation status, scientifically and effectively protect the biodiversity and natural ecosystem, and further provide basic information for species conservation, sustainable utilization of natural resources, ecological landscape planning and long-term dynamic monitoring, the research team of Nanjing Forestry University obtained substantial first-hand basic data of forest flora and vegetation based on the comprehensive research in Nanji Islands during 2017 and 2019, with the special fund support of “Forest plant resources survey of Nanji Islands” by Nanji Islands National Marine Nature Reserve Administration. In this book, we texted and photographed the main vascular plants of Nanji Islands based on the works in the last few years, combining with the historical materials such as Flora of Zhejiang, Flora of Wenzhou. In total, there were 457 species (containing a few of subspecies, varieties, forms and cultivars) of vascular plants belonging to 320 genera and 103 families, including pteridophytes 25 species of 18 genera and 11 families, gymnosperms 6 species of 5 genera and 4 families and angiosperms 426 species of 297 genera and 88 families. The family orders sorted according to PPG I (2016) in pteridophytes, Christenhusz (2011) in gymnosperms, and APG IV (2016) in angiosperms, and the genera and species listed by the Latin initials.

This book contains most of the local, exotic, invasive wild plants and a few cultivated plants (including cultivars) on Nanji Islands, which can be used as references for the departments of agriculture, forestry and environmental protection, etc., and also can be used as self-study textbooks and supplementary for plant amateurs, horticulturists, and faculties of plant researching and teaching. As one of the research cooperation achievements between Nanji Islands National Marine Nature Reserve Administration and Nanjing Forestry University, it is sincerely appreciated the supports and help from relevant departments, leaders, colleagues and friends during the preparation and publishing. Due to the compiling limitation of time and editor’s level, for the omissions and errors existing in the book, please point out without stint.

Editors

September 2021

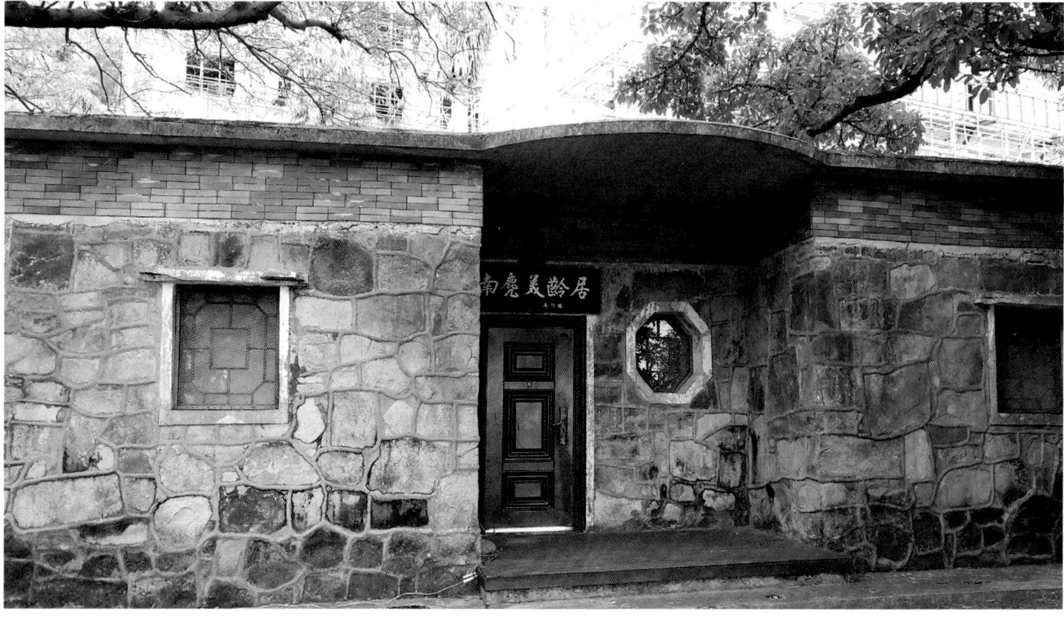

目　录

蕨类植物门（Pteridophyta）

裸子植物门（Gymnospermae）

被子植物门（Angiospermae）

蕨类植物门
PTERIDOPHYTA

P12. 里白科
Gleicheniaceae

芒萁
Dicranopteris pedata
(Dicranopteris dichotoma)
芒萁属 *Dicranopteris*

形态特征：植株高 0.5 ～ 3 m。根状茎横走，被浓密的深棕色毛。叶柄无毛；叶轴 5 ～ 8 回两叉分枝，基部节间被深棕色毛，后脱落；顶端芽密被锈色毛；苞片卵形到卵形长圆形；每分叉处两侧具一对托叶状羽片；叶片纸质，背面被白霜，正面黄绿色或绿色，在主脉和背面脉上被稀疏的棕色毛；主脉两面突出；每组小脉 3 ～ 5 条。孢子囊群在主脉两侧各 1 列；孢子囊 5 ～ 8。

分布：亚洲、澳大利亚和北美洲热带地区。广泛分布于我国秦淮以南地区。南麂主岛各处常见。

生境：灌木丛、森林、山谷、河边、山坡。

Description: Plants 0.5-3 m tall. Rhizomes creeping, covered with dense dark brown hairs. Stipe glabrous; rachis 5-8 times dichotomously branched, basal internode covered with dark brown hairs, glabrescent; apical buds covered with dense brown hairs; bracts ovate to ovate-oblong; rachises with a pair of lateral stipulelike pinnae at each dichotomy; lamina papery, glaucous abaxially, yellowish green or green adaxially, with sparse brown hairs on costae and veins abaxially; costae prominent on both surfaces; veins 3-5 in each group. Sori in 1 line on each side of costule; sporangia 5-8.

Distribution: South Asian to East Asia, Austrilia. Almost throughout China. Main island of Nanji, commonly.

Habitat: Thickets, forests, valleys, by rivers, hillsides.

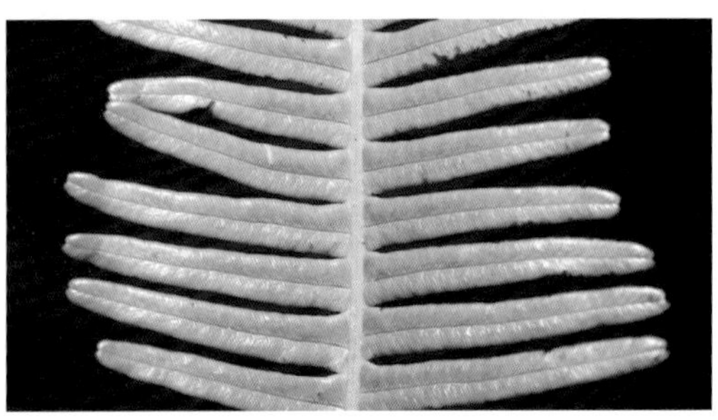

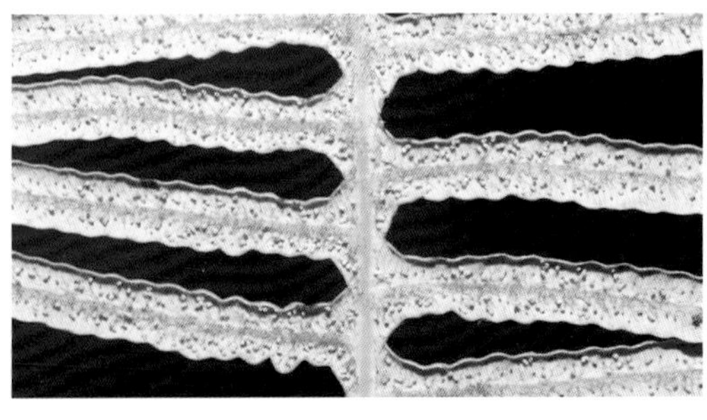

P13. 海金沙科 Lygodiaceae

海金沙（狭叶海金沙）
Lygodium japonicum（*Lygodium microstachyum*）
海金沙属 *Lygodium*

形态特征：多年生攀援草本，长 1 ~ 4 m。叶轴具窄边，羽片多数，对生于叶轴短距两侧；不育叶纸质，干后褐色，羽片尖三角形，两侧有窄边，二回羽状；能育羽片卵状三角形，长宽近相等，二回羽状。孢子囊穗长度过小羽片中央不育部分，排列稀疏，暗褐色，无毛。

分布：亚洲、澳大利亚和北美洲热带地区。广泛分布于我国秦淮以南地区。南麂主岛各处常见。

生境：攀援于次生植被上。

Description: Perennial climbing herbs, 1-4m long. Rachis with narrow sides, pinnae numerous, opposite on both sides of rachis with short spur; sterile fronds papery, brown after dry, pinnae sharp triangular, with narrow edge on both sides, bipinnate; fertile pinnae ovate triangular, nearly equal in length and width, bipinnate. Sori longer than central sterile part of pinnule, sparsely arranged, dark brown, glabrous.

Distribution: Asia, tropical Australia, North America. Widely distributed in the areas to the south of Qinling and Huaihe. Main island of Nanji, commonly.

Habitat: Climbing in secondary vegetation.

P29. 鳞始蕨科
Lindsaeaceae

阔片乌蕨
***Odontosoria biflora* (*Stenoloma biflorum*)**
乌蕨属 ***Odontosoria***

形态特征：植株高 30 cm。根状茎短而横走，密被鳞片。叶革质，近生，叶柄褐黄色，叶片三角状卵圆形，3～4回羽状，羽片 8～10 对，互生，披针形，下部 2～3 回羽状，小羽片近菱状长圆形，下部羽状分裂，裂片近扇形。孢子囊群生于 2～4 条细脉顶端，囊群盖基部或全部边缘着生，齿状或蚀刻状，孢子椭圆形，单沟。

分布：日本、菲律宾、太平洋岛屿。我国华东和华南地区。南麂主岛各处常见。

生境：海边石山上。

Description: Plants 30 cm tall. Rhizomes shortly creeping, densely scaly. Fronds leathery, proximal; stipe gramineous, lamina triangular-ovate, 3-4-pinnate, pinnae 8-10 pairs, alternate, lanceolate; 2-or 3-pinnate at base,pinnules nearly rhombic-oblong, proximal pinnatifid, lobes subflabellate. Sori terminal on 2-4 vein ends; indusia basally and entirely adnate laterally, denticulate to erose, spores ellipsoid, monolete.

Distribution: Japan, Philippines, Pacific islands. East China and South China. Main island of Nanji, commonly.

Habitat: On rocks along seashore.

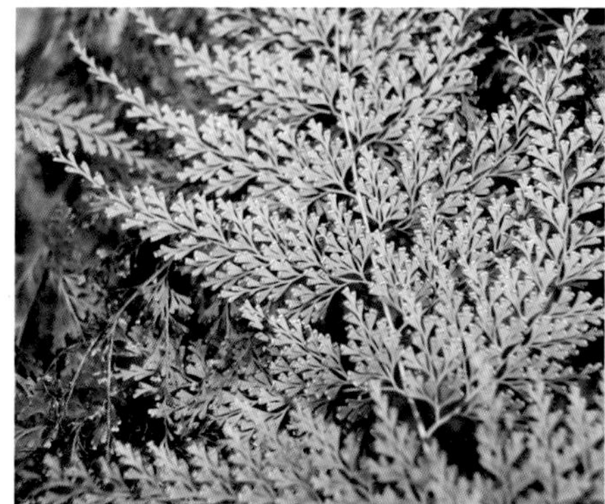

P30. 凤尾蕨科
Pteridaceae

扇叶铁线蕨
Adiantum flabellulatum
铁线蕨属 *Adiantum*

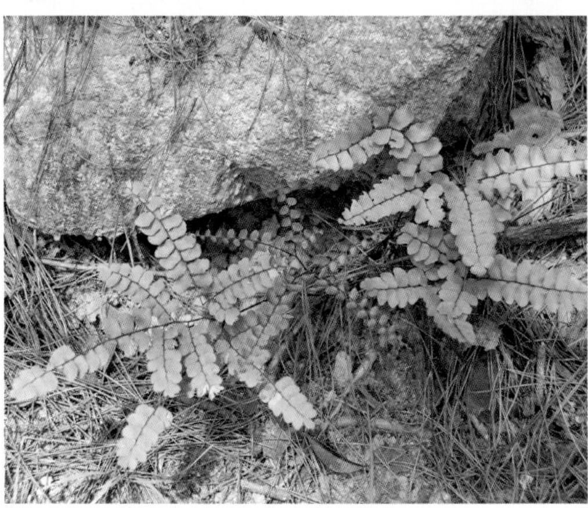

形态特征：植株高 20 ～ 45 cm。根茎短而直立，密被棕色鳞片有光泽，披针形。叶簇生；叶柄紫黑色，有光泽，上有纵沟，内有棕色短硬毛；叶片扇形，2 ～ 3 回不对称 2 叉分枝；羽片具小羽片 8 ～ 15 对，互生，水平开展，叶片近革质，绿色或暗褐色，两面无毛。孢子囊群每羽片 2 ～ 5 枚；假囊群盖黑褐色，半圆形或长圆形，全缘，宿存，孢子具不明显颗粒状纹饰。

分布：亚洲热带地区。我国华东、华南至西南地区。南麂主岛各处常见。

生境：开阔地的酸性红、黄壤土。

Description: Plants 20-45 cm tall. Rhizomes erect, short, scales dense, brown, glossy,lanceolate. Fronds clustered; stipe black-purple, glossy,adaxially grooved with short stiff brown hairs inside; lamina pedately 2-or 3-dichotomously branched, flabellate in outline;pinnules 8-15 pairs per pinna, alternate, horizontally spreading, blade thinly leathery, green or dark brown, both surfaces glabrous. Sori 2-5 per pinnule, false indusia dark brown, semi-orbicular or oblong, entire, persistent, perispore indistinctly granular.

Distribution: Tropical areas of Asia. East China, South China to Southwest China. Main island of Nanji, commonly.

Habitat: Acidic red and yellow soils in open areas.

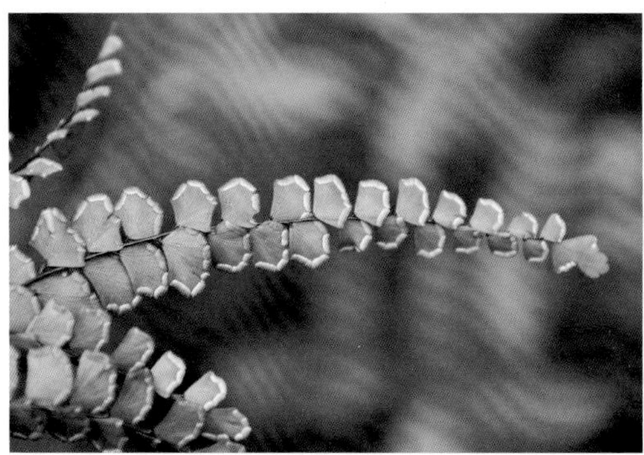

P30. 凤尾蕨科
Pteridaceae

野雉尾金粉蕨
Onychium japonicum
金粉蕨属 *Onychium*

形态特征：植株高约 60 cm。根茎长而横走，疏被鳞片，鳞片棕色或红棕色，披针形，筛孔明显。叶散生，柄基部褐棕色，叶片几和叶柄等长，卵状三角形或卵状披针形；叶干后坚纸质，灰绿色或绿色，羽轴坚挺。孢子囊群盖线形或短长圆形，膜质，灰白色，全缘。

分布：日本、泰国，大洋洲。广布我国长江以南地区。南麂主岛（关帝岙、国姓岙、门屿尾、三盘尾），常见。

生境：溪边、路旁、林缘。

Description: Plants ca. 60 cm tall. Rhizomes long creeping, sparsely scaly, scales brown or reddish brown, lanceolate, with conspicuous sieve pores. Fronds scattered, petiole base brown, lamina several as long as petiole, ovate-triangular or ovate-lanceolate; fronds firm papery after dry, gray-green or green, rachis firm. Sori indusia linear or shortly oblong, membranous, gray-white, margin entire.

Distribution: Japan, Thailand; Oceania.Widely to the south of Yangtze River. Main island (Guandiao, Menyuwei, Sanpanwei), commonly.

Habitat: Streams, roadsides, forest margins.

P30. 凤尾蕨科
Pteridaceae

刺齿半边旗（刺齿凤尾蕨）
Pteris dispar
凤尾蕨属 ***Pteris***

形态特征：植株高 30 ~ 90 cm。根茎斜生，被黑褐色鳞片。叶簇生，近二型；叶柄连同叶轴均栗色；叶片卵状长圆形，二回深羽裂或二回半边深羽裂；顶生羽片披针形，渐尖头，基部圆形；裂片宽披针形，略呈镰刀状；侧生羽片与顶生羽片同形，下部的有短柄，尾状渐尖头，裂片与顶生羽片同形同大；侧脉明显，2 叉，小脉达锯齿软骨质刺尖；叶干后草质。

分布：越南、马来西亚、菲律宾、日本。我国华东至西南地区。南麂主岛各处常见。

生境：山谷疏林。

Description: Plants 30-90 cm tall. Rhizome ascending, with brownish black scales. Fronds clustered, nearly dimorphic; petiole with all leaf axes chestnut colored; lamina ovate-oblong, 2-pinnatipartite or at one side deeply bipinnate-lobed; terminal pinnate lanceolate, acuminate, base rounded; lobes broadly lanceolate or linear-lanceolate, slightly falcate; lateral pinna homomorphic to terminal pinna, lower part with short handle, tail-like acuminate head, lobes homomorphic to terminal pinna; lateral veins conspicuous, 2-forked, venules up to serrate cartilaginous prickles; lamina herbaceous when dried.

Distribution: Japan, Malaysia, Philippines, Vietnam. East China to Southwest China. Main island of Nanji, commonly.

Habitat: Open forests along valleys.

P30. 凤尾蕨科
Pteridaceae

傅氏凤尾蕨
Pteris fauriei
凤尾蕨属 *Pteris*

形态特征： 植株高 50 ～ 90 cm。根茎短，直立或斜升，被深褐色鳞片。叶簇生，叶柄禾秆色，被暗褐色鳞片，光滑，上面具槽；叶片二回羽状深裂，卵形至卵状三角形；叶干后纸质，浅绿色至暗绿色，无毛。孢子囊群线形，沿裂片边缘延伸，仅裂片先端不育；囊群盖线形，灰棕色，膜质，全缘，宿存。

分布： 日本和越南北部。我国华东至西南地区。南麂主岛（关帝岙、门屿尾、三盘尾），偶见。

生境： 林下沟旁的酸性土。

Description: Plants 50-90 cm tall. Rhizome short, erect or obliquely ascending, covered with dark brown scales. Fronds clustered,stipe straw-colored, with dark brownish scales, glabrous, adaxially grooved; lamina 2-pinnatipartite,ovate to ovate-triangular; lamina light green to pale green, papery when dried, glabrous. Sori linear, extending along margin of lobes, sterile only at apex of lobes; indusia linear, gray-brown, membranous, margin entire, persistent.

Distribution: Japan, North Vietnam. East China to Southwest China. Main island (Guandiao, Menyuwei, Sanpanwei), occasionally.

Habitat: Acidic soils in forests along valleys.

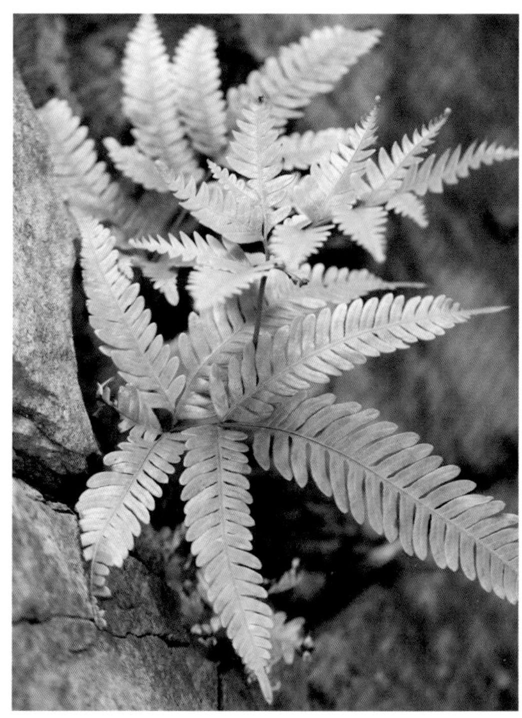

P30. 凤尾蕨科
Pteridaceae

井栏边草
Pteris multifida
凤尾蕨属 ***Pteris***

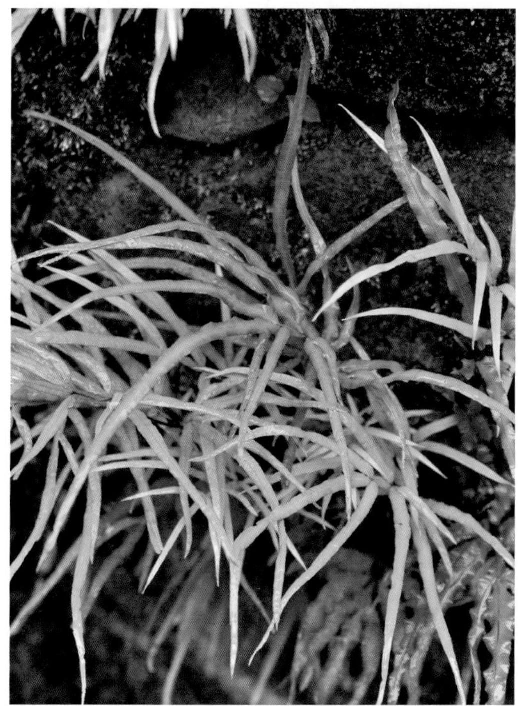

形态特征：植株高 20 ～ 45 cm，偶见 85 cm。根茎短而直立，被黑褐色鳞片。叶密而簇生，二型，不育叶柄较短，禾秆色或暗褐色，具禾秆色窄边；叶片卵状长圆形，尾状头，基部圆楔形，奇数一回羽状；能育叶柄较长，羽片 4 ～ 6(10) 对，线形，不育部分具锯齿；叶干后草质，暗绿色，无毛。

分布：东亚至东南亚。广布我国南北大多数地区。南麂主岛（美龄居、门屿尾、镇政府），偶见。

生境：墙壁、井边及石灰岩缝隙或灌丛下。

Description: Plants 20-45(85) cm tall. Rhizome short and erect, covered with dark brown scales. Fronds dense and clustered, dimorphic, sterile petiole short, culm color or dark brown, with narrow side of culm color; lamina ovate-oblong, caudate head, base rounded-cuneate, odd number back pinnate; fertile petiole longer, pinnate 4-6(10) pairs, linear, sterile part serrate; lamina herbaceous when dried, dark green, glabrous.

Distribution: East Asia to South east Asia. Widely distributed in most areas of south and north China. Main island (Meilingju, Menyuwei, Zhenzhengfu), occasionally.

Habitat: Walls, well margins and limestone crevices or under bushes.

P30. 凤尾蕨科
Pteridaceae

半边旗
Pteris semipinnata
凤尾蕨属 *Pteris*

形态特征：植株高 35 ～ 80 cm，偶见 1.2 m。根茎长而横走，被黑褐色鳞片。叶簇生，近一型；叶柄连同叶轴均栗红色；叶片长圆状披针形，奇数二回半边深羽裂，不育裂片有尖锯齿，能育裂片顶端有尖刺或具 2 ～ 3 尖齿；叶干后草质，灰绿色，无毛。

分布：亚洲热带地区。我国华东南部至西南东部地区。南麂主岛各处常见。

生境：疏林下阴处、溪边或岩石旁的酸性土。

Description: Plants 35-80(120) cm tall. Rhizome long creeping, apex with blackish brown scales. Fronds clustered, submonomorphic; stipe and rachis castaneous-reddish; lamina pinnate, oblong-lanceolate in outline, at one side deeply bipinnate-lobed; fertile segments entire except for with 1 spine or 2 or 3 acute teeth near apex, apex mucronate or obtuse; lamina herbaceous when dried, gray-green, glabrous.

Distribution: Tropical areas of Asia. Southern of Southeast China to eastern of Southwest China. Main island of Nanji, commonly.

Habitat: Acidic soil in open forests, by streams or rocks.

P30. 凤尾蕨科
Pteridaceae

蜈蚣凤尾蕨（蜈蚣草）
Pteris vittata
凤尾蕨属 ***Pteris***

形态特征：植株高 0.2 ～ 1.5 m。根茎短而直立，密被疏散的黄褐色鳞片。叶簇生，叶柄坚硬，深禾秆色或浅棕色，幼时密被鳞片；叶片倒披针状长圆形，长尾头，基部渐窄，奇数一回羽状；不育的叶缘有细锯齿；叶干后纸质或薄革质，绿色。孢子囊群线形，羽片边缘的边脉；囊群盖同形，全缘，膜质，灰白色。

分布：旧世界热带和亚热带地区。广布我国秦岭以南地区。南麂主岛（门屿尾、三盘尾），偶见。

生境：钙质土或石灰岩上，或石隙或墙壁上。

Description: Plants 0.2-1.5 m tall. Rhizomes short and erect, densely sparsely yellow-brown scales. Fronds clustered, stipe firm, dark straw color or light brown, densely scaly when young; lamina oblanceolate-oblong, long-tailed, base tapering, 1-odd-pinnated; sterile margin serrulate; lamina papery or thinly leathery when dried, green. Sori linear, attached to marginal veins of pinnate margin; indusia homomorphic, margin entire, membranous, gray-white.

Distribution: Tropics and subtropics of the Old World. Widely distributed to the south of Qinling Mountains. Main island (Menyuwei, Sanpanwei), occasionally.

Habitat: Calcareous soils, on limestone, also on stone and on walls.

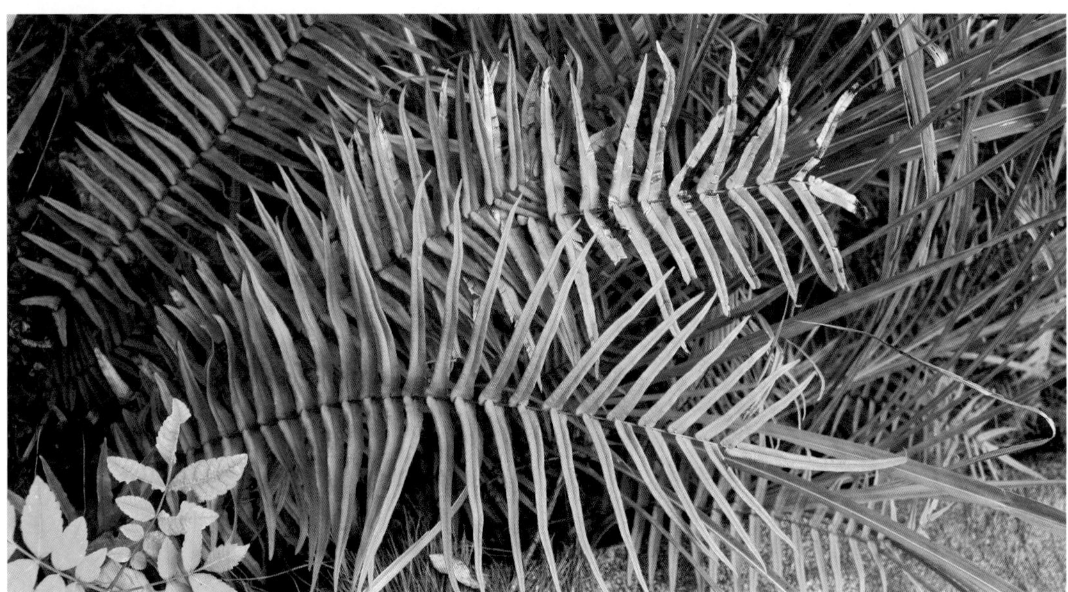

P31. 碗蕨科
Dennstaedtiaceae

姬蕨
Hypolepis punctata
姬蕨属 ***Hypolepis***

形态特征：植株高 1 m 以上。根状茎长而横走，密被淡棕色毛。叶疏生，暗褐色，粗糙有毛，叶片长卵状三角形，3 ~ 4 回羽状深裂，顶部为一回羽状；叶坚草质，干后黄绿色；叶轴、羽轴、小羽轴、叶柄淡褐色，有狭沟，粗糙，透明灰色节状毛。孢子囊群圆形或卵形，囊群盖由锯齿多少反卷而成，棕绿色，无毛。

分布：亚洲、大洋洲和美洲热带地区。我国长江以南各地。南麂主岛各处常见。

生境：溪边、密林。

Description: Plant to 1m tall. Rhizome long creeping, with pale brown hairs. Fronds sparsely, dark brown, scabrous hairy. Lamina long ovate-triangular, three to four-pinnate deep-divided, the top for a pinnate; bladeiridescent when dry, firmly herbaceous or papery,rachis,pinnate axis and stipe pale chestnut-brown,with narrow groove, rough, transparent gray node-like hair. Sori circular or ovate, indusia more or less convoluted by serrate, brown-green, glabrous.

Distribution: Tropical Asia, Oceania and America. To the south of Yangtze River. Main island of Nanji, commonly.

Habitat: Near streams, dense forests.

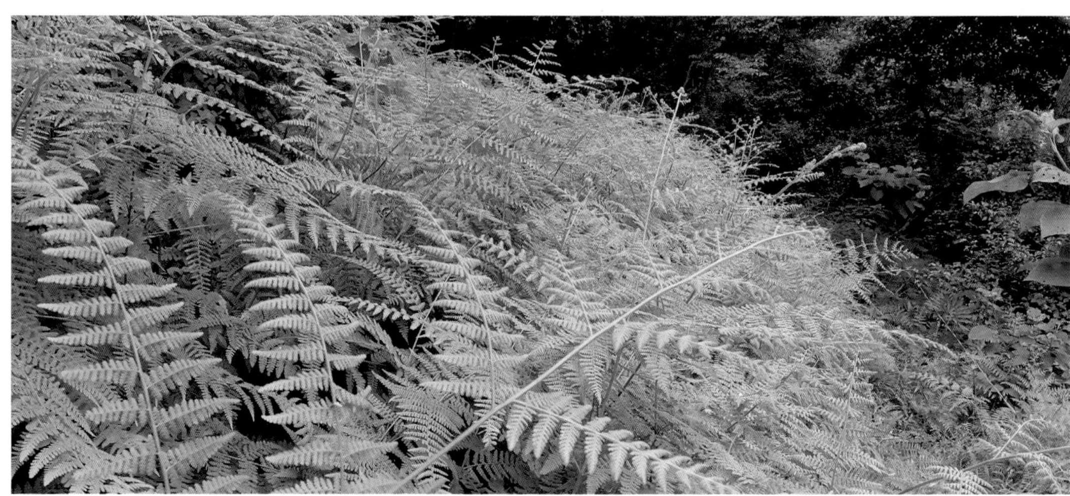

P31. 碗蕨科 Dennstaedtiaceae

蕨
Pteridium aquilinum var. *latiusculum*
蕨属 *Pteridium*

形态特征：植株高达 2 m。根茎长而横走，密被锈黄色柔毛。叶直立，叶柄褐色，木质；叶片 3 ~ 4 回羽状深裂，三角形至长圆状卵形，革质；叶轴淡褐色，羽片斜升或平展，卵状三角形至长圆形，小羽片线形至长圆形，除中脉和叶缘外无毛。孢子囊群伸长，外囊群盖膜质，具缘毛，内囊群盖退化，流苏状。

分布：日本、欧洲和北美洲。我国广泛分布，主产南方。南麂主岛各处常见。

生境：阳光充足的山坡和林缘。

Description: Plants up to 2 m tall. Rhizome long creeping, denselycastaneous hairy. Fronds erect, stipe brown, woody,lamina 3-or 4-pinnatepinnatifid,triangular to oblong-ovate in outline when pressed,leathery; rachis pale brown; pinnae ascending or horizontal,ovate-triangular to oblong; pinnules or segments linear to oblong, glabrous exceptfor pinnule margins and midvein. Sori elongate, outer indusia membranous, ciliate, inner indusia vestigial and fimbriate.

Distribution: Japan, Europe, North America. Throughout China, mostly in southern China. Main island of Nanji, commonly.

Habitat: Sunny slopes and forest margins.

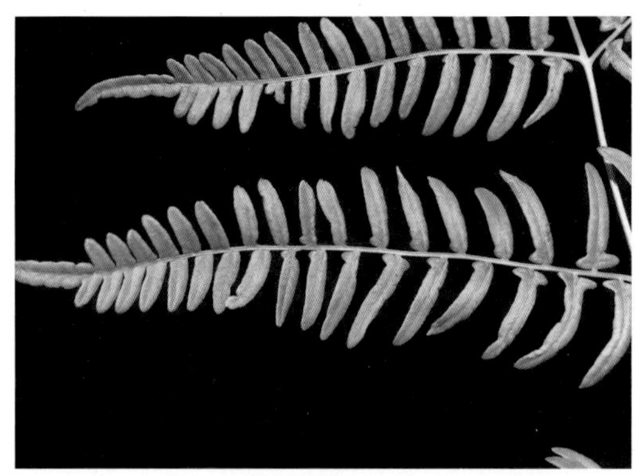

P40. 乌毛蕨科
Blechnaceae

珠芽狗脊（胎生狗脊蕨）
Woodwardia prolifera
狗脊属 *Woodwardia*

形态特征：植株高 0.7 ～ 2.3 m。根状茎匍匐，黑褐色，粗壮，密被鳞片；鳞片红褐色，披针形。叶近生，叶片二回羽状深裂，长圆状卵形或椭圆形，革质，无毛；羽片 5 ～ 9 对，羽状深裂，裂片 10 ～ 14 对，斜生，长圆状披针形，叶脉明显，羽片上面通常产生小珠芽。孢子囊群深陷于主脉两侧网眼内，新月形或椭圆形，囊群盖暗褐色，厚纸质。

分布：日本。我国华东至华南地区。南麂主岛（国姓岙、门屿尾、三盘尾），偶见。

生境：山坡、疏林下的开阔湿地或溪边。

Description: Plants 0.7-2.3 m tall. Rhizome decumbent, dark brown, stout, densely scaly; scales red-brown, lanceolate. Stipes close, lamina deeply bipinnatifid, oblong-ovate or elliptic, leathery, glabrous, pinnae 5-9 pairs, deeply pinnatifid, lobes 10-14 pairs, oblique, oblong-lanceolate; veins obvious. Leaf-bearing bulbils small and usually abundant on adaxial surfaces of pinna lobes. Sori occupying costular areoles, crescent-shaped or elliptic, sunken in rimmed depressions; indusia dark brown, thickly papery.

Distribution: Japan. East China to South China. Main island (Guoxingao, Menyuwei, Sanpanwei), occasionally.

Habitat: Mountain slopes, open and wet places in sparse forests, near streams.

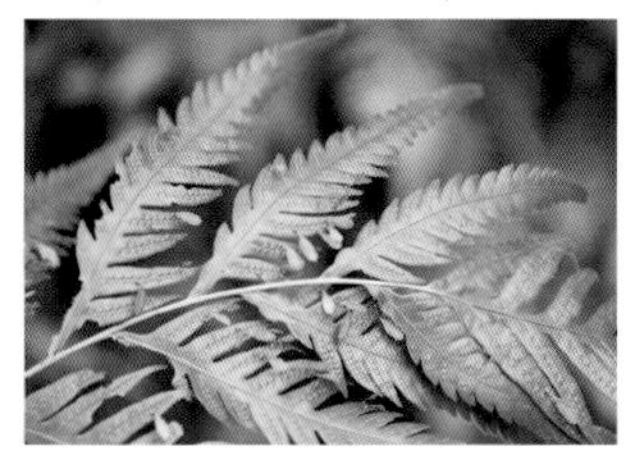

P42. 金星蕨科
Thelypteridaceae

渐尖毛蕨
Cyclosorus acuminatus
毛蕨属 *Cyclosorus*

形态特征：植株高 70 ～ 80 cm。根茎长而横走，顶端密被棕色披针形鳞片。叶疏生；叶柄禾秆色至褐色，叶片基部不变狭，先端尾状渐尖；侧生羽片 13 ～ 18 对，柄极短，中部羽片线状披针形至披针形；叶脉 7 ～ 9 对；叶片纸质或近革质，干后灰绿色，两面被短针毛。孢子囊群圆形，近边生，囊群盖被短毛或近无毛。

分布：日本、朝鲜、菲律宾。广布我国长江以南，东到台湾，北到陕西南部地区。南麂主岛各处常见。

生境：灌丛、草地、田边或路边的半开阔地。

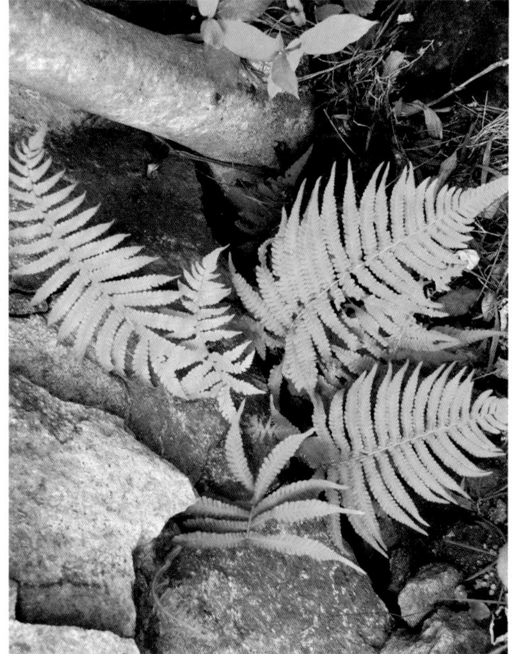

Description: Plants 70-80 cm tall. Rhizomes long creeping, apices including stipe bases with brown lanceolate scales. Fronds distant; stipes stramineous to brown; lamina bases not narrowed, apices caudate to acuminate; lateral pinnae 13-18 pairs, shortly stalked; middle pinnae linear-lanceolate to lanceolate;veinlets 7-9 pairs; lamina papery to subleathery, grayish green when dried, both surfaces with short acicular hairs. Sori orbicular, submarginal; indusia shortly hairy or subglabrous.

Distribution: Japan, Korea, Philippines. Widely distributed to the south of Yangtze River, east to Taiwan, north to the south of Shanxi, China. Main island of Nanji, commonly.

Habitat: Semi-open places in thickets, grasslands, farmland margins, roadsides.

P42. 金星蕨科
Thelypteridaceae

干旱毛蕨
Cyclosorus aridus
毛蕨属 *Cyclosorus*

形态特征：植株高 50 ~ 100 cm。根状茎长而横走，连同柄基部疏被棕色披针形鳞片。叶远生；叶片基部骤窄或渐窄，顶端尾状到渐尖；羽片 15 ~ 40 对，下端 2 ~ 10 对缩短；中部羽片线状披针形，裂片 20 ~ 40 对，三角形，全缘；叶片纸质至近革质，干后棕绿色或黄绿色，上面近无毛，下面沿叶脉疏生短针毛，有黄色或橙色棒状腺体。孢子囊群球形，囊群盖小；孢子具长翅或脊状褶皱。

分布：亚洲东部至太平洋岛屿。我国长江流域以南地区。南麂主岛各处常见。

生境：潮湿或半开阔处，通常成丛。

Description: Plants 50-100 cm tall. Rhizomes long creeping, including stipe bases with sparse brown lanceolate scales. Fronds distant; laminae base abruptly or gradually narrowed, apices caudate to acuminate; pinnae 15-40 pairs, proximal 2-10 pairs shortened; middle pinnae linear-lanceolate, segments 20-40 pairs on middle pinnae, triangular, entire, laminae papery to somewhat leathery, brownish green or yellowish green when dried, adaxially subglabrous, abaxial surface with short acicular hairs along costae and veins, also with yellow or orange clavate glands along veins. Sori orbicular, indusia glandular, sometimes hairy; spore with long wings or ridged folds.

Distribution: East Asia to Pacific islands. To the south of Yangtze River Basin. Main island of Nanji, commonly.

Habitat: Wet or semi-open places, usually among tall grasses.

P42. 金星蕨科
Thelypteridaceae

金星蕨
Parathelypteris glanduligera
金星蕨属 *Parathelypteris*

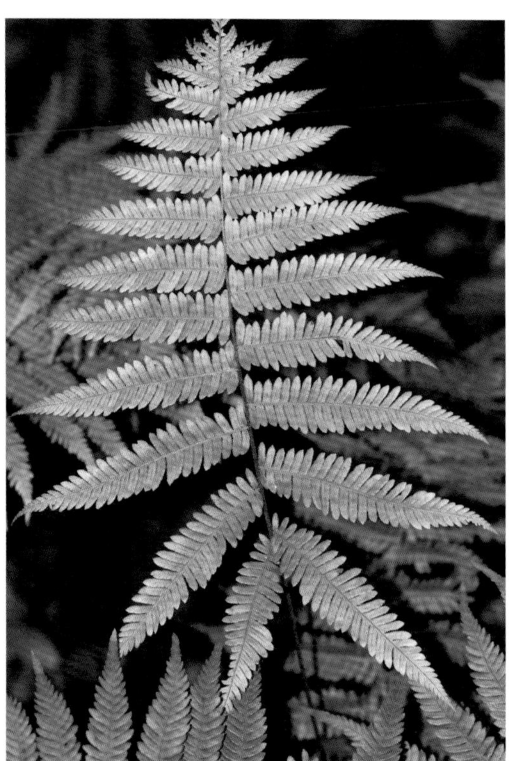

形态特征：植株高 35 ～ 50 cm，偶见 60 cm。根茎长而横走，顶端略被鳞片。叶近生，叶柄禾秆色，叶片披针形，二回羽状深裂；叶脉明显，侧脉单一，每裂片 5 ～ 7 对；叶草质，干后草绿；羽片背面近无毛，仅沿脉被稀疏的灰白色针状毛，正面沿脉被短针毛。孢子囊群圆形，每裂片 4 ～ 5 对，近叶缘；囊群盖圆肾形，棕色，背面疏被灰白色刚毛，宿存。

分布：日本、朝鲜、越南。分布于我国长江流域以南地区。南麂主岛各处常见。

生境：疏林下。

Description: Plants 35-50(60) cm tall. Rhizomes long creeping, apices with sparse lanceolate scales. Fronds approximate; stipes stramineous; lamina lanceolate, pinnate-pinnatifid; veins distinct, lateral veins simple, 5-7 pairs per segment;lamina herbaceous, when dry grass-green; pinnae abaxially subglabrous, except for sparse grayish white acicular hairs along costae and main veins, adaxially with short acicular hairs along costae. Sori orbicular, 4 or 5 pairs per segment, close to margins; indusia orbicular-reniform, brown, with sparse grayish white setae, persistent.

Distribution: Japan, Korea, Vietnam. To the south of the Yangtze River Basin. Main island of Nanji, commonly.

Habitat: Open forests.

P45. 鳞毛蕨科
Dryopteridaceae

中华复叶耳蕨
Arachniodes chinensis
复叶耳蕨属 *Arachniodes*

形态特征：植株高 40 ～ 65 cm。根状茎短匍匐，硬，密被黑褐色鳞片，线状披针形或钻形。叶密集，叶柄禾秆色，基部密被根状茎类似鳞片；叶片 2 ～ 3 回羽状，干时暗褐色或灰绿色，卵形或卵状长圆形；羽片 4 ～ 8 对，互生或下部 1(或 2)对有时对生，最下部羽片三角状披针形。孢子囊群生于细脉顶端，每节 1 ～ 8 对，囊群盖棕色，硬膜质，边缘具缘毛或撕裂状。

分布：东南亚至日本。我国长江流域以南。南麂主岛（国姓岙、美龄居、三盘尾），偶见。

生境：浓密森林、潮湿岩石或沟壑上的阴坡，常见于常绿阔叶林。

Description: Plants, 40-65 cm tall. Rhizome shortly creeping, stiff, densely scaly, cales (blackish) brown, linear-lanceolate or subulate. Fronds approximate, stipe stramineous, base densely scaly with scales similar to those on rhizome; lamina 2-or 3-pinnate, dark brown or dull green when dried, deltoid-ovate or ovate-oblong, papery or subleathery; pinnae 4-8 pairs, alternate or lower 1(or 2) pairs sometimes opposite, lowest pinna deltoid-lanceolate. Sori terminal on veinlets, 1-8 pairs per ultimate segment, indusia brown, firmly membranous, ciliate or lacerate on margin.

Distribution: Southeast Asia to Japan. To the south of Yangtze River Basin. Main island (Guoxingao, Meilingju, Sanpanwei), commonly.

Habitat: Shaded slopes in dense woods, on wet rocks or ravines, more often in evergreen broad-leaved forests.

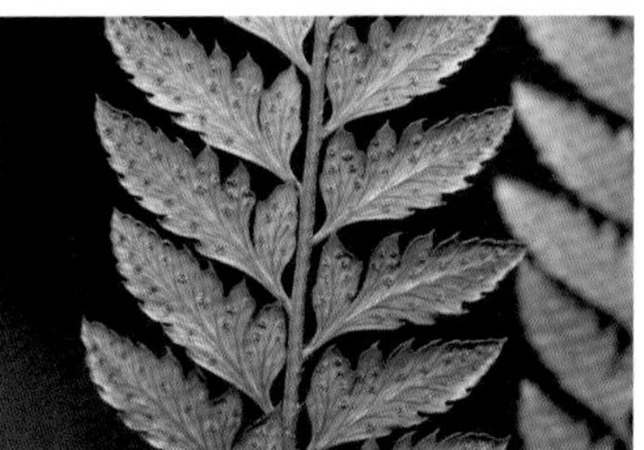

P45. 鳞毛蕨科 Dryopteridaceae

全缘贯众
Cyrtomium falcatum
贯众属 *Cyrtomium*

形态特征：高 30 ～ 40 cm。根茎直立，密被披针形棕色鳞片。叶簇生，革质，两面光滑；叶柄禾秆色，下部密被淡褐色的卵形鳞片；叶片宽披针形，一回奇数羽状；叶脉羽状，上面不显，下面微隆起，侧脉联成网状，在主脉两侧各有 3 ～ 4 行网眼。孢子囊群着生内藏小脉，密布羽片下面；囊群盖圆盾形，边缘具细齿。

分布：亚洲东部和太平洋岛屿。我国华东和华南地区。南麂主岛各处常见。

生境：海岸和低地森林。

Description: Plants 30-40 cm tall. Rhizome erect, densely covered with lanceolate brown scales. Fronds clustered,leathery, glabrous; stipe stramineous, lower portion densely scaly; scales pale brown, ovate; lamina broadly lanceolate,1-imparipinnate; venation pinnate, slightly raised abaxially, indistinct adaxially, veinlets anastomosing to form 3 or 4 rows of areoles. Sori throughout abaxial surface of pinnae; indusia margins slightly incised.

Distribution: Eastern Asia and Pacific islands. East and South China. Main island of Nanji, commonly.

Habitat: Coastal and lowland forests.

P45. 鳞毛蕨科
Dryopteridaceae

贯众
Cyrtomium fortunei
贯众属 *Cyrtomium*

形态特征：植株高 25 ~ 50 cm。根茎直立，密被棕色鳞片。叶簇生，纸质，叶柄禾秆色；叶片矩圆披针形，一回奇数羽状，侧生羽片 7 ~ 16 对，互生，披针形；顶生羽片狭卵形；羽状脉，腹面不明显，背面微凸起，小脉联结成 4 ~ 5 行网眼。孢子囊群遍布羽片背面；囊群盖圆形，盾状，全缘。

分布：亚洲东部。广布我国华北、西北和长江以南。南麂主岛（百亩山、大山、国姓岙），偶见。

生境：空旷地石灰岩缝或林下。

Description: Plants 25-50 cm tall. Rhizome erect, densely covered with brown scales. Fronds clustered, papery. Stipe stramineous, lamina oblong-lanceolate, 1-imparipinnate, lateral pinnae 7-16 pairs, alternate; terminal pinna ovate-lanceolate; venation pinnate, slightly raised abaxially, indistinct adaxially, veinlets anastomosing to form 4 or 5 rows of areoles . Sori throughout abaxial surface of pinnae; indusia grayish, margins entire.

Distribution: East Asia. Widely distributed in North China, Northwest China and the south of Yangtze River. Main island (Daimushan, Dashan, Guoxingao), occasionally.

Habitat: Limestone crevices in open areas or forests.

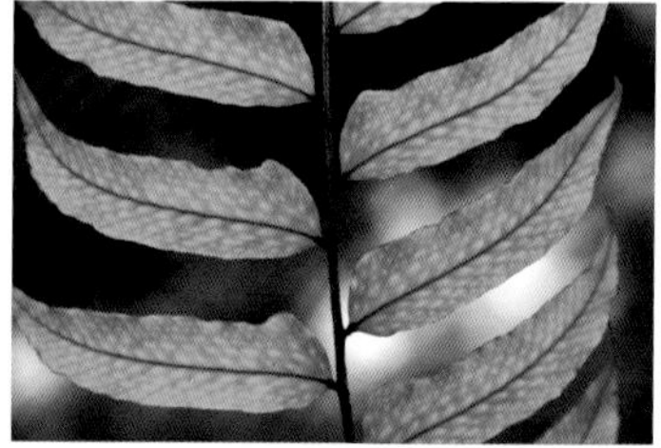

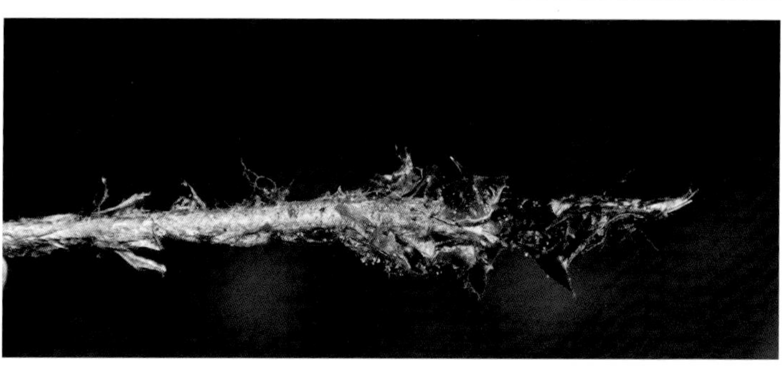

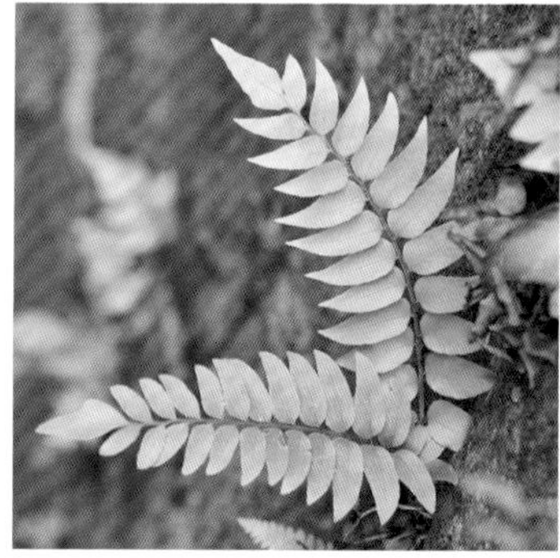

P45. 鳞毛蕨科 Dryopteridaceae

黑足鳞毛蕨
Dryopteris fuscipes
鳞毛蕨属 ***Dryopteris***

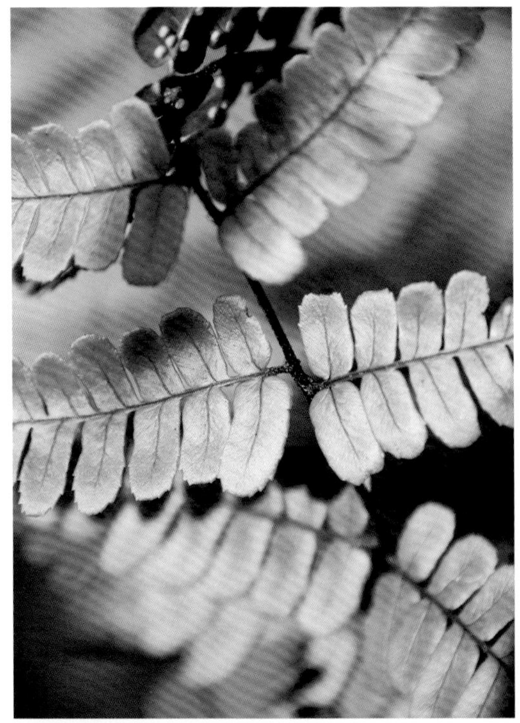

形态特征：植株高 40 ~ 92 cm。根茎斜升。叶簇生，下部叶柄深褐色，上部棕色，基部密被棕色披针形鳞片，叶簇生，叶片卵状披针形，二回羽状，草质，深绿色，羽片 10 ~ 15 对，披针形，下部羽片变短，上部变小，小羽片 10 ~ 12 对，三角状卵形，叶轴、羽轴和中脉上面均具浅沟。孢子囊群近主脉着生，在主脉两侧各排成 1 行；囊群盖肾形，全缘。

分布：日本、朝鲜、越南。我国长江以南地区。南麂主岛（百亩山、大山、打铁礁、国姓岙、三盘尾），偶见。

生境：亚热带常绿阔叶林。

Description: Plants 40-92 cm tall. Rhizome ascending. Fronds caespitose, stipes dark brown below, stramineous above, with densely brown, lanceolate scales at base. Lamina ovate-lanceolate, bipinnate, herbaceous, dark green, pinnae 10-15 pairs, lanceolate, lowest pairs shortened, upper pinnae becoming smaller, pinnules 10-12 pairs, deltoid-ovate,rachis, pinna rachis, and costa grooved adaxially. Sori usually in 1 row on either side of costa, slightly closer to costa; indusia reniform, entire.

Distribution: Japan, Korea, Vietnam. To the south of the Yangtze River. Main island (Baimushan, Dashan, Datiejiao, Guoxingao, Sanpanwei), occasionally.

Habitat: Subtropical broad-leaved evergreen forests.

P45. 鳞毛蕨科
Dryopteridaceae

太平鳞毛蕨
Dryopteris pacifica
鳞毛蕨属 *Dryopteris*

形态特征：植株高 60 ～ 80 cm。根状茎斜升，顶端密被黑色披针形鳞片，先端毛状。叶簇生，叶柄禾秆色，基部密被黑色披针形鳞片，向上鳞片变小，毛状。叶片五角状卵形，三回羽状，羽片 10 ～ 15 对，互生，小羽片 10 ～ 15 对，披针形，羽状全裂或羽状深裂；叶脉羽状，二叉或单一，背面可见。孢子囊略靠近边缘着生，囊群盖圆肾形，棕色，边缘啮蚀状。

分布：日本、朝鲜。我国华东地区。南麂主岛（百亩山、打铁礁、三盘尾）、大檑山屿，偶见。

生境：林下。

Description: Plants 60-80 cm tall. Rhizome ascending, apex densely covered with black, lanceolate scales; hairlike at apex. Fronds caespitose,stipes stramineous, base densely covered with scales like those on rhizome; upward scales smaller, hairlike; lamina pentagonal-ovate, tripinnate, pinnae 10-15 pairs, alternate, pinnules 10-15 pairs, lanceolate, pinnatipartite or pinnatisect; veins pinnate,veinlets forked or simple, visible abaxially. Sori slightly nearer to margin than to costa; indusia brown, reniform, erose.

Distribution: Japan, Korea. East China. Main island (Baimushan, Datiejiao, Sanpanwei), Daleishan Island, occasionally.

Habitat: Under forests.

P46. 肾蕨科 Nephrolepidaceae

肾蕨
Nephrolepis cordifolia
肾蕨属 *Nephrolepis*

形态特征：附生或土生，根状茎直立，被鳞片。叶簇生，柄暗褐色，略有光泽，上面有纵沟，下面圆形，密被淡棕色线形鳞片；叶片线状披针形；叶脉明显，侧脉纤细；叶坚草质，干后棕绿色，光滑。孢子囊群成 1 行位于主脉两侧，肾形，每组侧脉的上侧小脉顶端，位于从叶边至主脉的 1/3 处；囊群盖肾形，褐棕色，边缘色较淡，无毛。

分布：广布热带及亚热带地区。我国华东至西南地区。南麂主岛各处常见。

生境：溪边林下。

Description: Plants terrestrial or epiphytic, rhizome erect, scaled. Fronds fascicled, petioles dark brown, slightly shiny, longitudinally sulcate above, rounded below, densely covered with pale brown linear scales; lamina linear-lanceolate or narrowly lanceolate; veins conspicuous, lateral veins slender; lamina firmly herbaceous, brown-green after dry, smooth. Sori cluster into one row on both sides of the main vein, reniform, situated at the top of the superior venules of each group of lateral veins, 1/3 from the margins of the leaf to the main vein; indusia reniform, brown, margin paler, glabrous.

Distribution: Tropical and subtropical regions of the world. East China to Southwest China. Main island of Nanji, commonly.

Habitat: Forests, along streams.

P50. 骨碎补科
Davalliaceae

杯盖阴石蕨（圆盖阴石蕨）
Davallia griffithiana
(Humata griffithiana, Humata tyermannii)
骨碎补属 *Davallia*

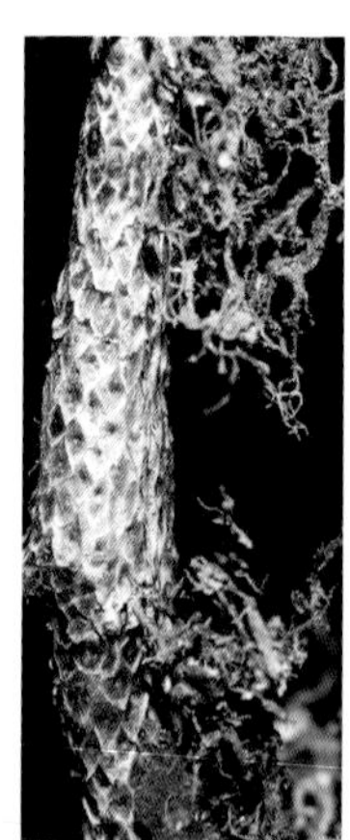
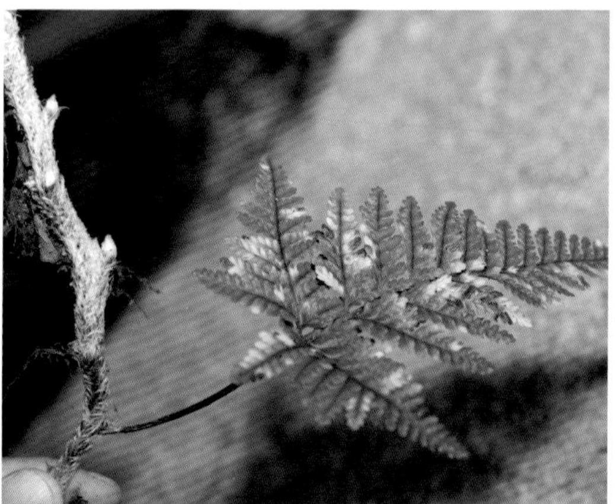

形态特征：植株高达40 cm。根状茎长而横走，密被鳞片。鳞片线状披针形，先端长渐尖，以红棕色的圆形基部盾状着生。叶片三角状卵形，基部和中部二回羽状或三回羽状。叶革质，干后上面浅褐色，下面棕色，无毛。孢子囊群生于裂片上侧小脉分叉点上，每裂片一枚或多枚。

分布：印度北部至越南。我国华东至西南地区。南麂主岛（大山、打铁礁、国姓岙、门屿尾、三盘尾），常见。

生境：森林潮湿处，林中树干或岩石上。

Description: Plants 40 cm tall. Rhizomes long creeping, densely scaly. Scales linear-lanceolate, apex long acuminate, basal peltate with reddish brown rounded. Lamina triangular-ovate, bipinnate or tripinnate toward base and in middle part. Leaves leathery, light brown above, brown below, glabrous after drying. Sori separate, borne several or single on a segment, at forking point of veins.

Distribution: Northern India to Vietnam. East China to Southwest China. Main island (Dashan, Datiejiao, Guoxingao, Menyuwei, Sanpanwei), commonly.

Habitat: Wet forests, climbing on tree trunks or rocks.

P51. 水龙骨科 Polypodiaceae

伏石蕨
Lemmaphyllum microphyllum
伏石蕨属 *Lemmaphyllum*

形态特征：植株高 6 cm。根茎细长，横走，淡绿色，疏被鳞片。叶疏生，二型；不育叶近无柄，近圆形，基部圆或宽楔形，全缘；能育叶柄长 3 ~ 8 mm，叶舌状或窄披针形，干后边缘反卷；叶脉网状，内藏小脉单一不分叉。孢子囊群线形，着生主脉与叶缘间，幼时被隔丝覆盖。

分布：日本、朝鲜南部和越南。我国华东至西南地区。南麂主岛各处常见。

生境：附生林中树干上或岩石上。

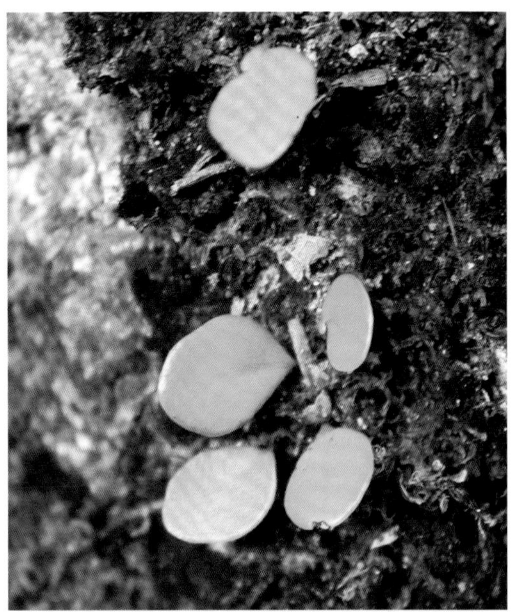

Description: Plants to 6 cm tall. Rhizome slender, transverse, pale green, sparsely scaly. Fronds remote, dimorphic; sterile fronds subsessile, lamina suborbicular, base rounded or broadly cuneate, margin entire; fertile petioles 3-8 mm long, lamina ligulate or narrow lanceolate, margin revolute when dried; veins reticulate, with simple included veinlets. Sori linear, between costa and margin, covered with paraphyses when young.

Distribution: Japan, South Korea, and Vietnam. East China to South west China. Main island of Nanji, commonly.

Habitat: On tree trunks in forests or on rocks.

P51. 水龙骨科
Polypodiaceae

江南星蕨
Lepisorus fortunei(Microsorum fortunei)
瓦韦属 *Lepisorus*

形态特征：附生，植株高 30 ~ 100 cm。根状茎长而横走，顶部被棕褐色，卵状三角形鳞片，易脱落。叶远生；叶柄上面有浅沟，基部疏被鳞片，向上近光滑；叶片线状披针形至披针形，顶端长渐尖，基部渐狭，下延于叶柄成狭翅，全缘，边缘软骨质；中脉明显，侧脉不明显。孢子囊群分离，大，圆形，沿中脉两侧排成整齐或者不规则的 1 列。

分布：分布于马来西亚、不丹、缅甸、越南。我国长江流域及以南。南麂主岛（国姓岙、门屿尾、三盘尾），偶见。

生境：附生于林下溪边岩石上或树干上。

Description: Epiphytic, 30 -100 cm tall. Rhizomes long creeping, apically covered with brown, oval triangular scales. Fronds distant; petiole shallow furrow above, base sparsely coated scales, upward nearly smooth; lamina linear-lanceolate to lanceolate, apically acuminate, base tapering, decurring to petiole and forming narrow wings, entire, margin cartilaginous; midrib obvious, lateral veins often obscure. Sori separate, orbicular, in 1 (irregular) row parallel to costa.

Distribution: Malaysia, Bhutan, Myanmar, Vietnam. Yangtze River Valley and south. Main island (Guoxingao, Menyuwei, Sanpanwei), occasionally.

Habitat: Epilithic or epiphytic often beside streams in forests.

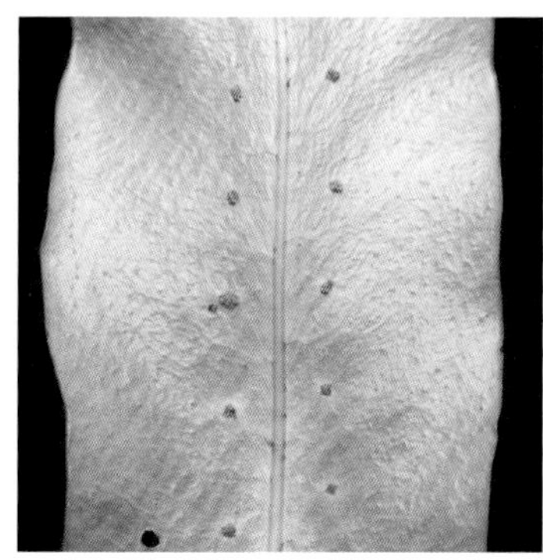

裸子植物门
GYMNOSPERMAE

G1. 苏铁科 Cycadaceae

苏铁
Cycas revoluta
苏铁属 *Cycas*

形态特征：茎干圆柱状，高可达 3 m，偶见 8 m，顶端密被厚绒毛。叶一回羽裂，羽片直或近镰刀状，革质，基部微扭曲，外侧下延，先端渐窄，具刺状尖头，下面疏被柔毛，边缘强烈反卷，横切面呈 V 字形。小孢子叶球卵状圆柱形，窄楔形，先端圆状截形，骤尖；大孢子叶密被灰黄色绒毛，不育顶片卵形或窄卵形。种子橘红色，倒卵状或长圆状，明显压扁。

花果期：花期 6 ~ 7 月，种子 9 ~ 10 月成熟。

分布：原产我国福建东部丘陵和海岛，现各地广为栽培。南麂主岛（大沙岙、三盘尾、镇政府），栽培。

生境：岛屿的山坡灌丛，大陆的疏林中。

Description: Stems terete, up to 3(8) m tall, apex tomentose. Leaves pinnate, pinnules straight or subsickle, leathery, slightly twisted at base, laterally decurtant, apex narrower, spiny apiculate, sparsely pubescent abaxially, margin strongly recurved, strongly "V"-shaped in cross section. Microsporophylls narrowly cuneate, oval cylindrical, cuspidate; megasporophylls densely grayish yellow tomentose, sterile blade ovate to narrowly. Seeds orange, obovoid or ellipsoid, compressed.

Flower and Fruit: Fl. Jun. -Jul., seed maturity in Sep. -Oct.

Distribution: Native in east Fujian, and some islands, usually cultivated as ornamental plants. Main island (Dashaao, Sanpanwei, Zhenzhengfu), cultivated.

Habitat: Thickets on hillsides on islands, sparse forests on mainland.

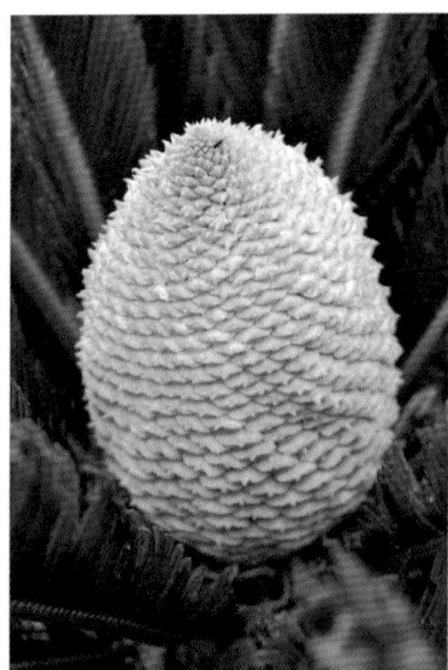

G7. 松科
Pinaceae

日本五针松（五针松）
Pinus parviflora
松属 ***Pinus***

形态特征：乔木，高达 25 m。幼树树皮淡灰色，光滑，大树树皮暗灰色，裂成鳞状块片。树冠圆锥形，一年生枝幼时绿色，后呈黄褐色，密生淡黄色柔毛；冬芽卵圆形，针叶 5 针一束，微弯曲，横切面三角形。球果卵圆形，无梗；种鳞宽倒卵状斜方形，鳞盾淡褐色，近菱形，鳞脐顶生，下凹，两侧边向外弯。种子为不规则倒卵圆形，近褐色，具黑色斑点，有翅。

花果期：花期 5 月，种子翌年 10 月成熟。

分布：原产日本。广泛种植于我国长江流域和山东，常用作园林树或盆景。南麂主岛（大沙岙、美龄居、镇政府），栽培。

生境：喜温，光照充足的山坡。

Description: Trees, up to 25 m tall. Bark pale gray, aging dull gray, smooth when young, furrowed with age into scaly plates. Crown conical, 1st-year branchlets initially green, aging yellow-brown, densely pale yellow pubescent. Winter buds ovoid, needles 5 per bundle, slightly curved, triangular in cross section. Seed cones ovoid, subsessile; seed scales obovate-rhombic, apophyses pale brown or dull gray-brown, almost rhombic, umbo terminal, sunken, margin recurved distally. Seeds irregularly obovoid, nearly brown, mottled with black, winged.

Flower and Fruit: Fl. May, seed maturity in Oct. of 2nd year.

Distribution: Native to Japan. Widely cultivated in Yangtze River Basin and Shandong, commonly used as a garden tree or for bonsai. Main island (Dashaao, Meilingju, Zhenzhengfu), cultivated.

Habitat: Temperate tree, sunny rocky slopes.

G7. 松科
Pinaceae

黑松
Pinus thunbergii
松属 *Pinus*

形态特征：常绿乔木，高达 30 m。幼树树皮暗灰色，老则灰黑色，树冠宽圆锥状或伞形，冬芽银白色。针叶 2 针一束，深绿色，有光泽，粗硬，边缘有细锯齿。雄球花淡红褐色，圆柱形，聚新枝下部；雌球花单生或 2 ~ 3 个聚新枝近顶端，直立，有梗，卵圆形，淡紫红色。球果成熟前绿色，熟时褐色，圆锥状卵圆形。种子倒卵状椭圆形，种翅灰褐色，有深色条纹。

花果期：花期 4 ~ 5 月，种子翌年 10 月成熟。

分布：原产日本、朝鲜。我国沿海地区城市、海滨及海岛有引种栽培。南麂各岛，常见。

生境：山坡砂石土中。

Description: Evergreen trees，up to 30 m tall. Young bark dark gray, old gray-black, crown broadly conical or umbrella-shaped, winter buds silvery white. Needles 2 per bundle, dark green, shiny, rough, the margins of a fine serrate. Male strobilus reddish brown, terete, aggregated at lower part of new branch; female strobilus solitary or 2-3 aggregated near apex of new branch, erect, pedicellate, ovoid, mauve or hazel. Cones green before maturity, brown at maturity, conical-ovoid or ovoid. Seeds obovate-elliptic, wings gray-brown, with dark stripes.

Flower and Fruit: Fl. Apr. -May, seed maturity in Oct. of 2nd year.

Distribution: Native to Japan, Korea. Cultivated in coastal areas, cities and islands in China. Throughout Nanji Islands, commonly.

Habitat: Sand and stone soil of the hillside.

G8. 南洋杉科 Araucariaceae

异叶南洋杉 *Araucaria cunninghamii*
南洋杉属 *Araucaria*

形态特征：乔木，树高 50 m，树干通直，树皮暗灰色，剥落。小枝平展或下垂，侧枝常成羽状排列，下垂。叶二型：幼树及侧生小枝的叶排列疏松，开展，钻形，光绿色，向上弯曲，通常两侧扁；大树及花果枝上的叶排列较密，微开展，基部宽，先端钝圆，中脉隆起或不明显。雄球花单生枝顶，雌球花近球形。种子椭圆形，稍扁，两侧具翅。

花果期：未见。

分布：原产澳大利亚。我国福建、广东、云南等地引种，长江以北常盆栽。南麂主岛（三盘尾、镇政府、兴岙），栽培。

生境：喜暖湿气候。

Description: Trees to 50 m tall, trunk straight, bark dark gray, flaking. Branchlets spreading horizontally or drooping, lateral branchlets usually pinnately arranged, drooping. Leaves dimorphic: those of young trees and lateral branchlets loosely arranged, spreading openly, subulate, bright green, usually laterally depressed, upcurved; those of mature trees and cone-bearing branchlets densely arranged, slightly openly spreading, widest at base, midvein obviously raised or not, apex obtuse. Pollen cones terminal, solitary, seed cones subglobose.Seeds ellipsoid, slightly flattened, with a lateral wing.

Flower and Fruit: Not seen.

Distribution: Native to Australia. Cultivated in Fujian, Guangdong and Yunnan, etc., potted in the north of the Yangtze River. Main island (Sanpanwei, Zhenzhengfu, Xingao), cultivated.

Habitat: Warm and humid climate.

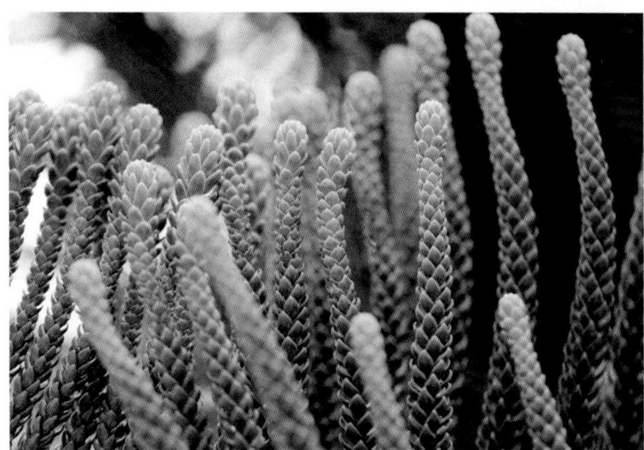

G11. 柏科
Cupressaceae

龙柏
Juniperus chinensis **'Kaizuca' (*Sabina chinensis* 'Kaizuca')**
刺柏属 *Juniperus*

形态特征：灌木或乔木，25 m，雌雄异株；树皮淡灰棕色，纵裂。树冠圆柱状，枝条向上直展，常有扭转上升之势；小枝密、在枝端成几相等长之密簇；鳞叶排列紧密，幼嫩时淡黄绿色，后呈翠绿色。球果蓝色，微被白粉，近球形，2～3种子。种子卵球形，稍扁平，脊状，先端钝。

花果期：花期4～5月，果期翌年10～11月。

分布：圆柏（*Juniperus chinensis*）的栽培品种，我国长江流域及华北各大城市庭园有栽培。南麂主岛（国姓岙、后隆、美龄居、三盘尾、镇政府），栽培。

生境：喜光树种，喜温凉、温暖气候及湿润土壤。

Description: Shrubs or trees, 25 m tall, dioecious; bark grayish brown, split. Crown cylindrical or pyramidal, branches curved and rotate up-spreading; branchlets density, clustered in apex of branchlets. Scalelike leaves closely arranged, light yellow-green when young, and emerald green when adult. Cones blue, usually glaucous, subglobose, 2-or 3-seeded. Seeds ovoid, slightly flat, ridged, apex blunt.

Flower and Fruit: Fl. Apr. -May, Fr. Oct. -Nov. of 2nd year.

Distribution: Cultiuar of *Juniperus chinensis*, cultivated in the Yangtze River Basin and North China. Main island (Guoxingao, Houlong, Meilingju, Sanpanwei, Zhenzhengfu), cultivated.

Habitat: Abundant light, warm and cool climate, and moist soil.

G11. 柏科
Cupressaceae

池杉
Taxodium distichum var. *imbricatum*
落羽杉属 *Taxodium*

形态特征：乔木，高达 25 m。树干基部膨大，通常有屈膝状的呼吸根。树皮褐色，纵裂，脱落。枝条向上伸展，树冠呈尖塔形。当年生小枝绿色，二年生小枝呈褐红色。叶钻形，在枝上螺旋状伸展，基部下延，向上渐窄，先端渐尖。球果圆球形，有短梗，向下斜垂，熟时褐黄色，种鳞木质，盾形。种子不规则三角形，微扁，红褐色，边缘有锐脊。

花果期：花期 3 ～ 4 月，球果 10 月成熟。

分布：原产北美洲东南部。我国中部地区有栽培，常用于造林。南麂主岛（美龄居、镇政府），栽培。

生境：沼泽、湿地。

Description: Trees, up to 25 m tall. Trunk swollen at base, bent breathing roots. Bark brown, longitudinal crack, flaking. Branches extend upward, crown steeple-shaped. Current branchlets green and biennial branchlets maroon. Leaves subulate, spirally spreading on branches, base decurbitate, tapering upward, apex acuminate. Cones globose or oblong-globose, short peduncle, downward sloping, brown-yellow when ripe; seed scales woody, peltate. Seeds irregularly triangular, slightly flat, reddish-brown, margin with acute ridges.

Flower and Fruit: Fl. Mar. -Apr. , seed maturity in Oct.

Distribution: Native to southeastern North America. Cultivated in central areas of China, usually affore station.Main island (Meilingju, Zhenzhengfu), cultivated.

Habitat: Marshy soils.

被子植物门

ANGIOSPERMAE

A11. 胡椒科
Piperaceae

山蒟
Piper hancei
胡椒属 *Piper*

形态特征：攀援藤本。茎、枝具细纵纹，节上生根。叶纸质近革质，卵状披针形，先端短尖，基部渐窄，叶脉5～7条，最上1对互生，离基1～3 cm，网脉明显。花单性，雌雄异株，穗状花序；雄花序长6～10 cm，总花梗与叶柄等长或略长，花序轴被毛，苞片近圆形，近无柄或具短柄，盾状，向轴面和柄上被柔毛；雌花序长约3 cm。核果球形，黄色。

花果期：花期3～8月。

分布：我国长江以南各地。南麂主岛各处常见。

生境：林中树上或石上。

Description:Climbers. Stems rooting at nodes, finely striated. Leaf blade papery to ± leathery, ovate-lanceolate or elliptic, apex acute or acuminate; base gradually tapered or cuneate, sometimes rounded; veins 5-7, apical pair arising 1-3 cm above base, alternate; reticulate veins usually conspicuous. Flowers unisexual, dioecism, spicate; male spikes 6-10 cm, peduncle as long as petioles or slightly longer, rachis pubescent, bracts suborbicular, ± sessile to shortly stalked, peltate, adaxially pilose; female spikes ca. 3 cm. Drupe globose, yellow.

Flower and Fruit: Fl. Mar. -Aug.

Distribution: To the south of the Yangtze River. Main island of Nanji, commonly.

Habitat: Forests, on trees or rocks.

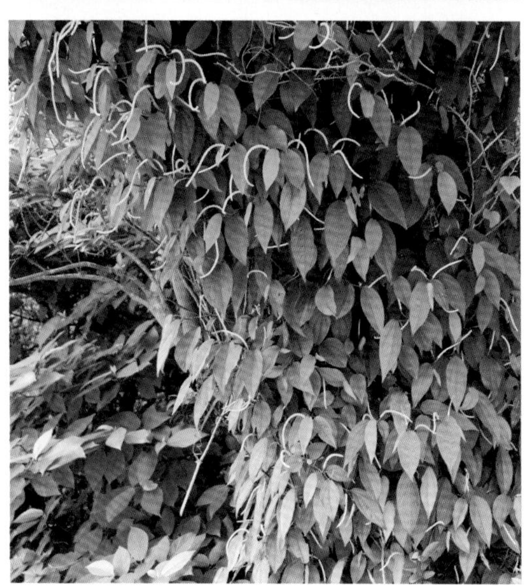

A11. 胡椒科
Piperaceae

风藤
Piper kadsura
胡椒属 *Piper*

形态特征：木质藤本。茎有纵棱，幼时被疏毛，节上生根。叶近革质，具均匀突起白色腺点，卵形或长卵形，基部心形至圆形；叶脉5条，基出或近基出，最外1对不甚显著。花单性，雌雄异株，穗状花序与叶对生；总花梗略短于叶柄，花序轴被微硬毛；苞片圆形，近无柄，盾状，边缘不整齐，腹面被白色柔毛。浆果球形，褐黄色。

花果期：花期3～8月。

分布：日本、朝鲜南部。我国华东沿海地区。南鹿主岛各处常见。

生境：低海拔林中，攀援于树上或石上。

Description: Climbers woody. Stems rooting at nodes, ridged, sparsely pubescent when young. Leaves ovate or long ovate, ± leathery, with uniformly scattered raised white glands base cordate to rounded; veins 5, emanating from or near the base; apical pair gracile, unapparent. Dioecism, spikes leaf-opposed base cordate to rounded; peduncle slightly shorter than petioles, rachis hispidulous; bracts orbicular, sessile, peltate, margin irregular, abaxially roughly white pubescent, sessile. Drupe brownish yellow, globose.

Flower and Fruit: Fl. Mar. -Aug.

Distribution: Japan, Korea. Coastal areas of East China.Main island of Nanji, commonly.

Habitat: Lowland forests, on trees or rocks.

A12. 马兜铃科
Aristolochiaceae

马兜铃
Aristolochia debilis
马兜铃属 *Aristolochia*

形态特征：草质藤本。茎圆柱形，光滑无毛，有腐肉味。叶纸质，卵状三角形、长圆状卵形或戟形，顶端钝圆或短渐尖，基部心形，两面无毛。花单生或 2 朵并生叶腋，花被筒基部球形，花药卵圆形，合蕊柱顶端 6 裂。蒴果近球形，具 6 棱。种子钝三角形，具白色膜质宽翅。

花果期：花期 7 ~ 8 月，果期 9 ~ 10 月。

分布：日本。我国黄河以南各地。南麂主岛（关帝岙、国姓岙、门屿尾、三盘尾），偶见。

生境：山谷、沟边、路旁阴湿处及山坡灌丛。

Description: Herbs twining. Stems terete, smooth, glabrous, with carrion smell. Leaf blade ovate or oblong-ovate to sagittate, apex acute or obtuse, base cordate, both surfaces glabrous. Flowers axillary, solitary or paired,perianth tubeglobose at base,anthers elliptic, gynostemium 6-lobed. Capsule subglobose, with 6 arris. Seeds obtusely deltoid, with white membranous wing.

Flower and Fruit: Fl. Jul. -Aug. , fr. Sep. -Oct.

Distribution: Japan. To the south of Yellow River. Main island (Guandiao, Guoxingao, Menyuwei, Sanpanwei), occasionally.

Habitat: Thickets, mountain slopes, moist valleys.

A25. 樟科
Lauraceae

樟（香樟）
Cinnamomum camphora
樟属 *Cinnamomum*

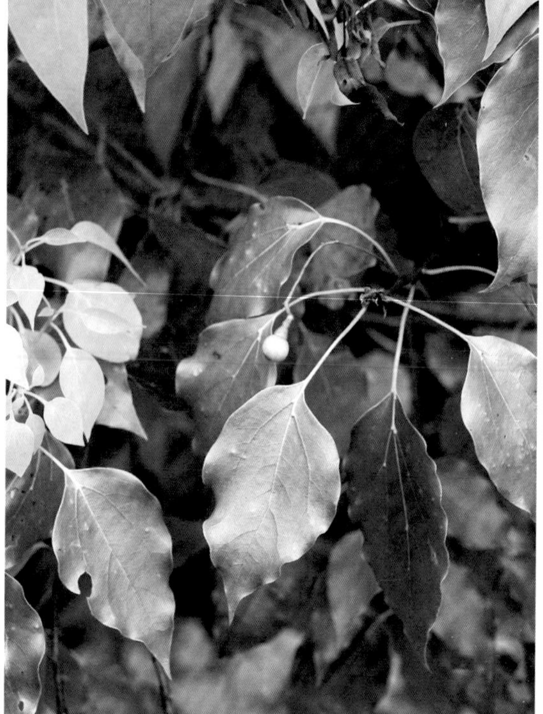

形态特征：常绿大乔木，高可达 30 m。树冠广卵形，整株有强烈的樟脑气味。树皮黄棕色，有不规则的纵裂。枝条圆柱形，淡褐色，无毛。叶互生，卵状椭圆形，边缘全缘，正面绿色，有光泽，背面黄绿色，晦暗。圆锥花序腋生，具梗；花绿白色，花被裂片椭圆形，花丝被短柔毛，子房球形，无毛。果卵球形，紫黑色。

花果期：花期 4 ~ 5 月，果期 8 ~ 11 月。

分布：日本、朝鲜、越南，世界各地广泛栽培。我国长江以南各地。南麂主岛偶见栽培。

生境：山坡或沟谷中。

Description: Evergreen trees, up to 30 m tall. Corona broadly ovate, strongly camphor-scented. Bark yellow-brown, irregularly and longitudinally fissured. Branchlets brownish, terete, glabrous. Leaves alternate, ovate-elliptic, entire, green or yellow-green and shiny adaxially, yellow-green or gray-green and glaucous abaxially. Panicle axillary with stalk; flowers green-white or yellowish, perianth lobes elliptic, filaments pubescent, ovary ovoid, glabrous. Fruit purple-black, ovoid or subglobose.

Flower and Fruit: Fl. Apr. -May, fr. Aug. -Nov.

Distribution: Japan, Korea, Vietnam, cultivated in many countries around the world. To the south of Yangtze River.Cultivated on main island of Nanji, occasionally.

Habitat: Slopes and valleys.

A25. 樟科
Lauraceae

天竺桂（浙江樟、普陀樟）
Cinnamomum japonicum（*Cinnamomum chekiangense*）
樟属 *Cinnamomum*

形态特征：常绿乔木，高 10 ~ 15 m。小枝圆柱形，红色或红褐色，无毛。叶卵状长圆形，革质，上面绿色，光亮，下面灰绿色，晦暗；叶柄带红褐色，无毛。圆锥花序腋生，无毛。花被片卵形，外面无毛，内面被柔毛；花丝被柔毛。果长圆形，果托浅波状，全缘或具圆齿。

花果期：花期 4 ~ 5 月，果期 7 ~ 9 月。

分布：东亚。我国华东地区。南麂主岛偶见栽培。

生境：低山或近海的常绿阔叶林中。

Description: Evergreen trees, 10-15 m. Branchlets terete, red or red-brown, glabrous. Leaves ovate-oblong or oblong-lanceolate, leathery, green and shiny adaxially, celadon and dark abaxially; petiole red-brown, glabrous. Panicle axillary, glabrous; perianth obconical, glabrous outside, villous inside; filaments villous. Fruit oblong, glabrous, perianth cup in fruit shallowly cupuliform, entire or shallowly dentate on margin.

Flower and Fruit: Fl. Apr. -May, fr. Jul. -Sep.

Distribution: East Asia. East China. Cultivated on main island of Nanji, occasionally.

Habitat: Evergreen broad-leaved forests on low hills, near seashores.

A25. 樟科
Lauraceae

红楠
Machilus thunbergii
润楠属 *Machilus*

形态特征：常绿乔木，高 10 ~ 15 m，偶见 20 m。树皮黄褐色。枝条多而伸展，紫褐色；老枝粗糙，嫩枝紫红色，小枝基部具环形芽鳞痕。叶倒卵形。花序顶生或在新枝上腋生，无毛；苞片卵形，被褐红色平伏绒毛；雄蕊花丝无毛。果扁球形，黑紫色，花序梗和果柄鲜红色。

花果期：花期 2 月，果期 6 ~ 8 月。

分布：日本、朝鲜。我国华东至华南地区。南麂各岛屿常见。

生境：山地或沟谷的阔叶林。

Description: Evergreen trees, usually 10-15(-20) m. Bark yellowish brown. Branchlets numerous and spread; older branchlets rough; young branchlets purple-brown when fresh with annular bud scale marks at base. Leaves obovate to obovate-lanceolate. Inflorescences terminal or axillary on new shoots, glabrous; bracts ovoid with maroon flat villi; filaments glabrous. Fruit compressed globose, dark purple; peduncle and fruiting pedicel reddish purple.

Flower and Fruit: Fl. Feb. , fr. Jun. -Aug.

Distribution: Japan and Korea. East China to South China.Throughout Nanji Islands, commonly.

Habitat: Mountain slopes or valleys, broad-leaved forests.

A28. 天南星科
Araceae

海芋
Alocasia odora
海芋属 *Alocasia*

形态特征：大型厚茎草本，高达2.5m，常绿，稍具白色乳汁。茎直立至斜升，基部具短匍匐茎。叶少数至多数簇生于枝顶；叶片盾形，心状箭形或心状卵形，基部边缘波状，先端短渐尖。花序2或3枚丛生叶基，佛焰苞基部绿色，卵圆形，檐部蕾时帽状，稍后反折，绿白色，阔长圆状披针形；肉穗花序短于佛焰苞，具短柄；附属物白色，狭圆锥形。浆果熟时猩红色，球形。

花果期：花期四季，但在密阴的林下常不开花。

分布：亚洲东南部。我国长江以南各地。南麂主岛（百亩山、美龄居），偶见。

生境：原始和次生热带雨林下，竹林灌丛，河岸，沼泽或石灰岩上。

Description: Pachycaul herbs, massive, to 2.5 m, evergreen, with slightly milky latex. Stem erect to decumbent, with short stolons arising from base. Leaves several to rather many together, clustered at tips of stems; leaf blade peltate, cordate-sagittate or cordate-ovate, basal margins undulate, apex shortly acuminate. Inflorescences 2 or 3 together among leaf bases, spathe proximal part green, ovoid, limb cowl-like at anthesis, later reflexed, greenish white, broadly oblong-lanceolate; spadix shorter than spathe, shortly stipitate; appendix white, narrowly conic. Berry ripening scarlet, globose.

Flower and Fruit: Fl. four seasons, not bloom under dense forests.

Distribution: Southeast Asia. To the south of Yangtze River.Main island (Baimushan, Meilingju), occasionally.

Habitat: Primary and secondary tropical rain forests, bamboo thickets, riverbanks, swamps, also on limestone.

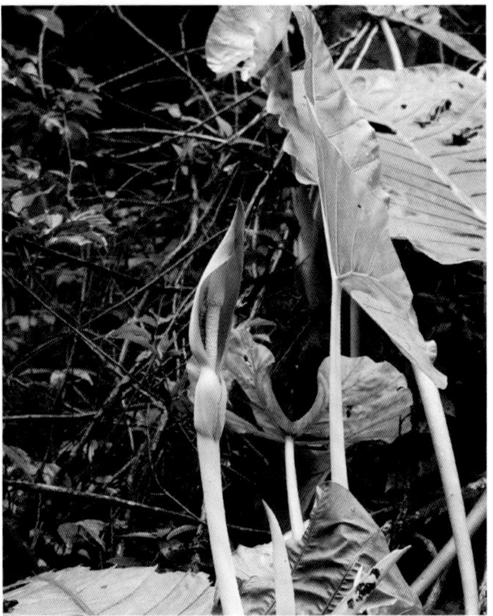

A28. 天南星科
Araceae

东亚魔芋（华东魔芋、疏毛磨芋）
***Amorphophallus kiusianus*（*Amorphophallus sinensis*）**
魔芋属 ***Amorphophallus***

形态特征：多年生草本。块茎扁球形，直径可达 20 cm。叶单生，小裂片长椭圆形。花序梗绿色，粗壮，光滑；子房近球形，淡绿，2 室，柱头头状，紫色；雄花小，黄色，无花丝，2 室。佛焰苞下部席卷，上部钟状展开，肉穗花序无梗。浆果红色，变蓝。

花果期：花期 5 ~ 6 月，果期 7 ~ 8 月。

分布：日本南部。我国华东至华南地区。南麂主岛（百亩山、门屿尾、三盘尾）、大檑山屿，偶见。

生境：荫蔽、半阴或全光地区，农田、次生林、竹阔混交林，果园等。

Description: Herbs perennial. Tuber depressed globose, to ca. 20 cm in diam. Leaf solitary, lobules long elliptic or lanceolate. Peduncle green, thick, smooth; ovary obovoid, pale green or yellow, 2-loculed; stigma head shape, purple; male flowers small, yellow, without filaments, 2-loculed. Spathe sweep lower, upper bell-shaped spread, spadix without stalk. Berries red to blue.

Flower and Fruit: Fl. May-Jun., fr. Jul.-Aug.

Distribution: South Japan. East China to South China. Main island (Baimushan, Menyuwei, Sanpanwei), Daleishan Island, occasionally.

Habitat: Shaded, semishaded, or sun-exposed places, plantations, secondary forests, mixed bamboo and broad-leaved forests, orchards.

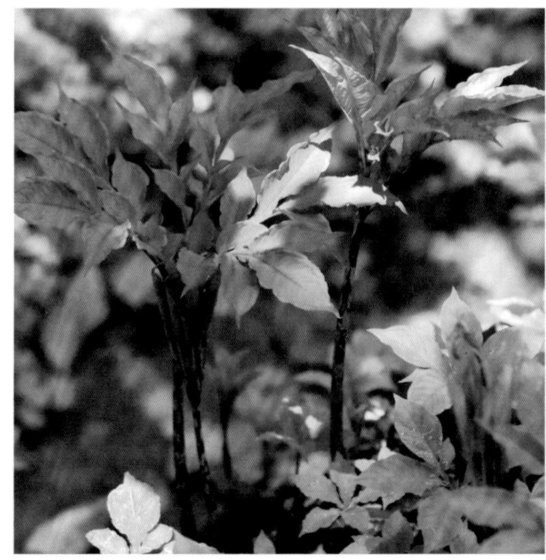

A28. 天南星科
Araceae

天南星
Arisaema heterophyllum
天南星属 *Arisaema*

形态特征：多年生草本。块茎扁球形。叶常单生，小叶 11 ~ 19，倒披针形，或线状长圆形，基部楔形，先端渐尖。佛焰苞管部外部有白霜，内灰绿色，圆筒状，喉部稍外卷，肉穗花序两性或雄性。浆果黄红或红色，圆柱形。种子 1 颗，黄色，具红色斑点。

花果期：花期 4 ~ 5 月，果期 7 ~ 9 月。

分布：日本、朝鲜。除西藏外几遍我国。南麂主岛各处、大檑山屿，常见。

生境：林下、灌丛或草地。

Description: Herbs perennial.Tuber depressed globose. Leaf usually solitary, leaflets 11-19, oblanceolate, or linear-oblong, base cuneate, apex acuminate. Spathe tube glaucous outside, whitish green inside, cylindric, throat slightly recurved, spadix bisexual or male. Berries yellowish red or red, cylindric. Seed 1, yellow, with red spots.

Flower and Fruit: Fl. Apr. -May, fr. Jul. -Sep.

Distribution: Japan, Korea. Almost throughout China, except Xizang. Main island and Daleishan Island, commonly.

Habitat: Forests, thickets, grasslands.

A28. 天南星科
Araceae

芋
Colocasia esculenta
芋属 *Colocasia*

形态特征：常绿多年生草本，常生多数小球茎。叶 2～3 枚或更多，叶柄绿色，叶片卵状，先端短尖，侧脉 4 对；花序柄常单生，佛焰苞长短不一，管部绿色，长卵形，檐部展开成舟状，边缘内卷。肉穗花序，雌花序圆锥状，中性花序细圆柱状，雄花序圆柱形，顶端骤狭，附属器钻形。

花果期：花期夏秋季。

分布：原产我国，热带和亚热带地区广泛栽培，时有逸生或归化。南麂主岛常见栽培。

生境：水田或森林、山谷、沼泽、荒地及水边的潮湿处。

Description: Evergreen perennial herbs, usually numerous corms. Leaves 2-3 or more, petiole green, leaf blade ovate, apex mucronate or short acuminate, lateral veins 4 pairs; peduncle usually solitary, length of the spathe different, tube green, long ovate, brim expands into a boat shape, involute edge. Spadix, female inflorescence paniculate, neutral inflorescence terete, male inflorescence cylindrical, narrow top, subulate appendix.

Flower and Fruit: Fl. summer-autumn.

Distribution: Native to China, widely cultivated in tropics and subtropics, usually wild or naturalized.Main island of Nanji, usually cultivated.

Habitat: Water fields, or wet places in forests, valleys, swamps, wastelands, and at watersides.

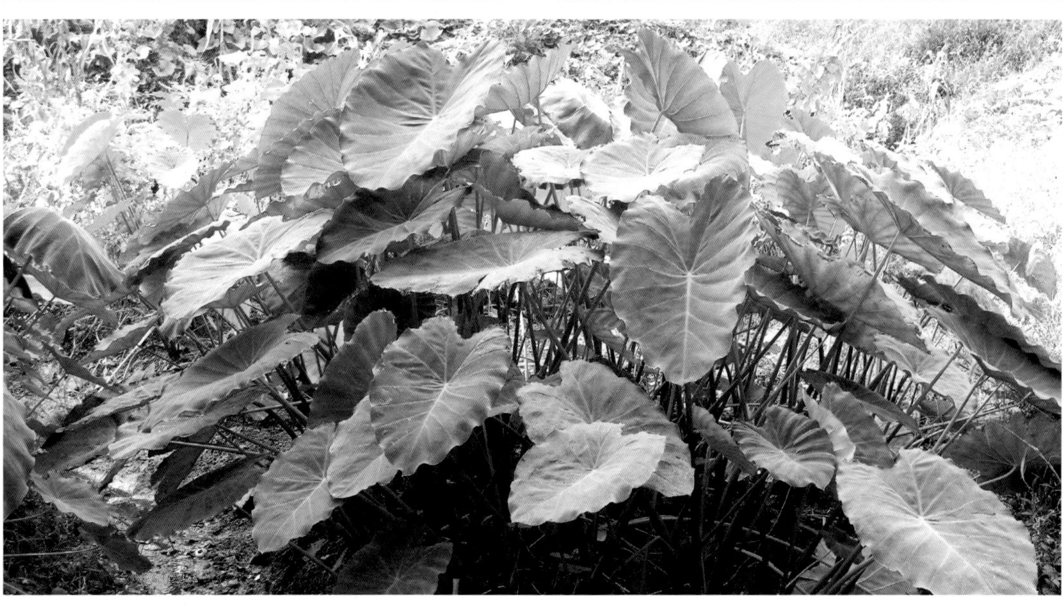

A28. 天南星科
Araceae

半夏
Pinellia ternata
半夏属 *Pinellia*

形态特征：多年生草本。块茎圆球形，具须根。叶2～5枚，叶柄基部具鞘，鞘内、叶柄中上部或叶片基部具珠芽；小叶3，长圆状椭圆形或披针形。佛焰苞绿色或灰绿色，微收缩，管部狭圆柱形，冠檐绿色，常具紫边；肉穗花序，附属器直立或“S”形，绿色至紫色。浆果黄绿色，卵圆形，花柱宿存。

花果期：花期5～7月，果期7～9月。

分布：日本、朝鲜。除内蒙古、青海、新疆和西藏外广布全国。南麂主岛（百亩山、门屿尾、三盘尾），偶见。

生境：草坡、次生林、荒地或耕地。

Description: Herbs perennial. Tubers globose, fibrous roots. Leaves 2-5, petiole base sheathing; bulbils present in sheath, at proximal or middle part of petiole, and at base of leaf blade; leaf blade 3-foliolate, oblong-elliptic or lanceolate. Spathe greenish or whitish green, slightly constricted, tube narrowly cylindric, limb green and usually violet at margin; spadix, appendix erect or sigmoid, green to violet. Berries yellowish green, ovoid, with persistent stigma and style.

Flower and Fruit: Fl. May-Jul. , fr. Jul. -Sep.

Distribution: Japan, Korea. Widely distributed in China, excluding Nei Mongol, Qinghai, Xinjiang, and Xizang. Main island (Baimushan, Menyuwei, Sanpanwei), occasionally.

Habitat: Grasslands, secondary forests, wastelands, cultivated lands.

A59. 菝葜科
Smilacaceae

圆锥菝葜
Smilax bracteata
菝葜属 *Smilax*

形态特征：攀援藤本。茎分枝，长可达 10 m，有时疏生刺，木质。叶纸质，叶片宽椭圆形到卵形椭圆形。圆锥花序通常具 3 ～ 6 个伞形花序，偶见 10 个，基部稍加厚，近球形；雄花花被片橄榄绿到暗红色，雌花花被片比雄花的小。浆果球形。

花果期：花期 11 月至翌年 2 月，果期 6 ～ 8 月。

分布：日本、菲律宾、越南和泰国。我国华东、华南至西南地区。南麂主岛（门屿尾、三盘尾），偶见。

生境：林地、灌丛或山坡荫蔽处。

Description: Vines climbing. Stem branched, woody, sometimes sparsely prickly, to 10 m. Leaf blade broadly elliptic to ovate-elliptic. Inflorescence a raceme of 3-6(-10) umbels, base slightly thickened, globose; male flowers tepals olive green to dark red, female flowers tepals smaller than male ones. Berries globose.

Flower and Fruit: Fl. Nov. -Feb. , fr. Jun. -Aug.

Distribution: Japan, Philippines, Vietnam and Thailand. East China, South China to South east China. Main island (Menyuwei, Sanpanwei), occasionally.

Habitat: Forests, thickets, shaded places on grassy slopes.

A59. 菝葜科
Smilacaceae

16. 菝葜
Smilax china
菝葜属 *Smilax*

形态特征：攀援灌木。茎圆柱状，多分枝，木质，疏生刺。叶柄长 0.5 ~ 1.5cm，占全长的 1/2 ~ 2/3 具翅，卷须常宿存。叶薄革质或坚纸质，椭圆形至圆形。伞形花单个腋生于幼叶，有花 10 ~ 25 朵，近球形，小苞片多而小；雄花黄绿色，雌花具 6 枚退化雄蕊。浆果熟时红色，球形，有粉霜。

花果期：花期 2 ~ 5 月，果期 9 ~ 11 月。

分布：东南亚。广布我国长江流域及其以南地区。南麂各岛屿常见。

生境：林下、灌丛、山坡或河谷溪边遮阴处。

Description: Vines climbing. Stem branched, terete, woody, sparsely prickly. Petiole 0.5-1.5 cm, narrowly winged for 1/2-2/3 its length, tendrils usually present. Leaf thinly leathery or hard papery, blade elliptic to orbicular. Umbel borne in axil of young leaf, of both sexes 10-25-flowered, subglobose, bracteoles many, small; male flowers tepals yellowish green, female flowers staminodes 6. Berries red, globose, minutely white powdery.

Flower and Fruit: Fl. Feb. -May , fr. Sep. -Nov.

Distribution: Southeast Asia. The Yangtze River Basin and to the south of it. Throughout Nanji Islands, commonly.

Habitat: Forests, thickets, hillsides, grassy slopes, shaded places along valleys or streams.

A59. 菝葜科
Smilacaceae

17. 小果菝葜
Smilax davidiana
菝葜属 *Smilax*

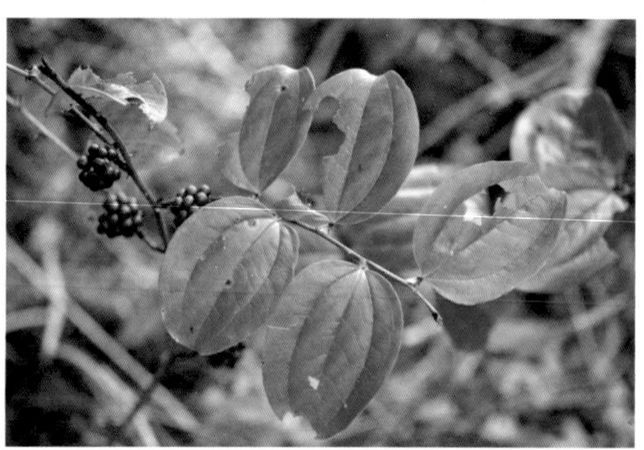

形态特征：攀援灌木。茎多分枝，圆柱形，稍木质，具疏刺。叶柄较短，5～7 mm，叶柄的 1/2～2/3 具翅，卷须很短。叶坚纸质，常椭圆形。伞形花序单个腋生于新枝幼叶，具花 3～13 朵，基部加厚，有时稍延长，具宿存小苞片；雄花黄绿色，雌花具 3 枚退化雄蕊。浆果球形，熟时暗红色。

花果期：花期 3～4 月，果期 10～11 月。

分布：日本。我国华东地区。南麂主岛各处常见。

生境：林下、灌丛，路旁阴湿处。

Description: Vines climbing. Stem branched, terete, slightly woody, sparsely prickly. Petiole usually 5-7 mm, winged for 1/2-2/3 its length; tendrils rather short. Leaf hard papery, blade usually elliptic. Umbel borne in axil of young leaf on new branchlets, of both sexes 3-13-flowered, base thickened, sometimes slightly elongate; bracteoles persistent; male flowers tepals yellowish green, female flowers staminodes 3. Berries globose, dark red when ripe.

Flower and Fruit: Fl. Mar. -Apr. , fr. Oct. -Nov.

Distribution: Japan. East China. Main island of Nanji, commonly.

Habitat: Forests, thickets, in wet places along the road.

A61. 兰科
Orchidaceae

叉唇角盘兰
Herminium lanceum
角盘兰属 ***Herminium***

形态特征：地生草本，块茎肉质，圆球形。茎无毛，直立而细长。叶互生，线状披针形，直立而伸展，先端急尖，基部抱茎。花序总状，花多数密生，苞片小，披针形，先端急尖，花小，黄绿色，花瓣直立线形，与中萼片相靠，唇瓣常下垂，基部扩大，凹陷，无距，在中部或中部以上呈叉状 3 裂，蕊喙小；柱头隆起。

花果期：花期 6 ~ 8 月。

分布：亚洲东部至南部。分布我国南北大部分地区。南麂主岛（百亩山），罕见。

生境：混交林、针叶林、竹林，或灌丛、草地。

Description: Terrestrial herbs, fleshy tubers, spherical or oval. Stem glabrous, erect and slender. Leaves alternate, linear-lanceolate, erect and extended, apex acute or acuminate, base amplexicaul. Inflorescences racemose, flowers mostly densely distributed, bracts small, lanceolate, apex sharp, flowers small, yellow-green or green, petals upright and linear, close to middle sepals, lips often drooping, bases enlarged, sunken and distanced, forked and 3-lobed in the middle or above, beak small; stigma bulge.

Flower and Fruit: Fl. Jun. -Aug.

Distribution: Eastern to southern Asia. Most areas in north and south China. Main island (Baimushan), rarely.

Habitat: Mixed forests, coniferous forests, bamboo forests, thickets, grasslands.

A72. 阿福花科
Asphodelaceae

山菅
Dianella ensifolia
山菅属 *Dianella*

形态特征：多年生草本，株高 1 ~ 2 m，具横走根状茎。叶狭条状披针形，基部收缩成鞘状，边缘及背部中脉有锯齿。顶端圆锥花序，花多生于侧枝上端。花梗稍弯曲，苞片小；花被片条状披针形，绿白色、淡黄色至青紫色，5 脉；花药条形，较花丝略长或近等长，花丝上部膨大。浆果近球形，深蓝色，具 5 ~ 6 颗种子。

花果期：花果期 3 ~ 8 月。

分布：热带亚洲、非洲和太平洋岛屿。我国华东、华南和西南等地。南麂各岛屿常见。

生境：森林、草坡。

Description: Herbs perennial,transverse rhizomes, 1-2 m tall. Leaf blades narrow, lanceolate, base shrinks into a sheath, edges and back midvein serrated. Apical panicle, flowers mostly born at the upper end of lateral branches. Pedicel slightly curved, bracts small; tepals strip-lanceolate, green-white, pale yellow to purplish blue, 5-veined; anthers strip-shaped, slightly longer or nearly as long as filaments, upper part of filaments enlarged. Berry nearly spherical, dark blue, 5-6 seeds.

Flower and Fruit: Fl. and fr. Mar. -Aug.

Distribution: Tropical Asia, Africa and Pacific islands. East China, South China and Southwest China. Throughout Nanji Islands, commonly.

Habitat: Forests, grassy slopes.

A72. 阿福花科
Asphodelaceae

萱草
Hemerocallis fulva
萱草属 *Hemerocallis*

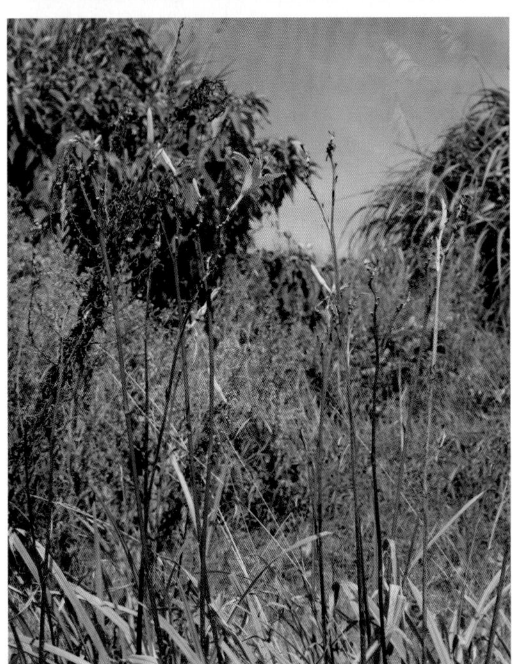

形态特征：多年生草本，具坚硬根状茎和肉质纤维根。叶条状披针形，基生成丛状，背面被白粉。花葶直立，中空，具不育苞片；螺旋状聚伞花序双生，2～5 花，无香味，晨开夜合，花被片单一，花被管较短，橙色至橙红色，裂片开展，有紫色或橙红色斑纹，边缘波状；花药紫黑色。蒴果椭圆形。

花果期：花期 6～11 月。

分布：朝鲜。我国大部分地区有分布。南麂主岛（关帝岙、三盘尾），栽培。

生境：林地、灌丛、草地或溪边。

Description: Herbs perennial, hard rhizomes, fleshy fibrous roots.Leaf blades strip-lanceolate, bases clumped, back covered with white powder. Helicoidal cymes double, 2-5-flowered; flowers unscented, strictly day opening, opening in morning and closing in evening of same day, perianth single, tube rather short, orange to reddish orange; segments spreading, with a purple or reddish orange patch, margin crinkly-undulate, anthers purplish black.Capsule ellipsoid.

Flower and Fruit: Fl. Jun. -Nov.

Distribution: Korea. Almost throughout China. Main island (Guandiao, Sanpanwei), cultivated.

Habitat: Forests, thickets, grasslands, streamsides.

A73. 石蒜科 Amaryllidaceae

薤白（小根蒜）

Allium macrostemon

葱属 *Allium*

形态特征：多年生草本，鳞茎单生，近球形，通常具珠芽在基部；鳞茎皮微黑，纸质或膜质，全缘。叶短于花茎，半圆柱形或具3角半圆柱形，背面具硬1棱，正面具沟。花茎圆柱状；佛焰苞2瓣裂，宿存；伞形花序半球形或球形，花朵繁密；花梗近等长，具小苞片；花被浅紫色或浅红。

花果期：花果期5～7月。

分布：东亚。我国大部分地区有分布。南麂主岛（关帝岙、三盘尾），常见。

生境：小山，山坡，山谷，平原。

Description: Herbs perennial,bulb solitary, subglobose, usually bulbels at base; tunic blackish, papery or membranous, entire. Leaves shorter than scape, semiterete or 3-angled-semiterete, abaxially strongly 1-angled, adaxially channeled. Scape terete; spathe 2-valved, persistent; umbel hemispheric to globose, densely many flowered; pedicels subequal, bracteolate; perianth pale purple to pale red.

Flower and Fruit: Fl. and fr. May -Jul.

Distribution: East Asia. Almost throughout China. Main island (Guandiao, Sanpanwei), commonly.

Habitat: Hills, slopes, valleys, plains.

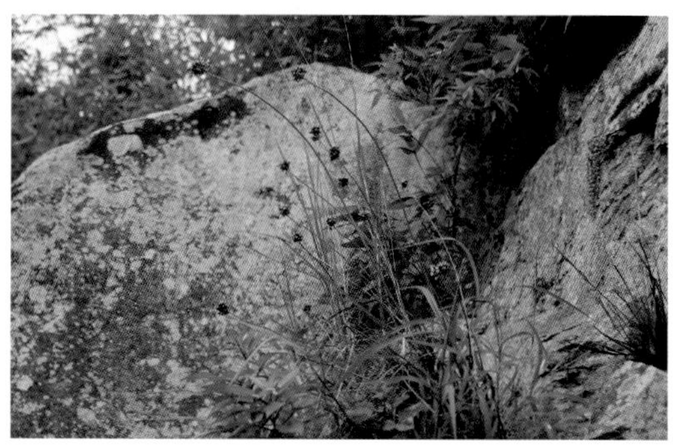

A73. 石蒜科
Amaryllidaceae

石蒜
Lycoris radiata
石蒜属 *Lycoris*

形态特征：多年生草本，鳞茎近球形。秋季出叶，深绿色，狭带状，中间有粉绿色带，顶端钝。伞形花序有花 4～7 朵，总苞片 2 枚，披针形；花被片鲜红色；管部绿色，花被裂片狭倒披针形，强度皱缩和反卷，雄蕊显著伸出于花被外。

花果期：花期 8～9 月，果期 10 月。

分布：日本、朝鲜、尼泊尔。我国黄河流域及其以南地区。南麂主岛（关帝岙、美龄居、门屿尾、三盘尾），常见。

生境：阴湿坡地，溪岸石质地。

Description: Herbs perennial, bulbs subglobose. Leaves appearing in autumn, dark green, narrowly ligulate,midvein pale, apex obtuse.Umbel 4-7-flowered; involucres 2, lanceolate; perianth bright red; tube green, lobes strongly recurved, narrowly oblanceolate, margin strongly undulate, stamens conspicuously exserted.

Flower and Fruit: Fl. Aug. -Sep. , fr. Oct.

Distribution: Japan, Korea, Nepal. The Yellow River Basin and to the south of it. Main island (Guandiao, Meilingju, Menyuwei, Sanpanwei), commonly.

Habitat: Shady and moist places on slopes, rocky places along stream banks.

A73. 石蒜科
Amaryllidaceae

换锦花
Lycoris sprengeri
石蒜属 *Lycoris*

形态特征：多年生草本，鳞茎卵球形。早春出叶，绿色，带状，先端钝。伞形花序有花 4 ~ 6 朵；总苞片 2 枚。花被片浅紫色，裂片通常在先端呈蓝色，倒披针形，边缘无波浪状；雄蕊近等长花被；花柱稍长于花被。蒴果具 3 角，室背开裂。种子黑色，近球形。

花果期：花期 8 ~ 9 月。

分布：我国华东和湖北地区。南麂主岛（关帝岙、美龄居、门屿尾、三盘尾），常见。

生境：疏林，竹林，山坡阴湿地。

Description: Herbs perennial, bulbs ovoid. Leaves appearing in early spring, green, ligulate, apex obtuse. Umbel 4-6-flowered; involucres 2. Perianth pale purple; lobes often bluish at apex, oblanceolate, margin not undulate; stamens nearly as long as perianth; style slightly longer than perianth. Capsule 3-angled, loculicidal. Seeds black, subglobose.

Flower and Fruit: Fl. Aug. -Sep.

Distribution: East China, Hubei. Main island (Guandiao, Meilingju, Menyuwei, Sanpanwei), commonly.

Habitat: Sparse forests, bamboo forests, shady and wet places on slopes.

A73. 石蒜科 Amaryllidaceae

水仙
Narcissus tazetta var. *chinensis*
水仙属 *Narcissus*

形态特征： 多年生草本，鳞茎卵球形。叶宽线形，平展，边缘全缘，先端钝。通常的花茎等长叶。伞形花序4～8花，总苞膜质；花芳香，花梗长短不一，花被管细，灰绿色，近三棱形，花被裂片6，裂片宽平展，白色，宽椭圆形到卵形，先端短锐尖；副花冠浅黄色，浅杯状，边缘不皱缩。蒴果室背开裂。

花果期： 花期2～3月。

分布： 原产亚洲东部海滨温暖地区，我国广泛栽培作观赏。我国福建东南部，浙江东部的沿海地区和近海岛屿有分布。大檑山屿、小檑山屿、竹屿。

生境： 沙地、荒地。

Description: Herbs perennial, bulbs ovoid. Leaves broadly linear, flat, margin entire, apex obtuse. Flowering stems usually equaling leaves. Umbels 4-8-flowered, involucres membranous; flowers fragrant, pedicels unequal, perianth tube glaucous, slender, nearly 3-angled; lobes 6, widely spreading, white, broadly elliptic to ovate, apex shortly acute; corona pale yellow, shallowly cupular, margin not undulate. Capsule loculicidal.

Flower and Fruit: Fl. Feb. -Mar.

Distribution: Native to warm coastal areas of East Asia, widely cultivated as an ornamental in China. Coastal areas and offshore islands of southeast Fujian, east Zhejiang.Daleishan Island, Xiaoleishan Island, Bamboo Island.

Habitat: Sandy places, wastelands.

A73. 石蒜科
Amaryllidaceae

葱莲（葱兰）
Zephyranthes candida
葱莲属 *Zephyranthes*

形态特征：多年生草本，鳞茎卵球形，具明显的颈部。叶亮绿色，圆柱状线形，肉质。花单生于花茎顶端，总苞红褐色；花被片白色，背面常带淡红色，裂片多少离生，近喉部常有很小的鳞片，先端圆钝至短尖；雄蕊 6，长约为花被的 1/2；花柱细长，柱头不明显 3 裂。蒴果近球形，3 瓣裂，室背开裂。

花果期：花期秋季。

分布：原产南美洲，广泛栽培作观赏。我国南方地区有归化。南麂主岛（美龄居、三盘尾、镇政府），常栽培。

生境：山坡、路旁。

Description: Herbs perennial, bulbs ovoid, neck distinct. Leaves bright green, terete-linear, fleshy. Flowers solitary, terminal, involucres red-brown; perianth white, often tinged with rose abaxially; lobes 6, ± free, usually with tiny scales near throat, apex obtuse to shortly acute; stamens 6, ca. 1/2 as long as perianth, style slender, stigma 3-notched. Capsule subglobose,3-valved, loculicidal.

Flower and Fruit: Fl. autumn.

Distribution: Native to South America, widely cultivated as an ornamental. Naturalized in south China. Main island (Meilingju, Sanpanwei, Zhenzhengfu), usually cultivated.

Habitat: Hillside, roadsides.

A74. 天门冬科
Asparagaceae

天门冬
Asparagus cochinchinensis
天门冬属 *Asparagus*

形态特征：多年生草本，雌雄异株。根膨大呈纺锤形。茎攀援，1 ～ 2 m，基部稍木质；分枝有棱或狭翅。叶状枝通常每 3 枚成簇，镰刀状，扁平或稍具三棱。茎上的鳞片状叶基部有时延伸硬刺，在分枝上的刺较短或不明显。雌雄花常成对，腋生于叶状枝上，近相等；花梗中部有关节。雄花花被片淡绿色，钟状。浆果绿色，有 1 ～ 2 颗种子。

花果期：花期 5 ～ 6 月，果期 9 月。

分布：日本、朝鲜、老挝、越南。分布于我国南北各地。南麂各岛屿常见。

生境：稀疏树林覆盖的山坡，路旁，废弃田野。

Description: Herbs perennial,dioecious. Roots with swollen, tuberous.Stems climbing, 1-2 m, slightly woody proximally; branches angled or narrowly winged. Cladodes usually in fascicles of 3, subfalcate, flat or slightly 3-angled. Leaf spur sometimes spinescent; spine minute or indistinct on branches. Inflorescences developing after cladodes, axillary; flowers of both sexes usually paired, subequal; pedicel articulate at middle; male flowers perianth greenish, campanulate. Berry green, 1-or 2-seeded.

Flower and Fruit: Fl. May -Jun. , fr. Sep.

Distribution: Japan, Korea, Laos, Vietnam. Throughout the north and south areas of China. Throughout Nanji Islands, commonly.

Habitat: Thinly forested slopes, roadsides, waste fields.

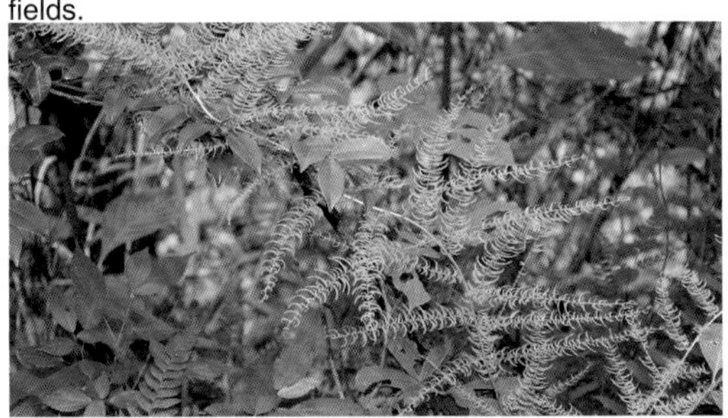

A74. 天门冬科
Asparagaceae

绵枣儿
Barnardia japonica（*Scilla scilloides*）
绵枣儿属 *Barnardia*

形态特征：多年生草本，鳞茎卵形或近球形，鳞茎皮黑褐色。叶基生，2～5枚，狭带状，柔软。总状花序具花多数，花葶长于叶；花被片紫红、粉红至白色，椭圆形或倒卵形，基部稍合生成盘状蒴果；近倒卵形。种子黑色。

花果期：花果期7～11月。

分布：日本、朝鲜、俄罗斯。分布于我国南北各地。南麂主岛（三盘尾），偶见。

生境：林缘、山腰，开阔的山坡、草地。

Description: Herbs perennial, bulb oval or subglobose, bulb skin dark brown. Leaves basal, 2-5, narrowly banded, soft. Racemes with many flowers, scape longer than leaves, perianth purplish red, pink to white, elliptic or obovate, slightly connate and discoid at base. Capsule subobovoid. Seeds black.

Flower and Fruit: Fl. and fr. Jul. -Nov.

Distribution: Japan, Korea and Russia. Throughout the north and south areas of China. Main island (Sanpanwei), occasionally.

Habitat: Forest margins, hillsides, open slopes, grasslands.

A74. 天门冬科
Asparagaceae

阔叶山麦冬
Liriope muscari (Liriope platyphylla)
山麦冬属 *Liriope*

形态特征：多年生草本，根近顶端有时膨大呈肉质块状，无匍匐茎。叶线形或更狭窄，硬。花序 8 ~ 45 cm，多花；苞片刚毛状；小苞片卵形；花 4 ~ 8 朵一簇，花梗近中部有节，花被片紫色或者紫丁香紫色，椭圆状长圆形。种子成熟时微黑紫色，球状。

花果期：花期 7 ~ 8 月，果期 9 ~ 10 月。

分布：日本。我国长江流域以南地区有分布。南麂主岛各处常见。

生境：森林，竹林，灌丛，在峡谷和山坡上里的在阴处和潮湿的地方。

Description: Herbs perennial, roots sometimes with fleshy, tuberous part near tip. Stolons absent. Leaves linear to narrowly so, stiff. Inflorescence 8-45 cm, many flowered; bracts setiform; bracteoles ovate; flowers in clusters of 4-8, pedicel articulate near middle, tepals purple or lilac-purple, elliptic-oblong. Seeds blackish purple at maturity, globose.

Flower and Fruit: Fl. Jul. -Aug. , fr. Sep. -Oct.

Distribution: Japan. To the south of Yangtze River Basin.Main island of Nanji, commonly.

Habitat: Forests, bamboo forests, scrub, shady and moist places in ravines and on slopes.

A74. 天门冬科
Asparagaceae

山麦冬
Liriope spicata
山麦冬属 *Liriope*

形态特征：多年生草本，根近顶端常膨大成纺锤形，肉质块根，匍匐茎纤细。叶片背面有白粉，狭线形，背面有5条明显的叶脉，基部被多条带褐色的鞘包围，边缘有细锯齿。花序多花，苞片披针形；花（2或）3～5簇，花梗上部有节，花被片略带紫色或者带蓝色，近长圆形。种子近球形。

花果期：花期5～7月，果期8～10月。

分布：日本、朝鲜、越南。我国大部分区域均有分布。南麂主岛各处常见。

生境：森林，草坡，山腰，潮湿的地方。

Description: Herbs perennial,roots usually, fusiform, fleshy, tuberous part near tip. Stolons creeping, slender. Leaves glaucous abaxially, narrowly linear, distinctly veins 5, abaxially, base surrounded by many brownish sheaths, margin serrulate. Inflorescence many flowered, bracts lanceolate; flowers in clusters of (2 or)3-5, pedicel articulate distally, tepals purplish or bluish, suboblong.Seeds subglobose.

Flower and Fruit: Fl. May-Jul. , fr. Aug.-Oct.

Distribution: Japan, Korea, Vietnam. Almost throughout China. Main island of Nanji, commonly.

Habitat: Forests, grassy slopes, hillsides, moist places.

A74. 天门冬科
Asparagaceae

麦冬（沿阶草）
Ophiopogon japonicus
沿阶草属 *Ophiopogon*

形态特征：植株有匍匐茎。根厚度适中，通常中部或端部具块茎状部分。叶基生，簇生，无梗，呈草状，有3～7条脉，边缘有细锯齿。花茎远短于叶，花序呈退化的圆锥花序，超过10朵花；苞片披针形；花单生或成对，通常有节；花梗近中部有节，花被片白色的或带紫色，披针形。种子球状。

花果期：花期5～8月，果期8～9月。

分布：日本、朝鲜。我国大部分区域均有分布。南麂主岛各处常见。

生境：山坡、沿着溪流、悬崖的森林、茂密的灌木。

Description: Plants stoloniferous. Roots moderately thick, usually with tuberous part near middle or tip. Leaves basal, tufted, sessile, grasslike,3-7-veined, margin serrulate. Scape much shorter than leaves, Inflo rescence a reduced panicle, several to more than 10 flowers; bracts lanceolate; flowers solitary or paired, usually nodding, pedicel articulate near middle, tepals white or purplish, lanceolate. Seeds globose.

Flower and Fruit: Fl. May-Aug. , fr. Aug. -Sep.

Distribution: Japan, Korea. Almost throughout China. Main island of Nanji, commonly.

Habitat: Forests, dense scrub in ravines, moist and shady places on slopes and along streams, cliffs.

A74. 天门冬科
Asparagaceae

凤尾丝兰（凤尾兰）
Yucca gloriosa
丝兰属 *Yucca*

形态特征：常绿灌木。茎高达 5 m，常分枝。叶线状披针形，先端长渐尖，坚硬刺状，全缘，稀具分离的纤维。圆锥花序高 1 ~ 1.5 m，常无毛；花下垂，白或淡黄白色，顶端常带紫红，花被片 6，卵状菱形，柱头 3 裂。果倒卵状长圆形，不裂。

花果期：花期春秋季。

分布：原产北美洲东部及东南部。我国大部分区域均有引种栽培。南麂主岛（大沙岙、后隆、三盘尾）、大檑山屿，栽培或野生。

生境：温暖湿润和阳光充足环境。

Description: Evergreen shrubs. Stem 5 m tall, branched. Leaves linear lanceolate, apex long acuminate, hard spiny, entire, sparsely with isolated fibers. Panicle 1-1.5 m, often glabrous; flowers pendulous, white or yellowish-white, apex often with purplish red, tepals 6, ovate-rhombus, stigma 3-lobed. Fruit obovate oblong, indehiscent.

Flower and Fruit: Fl. spring-autumn.

Distribution: Native to east and southeast North America. Cultivated in most areas of China. Main island (Dashaao, Houlong, Sanpanwei), Daleishan Island, cultivated or wild.

Habitat: Warm, humid and sunny environment.

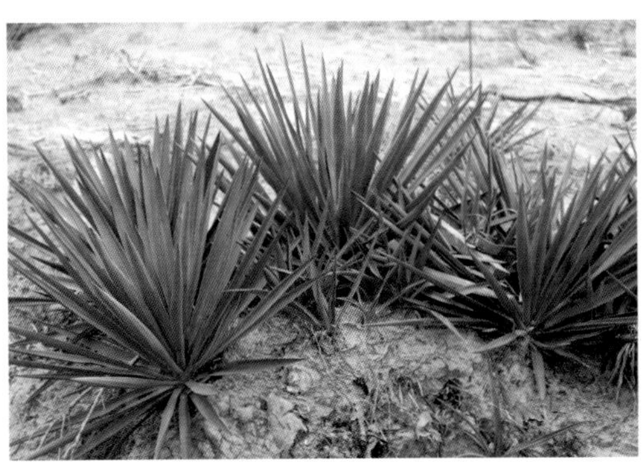

A76. 棕榈科
Arecaceae

棕榈
Trachycarpus fortunei
棕榈属 ***Trachycarpus***

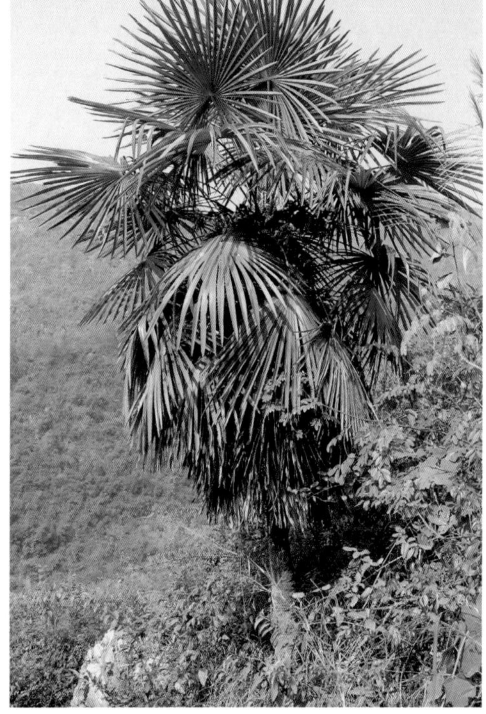

形态特征：常绿乔木，高 3 ~ 10 m 或更高，老叶柄基部不易脱落，被密集网状纤维。叶片圆形或者近圆形，深裂成线状剑形，2 裂或具 2 齿；叶柄两侧具细圆齿，顶端具戟突。花序粗壮多分枝，雌雄异株；雄花黄绿色，2 ~ 3 朵密生于小穗轴，花萼 3，卵状急尖，花瓣阔卵形；雌花 2 ~ 3，淡绿色，萼片阔卵形。果实阔肾形，熟时有白粉。

花果期：花期 4 月，果期 12 月。

分布：东南亚。我国长江以南各地，常栽培。南麂主岛（大沙岙、三盘尾、镇政府），栽培。

生境：罕见野生于疏林中。

Description: Evergreen trees, 3-10 m tall or more, old petiole base hard to fall off, densely reticulate fibers. Leaf blade orbicular or nearly orbicular, deeply divided into linear sword shape, 2-lobed or 2-toothed. Petiole with crenulate on both sides, apex with hastate. Inflorescences robust, much branched, dioecious; male flowers sessile, yellow-green, 2-3 dense in rachilla, calyx 3, ovate acute, petals broadly ovate, stamens 6; female flowers 2-3 on short tubercles, light green, sepals broadly ovate. Fruit kidney-shaped, with white powder when ripe.

Flower and Fruit: Fl. Apr. , fr. Dec.

Distribution: Southeast Asia. To the south of Yangtze River, commonly cultivated. Main island (Dashaao, Sanpanwei, Zhenzhengfu), cultivated.

Habitat: Rare in the wild sparse forest.

A76. 棕榈科
Arecaceae

大丝葵
Washingtonia robusta
丝葵属 ***Washingtonia***

形态特征：乔木状，高达 18 ～ 27 m。树干基部膨大，具明显的环状叶痕和不明显的纵向裂缝。叶片裂至基部 2/3 处，下部边缘被脱落性绒毛；叶柄粗壮，具粗壮钩刺。花序大型，长于叶，下垂，具多个大的分枝花序；花单生，花萼钟状，半裂成 3 个类似卵形的裂片；花冠长于花萼。果实椭圆形，亮黑色。种子卵球形。

花果期：果期 9 ～ 10 月。

分布：原产墨西哥西北部。我国南方地区有引种栽培。南麂主岛（大沙岙、三盘尾、镇政府），栽培。

生境：温暖湿润、阳光充足的环境。

Description: Treelike, 18 -27 m tall. Trunk base enlarged, with obvious annular leaf scars and inconspicuous longitudinal fissures. Leaf blade split to 2/3 of base, lower margin abscissive tomentose; petiole stout, with stout hook spines. Inflorescences large, longer than leaves, pendulous, with multiple large branched inflorescences; flowers solitary, calyx campanulate, semiflobed into 3 slightly ovate lobes; corolla longer than calyx. Fruit oval, bright black. Seeds ovoid.

Flower and Fruit: Fr. Sep. -Oct.

Distribution: Native to northwest of Mexico. Introduced and cultivated in southern China. Main island (Dashaao, Sanpanwei, Zhenzhengfu), cultivated.

Habitat: Warm, humid and sunny places.

A78. 鸭跖草科
Commelinaceae

饭包草
Commelina benghalensis
鸭跖草属 *Commelina*

形态特征： 多年生草本。匍匐茎多数，向上生长，铺散，分枝多数，疏生短柔毛。叶柄离生；叶片卵形，近无毛。总苞片与叶对生，常数个聚生枝顶；蝎尾状聚伞花序下部分枝具长梗和 1 ～ 3 外露的不育花，上部分枝长，花可育；萼片膜质，花瓣蓝色。蒴果椭圆形，3 瓣裂。种子黑色，圆筒状或半圆柱状的，具皱纹，不规则网状。

花果期： 花期夏秋季。

分布： 非洲和亚洲的热带、亚热带地区。广布我国南北各地。南麂主岛（关帝岙、后隆、门屿尾、三盘尾），常见。

生境： 潮湿的地方。

Description: Herbs perennial. Stems mostly creeping, ascending distally, diffuse, numerous branched, sparsely pubescent. Petiole distinct; leaf blade ovate, subglabrous. Involucral bracts borne opposite leaves, often several, aggregated at apex of branches; proximal branch of cincinni with elongate peduncle, and 1-3 exserted infertile flowers; distal branch longer,fertile flowers; sepals membranous, petals blue. Capsule ellipsoid, 3-valved. Seeds black, cylindric or semicylindric, rugose, irregularly reticulate.

Flower and Fruit: Fl. summer to autumn.

Distribution: Tropical and subtropical Africa and Asia. Throughout China. Main island (Guandiao, Houlong, Menyuwei, Sanpanwei), commonly.

Habitat: Wet places.

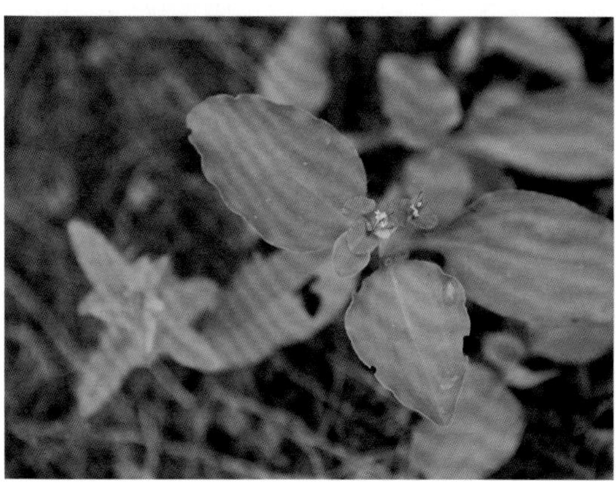

A78. 鸭跖草科
Commelinaceae

鸭跖草
Commelina communis
鸭跖草属 *Commelina*

形态特征：一年生披散草本。茎匍匐生根，多分枝。叶鞘无毛；叶披针形至卵状披针形，无毛。总苞片具长柄，与叶对生，心形，折叠；蝎尾状聚伞花序的下部分枝具1～2朵雄花，上部分枝具3～4朵两性花；萼片膜质，花瓣深蓝色。蒴果椭圆形，2瓣裂，有种子4颗。种子棕黄色，半椭圆形，有不规则窝孔。

花果期：花期夏秋季。

分布：东南亚至俄罗斯远东。我国除青海、新疆和西藏以外广泛分布。南麂主岛各处常见。

生境：潮湿的地方。

Description: Herbs annual, spreading. Stems creeping, numerous branched. Leaf sheaths glabrous; leaf blade lanceolate to ovate-lanceolate, glabrous. Involucral bracts borne opposite leaves, with long stalk, cordate, folded; proximal branch of cincinni with 1 or 2 male flowers, distal branch with 3 or 4 bisexual flowers; sepals membranous, petals dark blue. Capsule ellipsoid, 2-valved, seeds 4. Seeds brown-yellow, semiellipsoid, irregularly pitted.

Flower and Fruit:Fl. summer-autumn.

Distribution: Southeast Asia to Russian (Far East). Throughout China except for Qinghai, Xinjiang, and Xizang. Main island of Nanji, commonly.

Habitat: Humid places.

A78. 鸭跖草科
Commelinaceae

裸花水竹叶
Murdannia nudiflora
水竹叶属 *Murdannia*

形态特征：多年生草本，根纤细，无毛或被长绒毛。叶茎生，偶有 1 ~ 2 条形基生叶，叶片禾叶状或披针形，两面无毛或疏生刚毛。蝎尾状聚伞花序数个排成顶生圆锥花序，或仅单个，聚伞花序具数朵花密集排列，具纤细总梗，苞片早落，萼片草质，卵状椭圆形，浅舟状，花瓣紫色，能育雄蕊 2 枚，不育雄蕊 2 ~ 4 枚，花丝下部有须毛。蒴果卵圆状三棱形。种子有深窝孔。

花果期：花果期（6）8 ~ 9（10）月。

分布：亚洲东南部至印度洋、太平洋岛屿。我国华北以南各地。南麂主岛（三盘尾），偶见。

生境：水边潮湿处，少见于草丛中。

Description: Herbs perennial, roots slender, glabrous or tomentose. Leaves cauline, occasionally 1-2 strip-shaped basal leaves, leaf blades graminiform or lanceolate, both surfaces glabrous or sparsely setose. Several scorpioid cymes, arranged in terminal panicles, or only single, cymes with several flowers densely arranged, slender pedicels, bracts caducous, sepals herbaceous, ovate-elliptic, shallow boat-shaped, petals purple, fertile stamens 2, sterile stamens 2-4, filaments with hairs below. Capsule ovoid-trigonal. Seeds with deep pits.

Flower and Fruit: Fl. and fr. (Jun.) Aug. -Sep. (Oct.)

Distribution: Southeastern Asia to Indian Ocean and Pacific Islands. To the south of North China. Main island (Sanpanwei), occasionally.

Habitat: Wet places by water, rarely among grass.

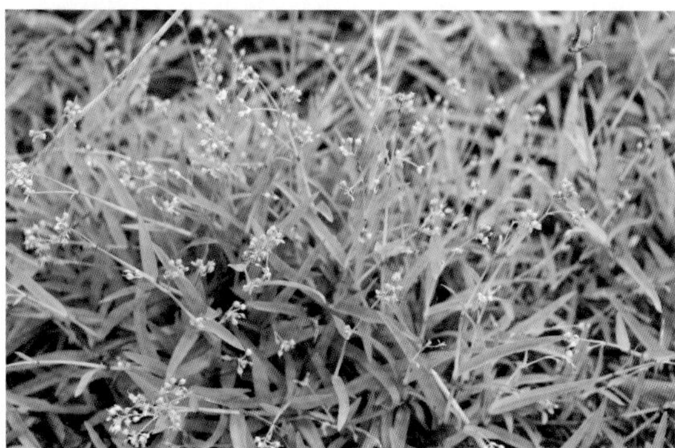

A86. 美人蕉科
Cannaceae

兰花美人蕉
Canna orchioides
美人蕉属 *Canna*

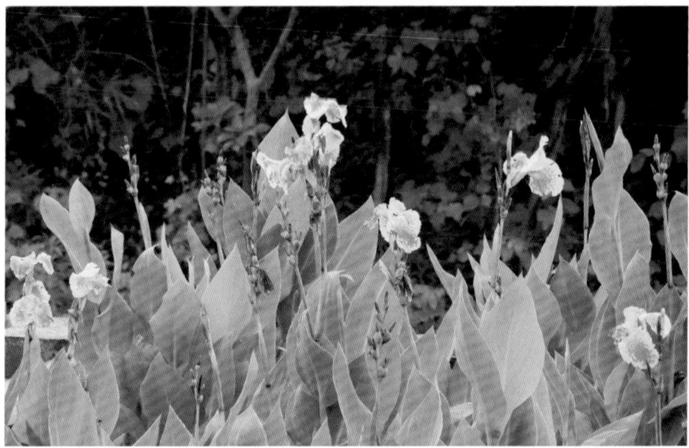

形态特征：多年生草本，植株高 1 ～ 1.5 m；茎绿色。叶片椭圆形至椭圆状披针形，顶端具短尖头，基部渐狭，下延，绿色。总状花序通常不分枝，花大，花萼长圆形，花冠裂片披针形，浅紫色，在开花后一日内即反卷下向；退化雄蕊倒卵状披针形，质薄而柔，纸质，鲜黄至深红，具红色条纹或斑点；子房长圆形；花柱狭带形。蒴果卵状三角形。种子具凹尖。

花果期：花期夏秋季。

分布：原产欧洲。我国各大城市公园栽培观赏。南麂主岛（大沙岙、后隆、美龄居、门屿尾、三盘尾、镇政府），栽培。

生境：温暖湿润气候。

Description: Herbs perennial, 1-1.5 m tall; stems green. Leaf blade elliptic to elliptic lanceolate, apex mucronate, base tapering, decurrent, green. Racemes usually unbranched, flowers large, calyx oblong, corolla lobes lanceolate, light purple, retrograde downward within one day after flowering; staminodes obovate-lanceolate, thin and soft, papery, bright yellow to deep red, with red stripes or splashes; ovary oblong; style narrowly zonate. Capsule ovoid-trigonal. Seeds with deep pits.

Flower and Fruit: Fl. and Fr. summer -autumn.

Distribution: Native to Europe. Cultivated as ornamental in city parks. Main island (Dashaao, Houlong, Meilingju, Menyuwei, Sanpanwei, Zhenzhengfu), cultivated.

Habitat: warm and humidclimate.

A89. 姜科
Zingiberaceae

艳山姜
Alpinia zerumbet
山姜属 ***Alpinia***

形态特征：多年生草本，具根状茎。叶披针形，顶端具旋卷小尖头，边缘具短柔毛。圆锥花序下垂，呈总状花序式，花序轴被绒毛，分枝上具花 1 ～ 3 朵；小苞片白色，顶端粉红，无毛；小花梗极短，花萼钟形，白色，一侧开裂；花冠管较花萼短，裂片长圆形，乳白色；侧生退化雄蕊钻状；唇瓣匙状宽卵形，先端皱波状，黄色，有紫红色条纹。蒴果卵圆形，被疏毛，条纹明显，熟时朱红色。

花果期：花期 4 ～ 6 月，果期 7 ～ 10 月。

分布：亚洲热带地区广布。我国东南部至西南部地区。南麂主岛各处常见。

生境：山坡、路边或沟边草地。

Description: Herbs perennial, rhizomes. Leaves lanceolate, apex with convoluted cusps, margin pubescent. Panicle drooping, raceme-shaped,downy inflorescence axis, flowers 1-3 on branches; bracteoles white, tip pink, glabrous; pedicel short, calyx bell-shaped, white, one side cracked; corolla tube shorter than calyx, lobes oblong and milky white; lateral staminodes subulate; labellam spoon-shaped wide ovate, apex crisped, yellow, purplish red stripes. Capsule oval, sparsely hirsute, exposed stripes, vermilion when ripe.

Flower and Fruit: Fl. Apr. -Jun. , fr. Jul. -Oct.

Distribution: Tropical Asia. Southeast to southwest China.Main island of Nanji, commonly.

Habitat: Grasslands of slopes, roadsides, and ditches.

A97. 灯心草科
Juncaceae

笄石菖
Juncus prismatocarpus
灯心草属 *Juncus*

形态特征：多年生草本，具根状茎和多数黄褐色须根。基生叶少，茎生叶 2 ~ 4 枚，叶片线形，圆柱形或压扁，具横隔。花序顶生，有分枝，头状花序 5 ~ 10，球状至半球形，具花 8 ~ 15 朵；叶状苞片线形，短于花序，苞片多枚，宽卵形，膜质，花被片线状披针形或窄披针形，绿或淡红褐色，背面有纵脉。蒴果三棱状圆锥形，具短尖头，淡褐或黄褐色。

花果期：花期 3 ~ 6 月，果期 5 ~ 11 月。

分布：亚洲东部至东南部，大洋洲。我国华北以南各地。南麂主岛（三盘尾），偶见。

生境：疏林的潮湿草地，灌丛，田野，沼泽，河岸，溪边。

Description: Herbs perennial, rhizomes, many yellowish-brown fibrous roots. Basal leaves few, cauline leaves 2 to 4, blade linear, terete to compressed,septate.Inflorescences terminal, sparingly or much branched;heads 5-10, globose to hemispheric, 8-15-flowered; leaflike bracts linear, shorter than inflorescence, bracts numerous, broadly ovate, membranous, perianth segments linear-lanceolate or narrowly lanceolate, green or pale red-brown, abaxially with longitudinal veins. Capsule trigonous-conic, mucronate, pale brown or yellow-brown.

Flower and Fruit: Fl. Mar. -Jun. , fr. May-Nov.

Distribution: Eastern to southern Asia, Oceania. To the south of North China. Main island (Sanpanwei), occasionally.

Habitat: Wet grasslands in sparse forests, thickets, fields, marshy places, swampy river banks, streamsides.

A97. 灯心草科
Juncaceae

星花灯心草
Juncus diastrophanthus
灯心草属 *Juncus*

形态特征：多年生草本，簇状，高 15 ～ 25 cm。茎稍扁，两侧有狭翅。基生叶很少，茎生叶 1 ～ 3；叶鞘耳稍钝；叶片线形，扁状，先端渐尖。花序顶生，通常 2 ～ 3 分枝；总苞片成薄片状。花被片狭披针形，边缘膜质，先端具刚毛状尖端。蒴果黄绿色到栗棕色，有光泽，三角状长圆形，先端尖。种子倒卵状椭圆形，具条纹，两端短尖。

花果期：花期 4 ～ 6 月，果期 4 ～ 10 月。

分布：日本、朝鲜、印度。我国秦淮以南地区。南麂主岛（三盘尾），偶见。

生境：疏林下的潮湿地，溪边，田野。

Description:Herbs perennial, tufted, 15-25 cm tall. Stems slightly compressed, very narrowly winged on both sides. Basal leaves few, cauline leaves 1-3; leaf sheath auricles somewhat obtuse; leaf blade linear, compressed, apex acuminate. Inflorescences terminal, usually 2 or 3 branched; involucral bract leaflike. Perianth segments narrowly lanceolate, margin scarious, apex with setaceous tip. Capsule yellowish green to chestnut brown, shiny, trigonous oblong, apex acute. Seeds obovate-ellipsoid, striate, both ends mucronate.

Flower and Fruit: Fl. Apr. -Jun. , fr. Apr. -Oct.

Distribution: Japan, Korea, India. To the south of Qinling and Huaihe. Main island (Sanpanwei), occasionally.

Habitat: Moist places in sparse forests, streamsides, fields.

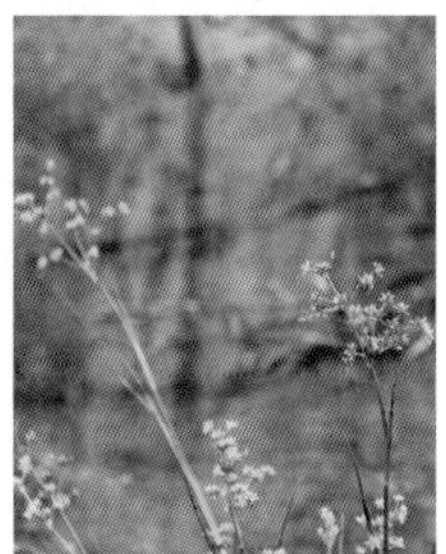

A98. 莎草科
Cyperaceae

青绿薹草
Carex breviculmis
薹草属 *Carex*

形态特征：多年生草本。秆丛生，纤细，三棱形。叶短于秆，平张，边缘粗糙，质硬。苞片最下部的叶状，长于花序。小穗 2 ~ 5 个，顶生小穗雄性，侧生小穗雌性，具稍密生的花。雄花鳞片膜质，黄白色；雌花鳞片膜质，苍白色，向顶端延伸成长芒。果囊近等长于鳞片，倒卵形，淡绿色，具多条脉；小坚果紧包于果囊中，栗色，顶端缢缩成环盘。

花果期：花果期 3 ~ 6 月。

分布：俄罗斯、朝鲜、日本、印度、缅甸。我国南北各地均有分布。南麂各岛屿常见。

生境：山坡草地、路边、山谷沟边。

Description: Herbs perennial. Culms tufted, slender, trigonous. Leaves shorter than culm, flat, margin rough, hard. Bracts lowermost leaflike, longer than inflorescence. Spikes 2-5, terminal spikelet male, lateral spikelet female, densely flowered. Male glumes membranous, yellow-white; female glumes membranous, pale, excurrent into an awn. Utricles pale green, nearly equaling glume, obovate, green, several veined; nutlet chestnut-colored, tightly enveloped,apex discoid-annulate.

Flower and Fruit: Fl. and fr. Mar. -Jun.

Distribution: Russia, Korea, Japan, India, Myanmar. Throughout the north and south areas of China. Throughout Nanji Islands, commonly.

Habitat: Grasslands on mountain slopes, waysides, ditch sides in valleys.

A98. 莎草科
Cyperaceae

褐果薹草（栗褐薹草）
Carex brunnea
薹草属 *Carex*

形态特征： 多年生草本。秆密丛生，细长。叶长于或短于秆，两面及边缘均粗糙，具鞘。小穗几个至十几个，常 1 ~ 2 个出自同一苞片鞘内，全部为雄雌顺序，雄花较雌花短很多，圆柱形，具多数密生的花，具柄。果囊近直立，长于鳞片，椭圆形或近圆形，扁平凸状，膜质，褐色，两面均被白色短硬毛，顶端急狭成短喙，喙顶二齿；小坚果紧包于果囊内，近圆形。

花果期： 花果期 6 ~ 10 月。

分布： 亚洲东部至东南部。我国长江流域以南及陕西、甘肃等地。南麂主岛（后隆、门屿尾、三盘尾）、大檑山屿，常见。

生境： 山坡、山谷林下或灌丛、河边、路边阴处或水边阳处。

Description: Herbs perennial. Culms densely tufted, slender. Leaves longer or shorter than culm, scabrid on both surfaces and margins, sheathing. Spikes several to more than 10, usually 1-2 in an involucral bract sheath, all male and female in order, male flower part much shorter than female part, cylindric, densely many flowered, stalked.Fruit sac nearly erect, longer than scales, elliptic or suborbicular, flat and convex, membranous, brown, both surfaces covered with white short bristles, tip sharply narrowed into a short beak, beak tip bidentate; nutlets tightly enclosed in fruit sac, suborbicular.

Flower and Fruit: Fl. and fr. Jun. -Oct.

Distribution: Eastern to southwestern Asia. To the south of Yangtze River Basin and Shanxi, Gansu. Main island (Houlong,Menyuwei, Sanpanwei), Daleishan Island, commonly.

Habitat: Mountain slopes, forests or among shrubs in valleys, riversides, shady places at roadsides, sunny places at watersides.

A98. 莎草科
Cyperaceae

蕨状薹草
Carex filicina
薹草属 *Carex*

形态特征：多年生草本。秆密丛生，锐三棱形，无毛。叶长于秆，少有短于秆，下面粗糙，上面光滑或两面均光滑。苞片叶状，具长鞘。圆锥花序复出，4～8分枝，三角状卵形，单生，稀双生；小苞片鳞片状。小穗多数，两性，长圆形或长圆状圆柱形。果囊椭圆形或狭椭圆形，三棱形，膜质，无毛，上部收缩成喙，喙口斜截形；小坚果椭圆形，三棱形，成熟时黄褐色。

花果期：花果期5～11月。

分布：东南亚。我国长江以南各地。南麂主岛（百亩山、国姓岙、门屿尾），偶见。

生境：林间或林边湿润草地。

Description: Herbs perennial. Culms densely tufted, sharply trigonous, glabrous. Leaves longer than culm, rarely shorter than culm, rough below, smooth above or smooth on both sides. Involucral bracts leafy, long sheathed. Panicle compound 4-8 branched, outline triangular-ovate, single, sparse twin; bracteoles scalelike.Spikes numerous, bisexual, oblong or oblong-cylindric. Fruit sac elliptic or narrowly elliptic, triangular, membranous, glabrous, contracted into a long beak slightly curved out to slightly curved down in the upper part, beak mouth obliquely truncated; nutlets ellipsoid, trigonous, yellow-brown at maturity.

Flower and Fruit: Fl. and fr. May-Nov.

Distribution: Southeast Asia. To the south of Yangtze River. Main island (Baimushan, Guoxingao, Menyuwei), occasionally.

Habitat: Wet grasslands in or near forests.

A98. 莎草科
Cyperaceae

矮生薹草
Carex pumila
薹草属 *Carex*

形态特征：多年生草本。根状茎具细长匍匐茎。茎散生成丛，高 10 ~ 30 cm，三棱形。叶片平展或有褶皱，硬，叶脉及边缘粗糙，具鞘。叶状的总苞片，长于茎，具短鞘。穗状花序顶部 2 ~ 3 枚雄性，棒状，其余雌性，长圆形，小花松散，常有短花序梗。雌颖片褐色，宽卵形，膜质，3 脉，短尖或具芒。胞果淡黄，卵形，先端具喙；小坚果三棱形。

花果期：花果期 4 ~ 6 月。

分布：日本、朝鲜北部、俄罗斯远东。我国东部沿海地区。南麂主岛（百亩山、大沙岙、三盘尾），常见。

生境：海边沙地。

Description: Herbs perennial. Rhizome with slender stolons. Culms laxly tufted, 10-30 cm tall, triquetrous. Leaf blades flat or sometimes plicate, stiff, scabrous on veins and margins, sheathed. Involucral bracts leafy, longer than culm, shortly sheathed. Terminal 2 or 3 spikes male, clavate, shortly pedunculate; remaining spikes female, oblong, slightly laxly many flowered, usually shortly pedunculate. Female glumes brownish, broadly ovate, membranous, veins 3, apex mucronate or aristate. Utricles yellowish, ovate , apex beak; nutlets trigonous.

Flower and Fruit: Fl. and fr. Apr. -Jun.

Distribution: Japan, North Korea, Russia Far East. Coastal areas of eastern China. Main island (Baimushan, Dashaao, Sanpanwei), commonly.

Habitat: Seaside sands.

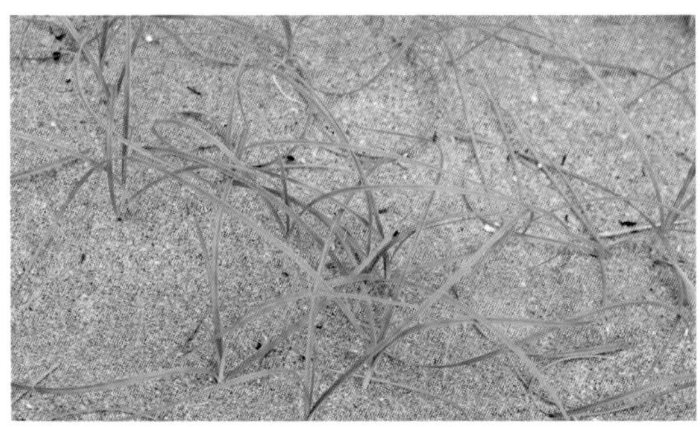

A98. 莎草科
Cyperaceae

健壮薹草
Carex wahuensis subsp. *robusta*
薹草属 *Carex*

形态特征：多年生草本，根状茎纤维状，浓密，灰色。茎丛生，高达 100 cm。叶基生，叶片宽线形，扁平，革质，鞘短。顶生穗状花序雄性，狭棍棒状或近圆筒状，侧生穗状花序通常雌性；小穗 3 ~ 6 个，圆柱形，多花，苞片叶状，具长鞘。胞果等长颖片，卵形球状，无毛，多脉；小坚果阔倒卵形，先端具喙。

花果期：花期 2 月。

分布：日本、韩国。我国浙江、台湾、香港等地海岛。南鹿各岛屿常见。

生境：海岸，稳定的沙土和岩石缝隙、坡地。

Description: Herbs perennial, rhizomes fibrous, density, gray. Culms densely tufted, up to 100 cm tall. Leaves basal, blades wide linear, flat, leathery, short sheaths. Terminal spikes male, narrowly clavelike or subcylindric, lateral spikes usually female; spikelets 3-6, cylindric, floriferous, bracts leaflike, with long sheaths. Utricles equal to glumes, ovate globose, glabrous, veined; nutlets broadly obovoid,apex beaked.

Flower and Fruit: Fl. Feb.

Distribution: Japan, South Korea. Islands of Zhejiang, Taiwan and Hong Kong, China. Throughout Nanji Islands, commonly.

Habitat: Seashore, stable crevices of sand and rocks, slopes.

A98. 莎草科
Cyperaceae

风车草
***Cyperus involucratus* (*Cyperus alternifolius* subsp. *flabelliformis*)**
莎草属 *Cyperus*

形态特征： 多年生草本，高 30 ～ 150 cm。茎稍粗壮，钝三棱形，基部具无叶的鞘。基生叶鞘黄褐色和干状；顶端叶鞘淡绿色。总苞片叶状，边缘粗糙，基部水平，顶部弯曲并下垂；长侧枝聚伞花序复出，多数辐射状；小穗簇生枝顶，狭卵形，扁平。成熟时小坚果棕色，宽椭圆形。

花果期： 花果期 5 ～ 12 月。

分布： 原产东非和阿拉伯半岛。我国常作观赏栽培，偶有逸生归化。南麂主岛（大沙岙、镇政府），栽培。

生境： 河流边缘、湿地灌丛和干扰地。

Description: Herbs perennial, 30-150 cm tall. Culms slightly stout, obtusely angles 3, base with bladeless sheaths. Basal leaf sheaths yellowish brown and cataphylloid; apical leaf sheaths pale green. Involucres leaflike, margin scabridulous, basally horizontal, apically flexuose and drooping. Inflorescence a decompound anthela, rays numerous; spikelets clustered at apex of raylets, narrowly ovoid, flattened. Nutlet brown at maturity, broadly ellipsoid.

Flower and Fruit: Fl. and fr. May-Dec.

Distribution: Native to East Africa and Arabian Peninsula. Cultivated as an ornamental in China, sometimes escaped and naturalized. Main island (Dashaao, Zhenzhengfu), cultivated.

Habitat: Along streams and in wet thickets and disturbed areas.

A98. 莎草科
Cyperaceae

扁穗莎草
Cyperus compressus
莎草属 *Cyperus*

形态特征：一年生草本；根为须根。秆稍纤细，锐三棱形，基部具较多叶。叶短于秆，或与秆等长；叶鞘紫褐色。总苞片3～5枚，叶状，较花序长；长侧枝聚伞花序简单，伞辐2～7，穗状花序近于头状；花序轴短，小穗排列紧密，斜展，线状披针形，近于四棱形，鳞片顶端具长芒。小坚果倒卵形，三棱形，侧面凹陷，深棕色，表面具密的细点。

花果期：花果期7～12月。

分布：热带亚热带地区。我国南北各地多有分布。南麂主岛各处常见。

生境：草地、海岸，小路、湖边、森林、石缝，潮湿的砂质河岸、开阔田野和田边。

Description: Herbs annual; root fibrous. Culms slightly slender, sharply trigonous, base with many leaves. Leaves shorter than culm or equaling culm; leaf sheath purplish brown. Involucral bracts 3-5, leaflike, longer than inflorescence,inflorescense a simple anthla,ray 2-7, spikes nearly capitate, spikelets closely arranged, obliquely spreading, linear-lanceolate, nearly quadrangular. Nutlets obovate, trigonous, laterally impressed, dark brown, surface densely punctate.

Flower and Fruit: Fl. and fr. Jul. -Dec.

Distribution: Tropical and subtropical areas. Most areas throughout north and south China. Main island of Nanji, commonly.

Habitat: Grasslands, seashores, along trails, lake margins, forests, stony clefts, wet sandy riverbanks, open fields, paddy field margins.

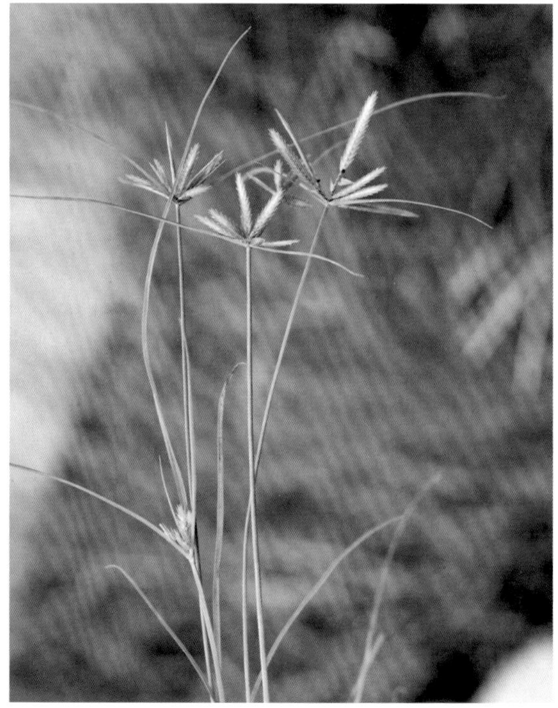

A98. 莎草科
Cyperaceae

砖子苗
Cyperus cyperoides（*scirpus cyperoides*）
莎草属 *Cyperus*

形态特征：多年生草本，根状茎短。秆疏丛生，锐三棱形，具稍多叶。叶短于秆或与秆等长，边缘不粗糙；叶鞘褐色或红棕色。叶状苞片 5 ～ 8 枚；长侧枝聚伞花序简单；穗状花序圆筒形或长圆形，小穗多数，密生，平展或稍俯垂，线状披针形，小穗轴具宽翅，翅披针形，白色透明。小坚果狭长圆形，三棱形，表面具微突起细点。

花果期：花果期 4 ～ 10 月。

分布：热带和亚热带地区。我国长江流域以南地区。南麂主岛各处常见。

生境：山坡阳处、路旁草地、溪边以及松林下。

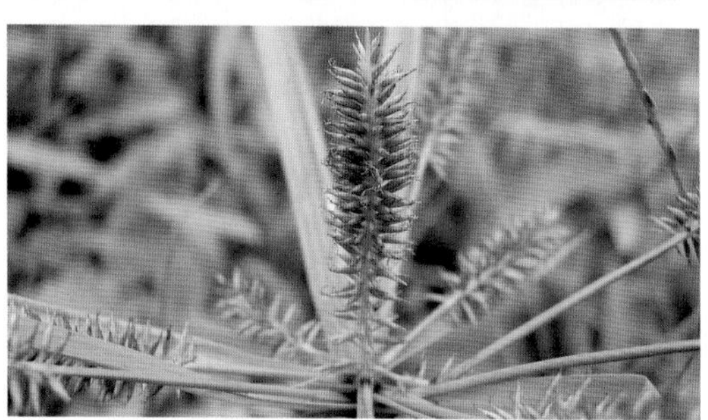

Description: Herbs perennial, rhizome short. Culms laxly tufted, acutely trigonous, slightly leafy. Leaves shorter than to equaling culm, margin not scabrous; leaf sheaths brown or reddish brown. Leaflike bracts 5-8; inflorescences long lateral branches, cymes simple; spikes cylindric or oblong, spikelets numerous, densely, spreading or slightly prostrate, linear-lanceolate; rachis broadly winged, wings lanceolate, white transparent. Nutlets narrowly oblong, trigonal, minutely punctate.

Flower and Fruit: Fl. and fr. Apr. -Oct.

Distribution: Tropical and subtropical areas. To the south of Yangtze River Basin. Main island of Nanji, commonly.

Habitat: Sunny slopes, roadside grasslands, streams and pine forests.

A98. 莎草科
Cyperaceae

异型莎草
Cyperus difformis
莎草属 *Cyperus*

形态特征：一年生草本，根为须根。秆丛生，稍粗或细弱，扁三棱形，平滑。叶短于秆；叶鞘稍长，褐色。苞片2枚，叶状，长于花序；长侧枝聚伞，花序简单，少数为复出；头状花序球形，具极多数小穗，小穗密聚，披针形或线形，具8～28朵花；小穗轴无翅。小坚果倒卵状椭圆形，三棱形，几与鳞片等长，淡黄色。

花果期：花果期7～10月。

分布：亚洲、非洲、中美洲。广布我国各地。南麂主岛（大沙岙、国姓岙、后隆、门屿尾、三盘尾），常见。

生境：稻田中或水边潮湿处。

Description: Herbs annual, roots fibrous. Culms tufted, slightly thick or thin, flat triangular, smooth. Leaves shorter than culm; leaf sheaths slightly longer, brown. Bracts 2, leaflike, longer than inflorescence; inflorescence long lateral branches, spike simple, a few recurrent; head globose, with very many spikelets, spikelets dense, lanceolate or linear, flowers 8-28; rachilla wingless. Nutlets obovate-elliptic, triangular, almost as long as scales, yellowish.

Flower and Fruit: Fl. and fr. Jul. -Oct.

Distribution: Asia, Africa, Central America.Throughout China. Main island (Dashaao, Guoxingao, Houlong, Menyuwei, Sanpanwei), commonly.

Habitat: Rice paddies or wet places near water.

A98. 莎草科
Cyperaceae

碎米莎草
Cyperus iria
莎草属 *Cyperus*

形态特征：一年生草本，无根状茎，具须根。秆丛生，细弱或稍粗壮，扁三棱形，基部具少数叶，叶短于秆。叶状苞片 3 ～ 5 枚；穗状花序卵形或长圆状卵形，小穗排列松散，斜展开，长圆形、披针形或线状披针形，压扁。小坚果倒卵形或椭圆形，三棱形，褐色，具密集的微突细点。

花果期：花果期 6 ～ 10 月。

分布：热带亚热带地区。我国南北各地广布。南麂主岛各处常见。

生境：田间、山坡、路旁阴湿处，为一种常见的杂草。

Description: Herbs annual, rhizomatous, roots fibrous. Culms tufted, slender or slightly stout, compressed triangular, base few leaved, leaves shorter than culms. Leaflike bracts 3-5; spikes ovate or oblong-ovate, spikelets laxly arranged, obliquely spreading, oblong, lanceolate or linear-lanceolate, compressed. Nutlets obovate or elliptic, trigonal, brown, densely prominently puncticulate.

Flower and Fruit: Fl. and fr. Jun. -Oct.

Distribution: Tropical and subtropical areas. Throughout the north and south areas of China. Main island of Nanji, commonly.

Habitat: Common weed in fields, hillsides and roadside damp places.

A98. 莎草科
Cyperaceae

香附子
Cyperus rotundus
莎草属 *Cyperus*

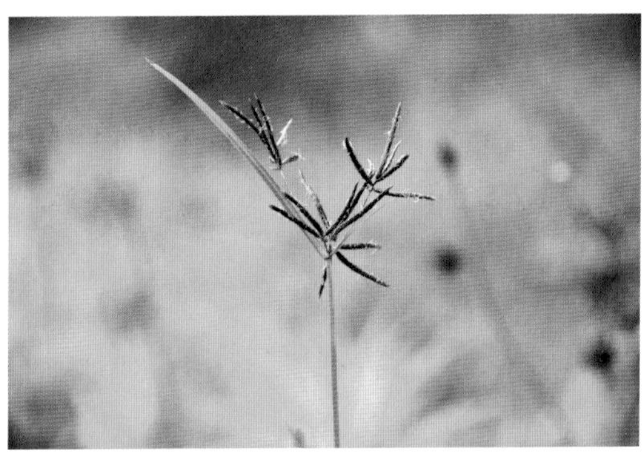

形态特征：多年生草本，匍匐根状茎长，具椭圆形块茎。秆稍细弱，锐三棱形，基部呈块茎状。叶较多，短于秆；鞘棕色，常裂成纤维状。叶状苞片 2 ~ 3 枚，偶见 5 枚；长侧枝聚伞花序简单或复出，穗状花序陀螺形，稍疏松；小穗斜展开，线形，小穗轴具较宽的、白色透明的翅。小坚果长圆状倒卵形，三棱形，具细点。

花果期：花果期 5 ~ 11 月。

分布：世界广布。我国南北各地广布。南麂主岛各处常见。

生境：山坡荒地草丛中或水边潮湿处。

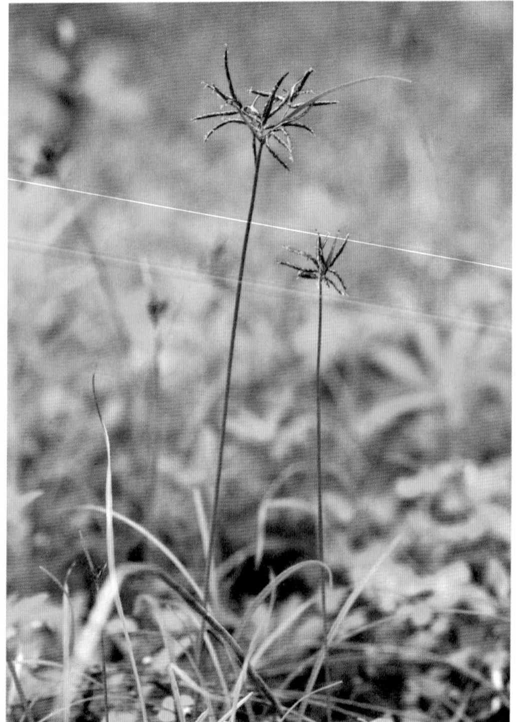

Description: Herbs perennial, prostrate rhizomes long, elliptic tubers. Culm slightly thin, weak, sharply triangular, tuberous base. Leaves more numerous, shorter than culm; sheath brown, often fissured into fibrous. Leaflike bracts 2-3(5); inflorescences long lateral branches, simple or regurgitated, spikes obdeltoid, slightly lax; spikelets obliquely spreading, linear, rachis with wider, white hyaline wings.Nutlets oblong-obovate, trigonal, minutely punctate.

Flower and Fruit: Fl. and fr. May-Nov.

Distribution: Cosmopolitan.Throughout the north and south areas of China. Main island of Nanji, commonly.

Habitat: Grasslands, wet or dry areas on mountain slopes.

A98. 莎草科
Cyperaceae

水虱草（日照飘拂草）
Fimbristylis littoralis（*Fimbristylis miliacea*）
飘拂草属 *Fimbristylis*

形态特征：一年生或短暂多年生草本无根状茎。秆丛生，扁四棱形，具纵槽，基部包着 1 ～ 3 个无叶片的鞘；叶边有稀疏细齿；鞘侧扁，背面锐龙骨状，前面具膜质边，鞘口斜裂，无叶舌。苞片 2 ～ 4 枚，刚毛状；长侧枝聚伞花序复出，伞辐 3 ～ 6，小穗单生，球形或近球形。小坚果倒卵形或宽倒卵形，钝三棱形，具疣状突起和横向长圆形网纹。

花果期：花果期 5 ～ 10 月。

分布：亚洲、大洋洲等。广布我国大部分省区。南麂主岛（国姓岙、门屿尾、三盘尾），偶见。

生境：溪边、沼泽地、水田及潮湿的山坡、路旁和草地。

Description: Herbs annual or short-lived perennial rhizomes absent. Culms tufted, compressed quadrangular, longitudinally grooved, with 1-3 bladeless sheaths at base; leaves margin sparsely denticulate; sheath laterally compressed, abaxially sharply keeled, anterior with membranous margin, sheath mouth obliquely split, ligule absent. Bracts 2-4, setaceous; inflorescences long lateral branches compound,rays 3-6, spikelets solitary, globose or subglobose. Nutlets obovate or broadly obovate, obtusely trigonous, verrucose and horizontally oblong reticulate.

Flower and Fruit: Fl. and fr. May-Oct.

Distribution: Asia, Australia. Most areas throughout China. Main island (Guoxingao, Menyuwei, Sanpanwei), occasionally.

Habitat: Stream sides, marshes, paddy fields and wet slopes, roadsides and grasslands.

A98. 莎草科
Cyperaceae

独穗飘拂草
Fimbristylis ovata
飘拂草属 *Fimbristylis*

形态特征： 多年生草本，根状茎短。秆丛生，纤细。叶片狭窄。苞片 1 ~ 3 枚，颖片状，但有时为叶状。花序退化为单个小穗，顶生，卵形、椭圆形或长圆状卵形，稍扁。颖片黄绿色，宽卵形至卵形，革质；雄蕊 3；花柱三棱形，基部膨大，柱头 3。小坚果倒卵状三棱形，表面具明显的疣状突起。

花果期： 花果期 6 ~ 9 月。

分布： 世界热带和温暖地区。我国华东至西南等地区。南麂主岛（关帝岙、门屿尾、三盘尾），偶见。

生境： 草地、路边、荒地，开阔湿润的山边、溪边，光照充足的干草坡。

Description: Herbs perennial.Rhizome short. Culms tufted, slender. Leaf blade narrow. Involucral bracts 1-3, glumelike but basal sometimes leaflike. Inflorescences reduced to a single terminal spikelet, ovoid, ellipsoid, or oblong-ovoid, slightly compressed. Glumes yellowish green, broadly ovate to ovate, leathery; stamens 3; style 3-sided, basally inflated, stigmas 3. Nutlets obovoid, 3-sided, prominently verruculose.

Flower and Fruit: Fl. and fr. Jun. -Sep.

Distribution: Tropical and temperate areas. East China to Southwest China. Main island (Guandiao, Menyuwei, Sanpanwei), occasionally.

Habitat: Grasslands, roadsides, waste fields, open moist hillsides, streamsides, sunny dry slopes, grassy slopes.

A98. 莎草科
Cyperaceae

锈鳞飘拂草
Fimbristylis sieboldii
飘拂草属 *Fimbristylis*

形态特征：多年生草本。根状茎短，呈水平匍匐状，木质。茎灰绿色，簇状，纤细，扁三棱形，光滑。茎顶1叶，线形，折叠。总苞片近直立，基部稍宽；花序简单或较少呈复合状；小穗单生，长圆形，浓密多花，先端锐尖或很少钝。小坚果成熟时棕色到黑棕色，具短柄，倒卵球形到宽倒卵球形，扁平双凸，近光滑。

花果期：花果期6～8月。

分布：日本、朝鲜。我国长江流域以南地区。南麂主岛（打铁礁、东方岙、关帝岙、国姓岙、后隆、门屿尾、三盘尾），常见。

生境：海岸，海边潮湿且阳光充足的地方、盐碱的沼泽。

Description: Herbs perennial, rhizomes short or well developed and horizontally creeping, woody. Culms grayish green, tufted, slender, flatly 3-angled, smooth. Leaves apically on culm with a blade; leaf blade often linear, folded. Involucral bracts suberect, base slightly broader; inflorescence a simple or rarely subcompound anthela; spikelets solitary, oblong, densely many flowered, apex acute to rarely obtuse. Nutlet brown to blackish brown when mature, shortly stipitate, obovoid to broadly obovoid, flatly biconvex, subsmooth.

Flower and Fruit: Fl. and fr. Jun. -Aug.

Distribution: Japan, Korea. To the south and near the Yangtze River Basin in China. Main island (Datiejiao, Dongfangao, Guandiao, Guoxingao, Houlong, Menyuwei, Sanpanwei), commonly.

Habitat: Seashores, sunny wet places at seashores, salty marshes.

A98. 莎草科
Cyperaceae

双穗飘拂草
Fimbristylis subbispicata
飘拂草属 *Fimbristylis*

形态特征：一年生草本，无根状茎。秆丛生，细弱，扁三棱形，具多条纵槽，基部具少数叶。叶短于秆。苞片无或只有1枚，小穗通常1个，顶生，罕有2个，卵形、长圆状卵形或长圆状披针形，圆柱状，具多数花；鳞片螺旋状排列，膜质，卵形、宽卵形或近于椭圆形；雄蕊3；花柱长而扁平，具缘毛，柱头2。小坚果圆倒卵形，扁双凸状，褐色，基部具柄，表面具六角形网纹，稍有光泽。

花果期：花期6～8月，果期9～10月。

分布：日本、朝鲜、越南。分布我国南北大部分地区。南麂主岛（三盘尾），偶见。

生境：山坡、山谷空地、沼泽地、溪边、沟旁近水处，海边、盐沼地。

Description: Herbs annual,rhizomes absent. Culms tufted, slender, flat triangular, with many longitudinal grooves, with few leaves at base. Leaves shorter than culms. Bract absent or only 1, spikelet 1, terminal, rarely 2, ovate, oblong-ovate or oblong-lanceolate, terete, many flowered; scales spirally arranged, membranous, ovate, broadly ovate or nearly elliptic; stamens 3; style long and flattened,ciliate, stigmas 2. Nutlets obovate, flat biconvex, brown, base stipe, surface with hexagonal reticulate, slightly shiny.

Flower and Fruit: Fl. Jun. -Aug., fr. Sep. -Oct.

Distribution: Japan, Korea, Vietnam. Most areas in north and south of China. Main island (Sanpanwei), occasionally.

Habitat: Slopes, valleys, marshes, streamsides, watersides, seashores, salt marshes.

A98. 莎草科
Cyperaceae

短叶水蜈蚣
Kyllinga brevifolia
水蜈蚣属 ***Kyllinga***

形态特征：多年生草本，具匍匐根状茎。秆散生，纤细，扁三棱形，下部具叶。叶常与秆等长；叶片线形；叶鞘通常淡红色。苞片通常3，叶状，开展；穗状花序单一，近球形或卵状球形；密生多数小穗，小穗基部具关节，长圆形或长圆状披针形；具两性花，鳞片卵形膜质，雄蕊3，柱头2。小坚果倒卵状长圆形，褐色，扁双凸状。

花果期：花果期6～10月。

分布：热带亚热带地区。分布于我国各地。南麂主岛各处常见。

生境：山坡荒地、路旁草丛中，湿润的路边、沟边、海边沙滩和田缘。

Description: Herbs perennial, rhizomes creeping. Culms scattered, slender, compressed trigonous, proximally with leaves. Leaves often as long as culm; blade linear; leaf sheaths usually reddish. Bracts 3, leaflike, spreading; spikes simple, subglobose or ovate-globose; densely numerous spikelets, jointed at base, oblong or oblong-lanceolate; hermaphrodite flower; scales ovate, membranous, stamens 3, stigma 2. Nutlets obovate-oblong, brown, compressed biconvex.

Flower and Fruit: Fl. and fr. Jun. -Oct.

Distribution: Tropical and subtropical areas. Throughout China. Main island of Nanji, commonly.

Habitat: Grasslands of mountain slopes, roadsides, wet places along trails, ditch margins, and seashores, paddy field margins.

A98. 莎草科
Cyperaceae

球穗扁莎
Pycreus flavidus
扁莎属 ***Pycreus***

形态特征：多年生草本。秆丛生，细弱，钝三棱形。叶少，短于秆；叶鞘长，叶片对折或平展。苞片 2 ~ 4，长于花序纤细。长侧枝聚伞花序简单，伞辐 1 ~ 6；每枝具多个小穗；小穗密聚于辐射枝顶呈球形，多花；颖片两侧黄褐色、红褐色或为暗紫红色，疏松或紧密，长圆状卵形，3 脉。小坚果褐色或暗褐色，倒卵形，稍扁，双凸状，具明显细点。

花果期：花果期 6 ~ 11 月。

分布：世界热带和温带地区。我国南北大部分地区有分布。南麂主岛（大沙岙、三盘尾），偶见。

生境：田边、沟旁潮湿处、溪边湿润的沙土上。

Description: Herbs perennial. Culms tufted, slender, obtusely trigonous. Leaves few, shorter than culms; leaf sheath long, leaf blade folded or flat. Involucral bracts 2-4, longer than inflorescence, slender. Simple anthelate with 1-6 ray, each with many spikelets; spikelets congested at apex of rays, multi-flowered; glumes yellowish brown, reddish brown, dark purplish red, dark grayish brown, or purplish brown on both surfaces, lax or dense, oblong-ovate, 3-veined. Nutlet brown to dark grayish brown, obovoid, slightly compressed, biconvex, prominently puncticulate.

Flower and Fruit: Fl. and fr. Jun. -Nov.

Distribution: Tropical and temperate areas. Most areas in north and south of China. Main island (Dashaao, Sanpanwei), occasionally.

Habitat: Shady wet or sandy places of water margins, ditch margins, paddy field margins.

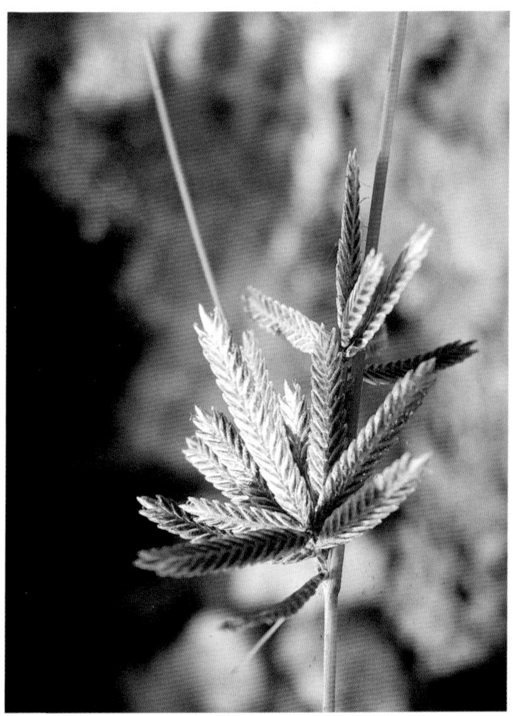

A98. 莎草科
Cyperaceae

多枝扁莎（多穗扁莎）
Pycreus polystachyos
扁莎属 *Pycreus*

形态特征： 一年生或短暂多年生草本。秆密丛生，扁三棱形，坚挺，光滑。叶短于秆，稍硬。苞片 4 ~ 6 枚，叶状；长侧枝聚伞花序简单或头状，伞辐 5 ~ 8，具多数小穗；小穗紧密排列成球形，6 ~ 30 朵花；小穗轴弯曲，具狭翅；颖片密覆瓦状排列，膜质，卵状长圆形，三脉。小坚果近长圆形或卵状长圆形，双凸状，顶端具短尖，表面具微突的细点。

花果期： 花果期 5 ~ 10 月。

分布： 热带亚热带地区。我国华东至华南。南麂主岛（打铁礁、东方岙、关帝岙、门屿尾、三盘尾），常见。

生境： 稻田旁、阴湿的沙土、海岸、水边沙地。

Description: Herbs annual or short-lived perennial. Culms densely tufted, compressed 3-angled, stiff, smooth. Leaves shorter than culm, slightly rigid. Involucral bracts 4-6, leaflike; simple or almost capitate anthela; rays 5-8, each with many spikelets; spikelets congested into a globose spike, 6-30-flowered or more; rachilla flexuose, narrowly winged; glumes densely imbricate, membranous, ovate-oblong, 3-veined. Nutlets suboblong or ovate-oblong, biconvex, apically mucronate, minutely punctate.

Flower and Fruit: Fl. and fr. May. -Oct.

Distribution: Tropical and subtropical areas. East to South China. Main island (Datiejiao, Dongfangao, Guandiao, Menyuwei, Sanpanwei), commonly.

Habitat: Shady areas in wet sand, paddy field margins, wet places, sandy areas at seashores, water margins.

A98. 莎草科
Cyperaceae

刺子莞
Rhynchospora rubra
刺子莞属 *Rhynchospora*

形态特征：一年生或短暂多年生草本。秆丛生，直立，圆柱状，基部不具无叶片的鞘。叶基生，叶片狭长，钻状线形，纸质。苞片 4 ~ 10 枚，叶状；头状花序顶生，具多数小穗；小穗钻状披针形，有 2 ~ 4 朵花。小坚果宽或狭倒卵形，双凸状，近顶端被短柔毛，上部边缘具细缘毛，成熟后为黑褐色，表面具细点。

花果期：花果期 5 ~ 11 月。

分布：亚洲、非洲、大洋洲的热带地区。我国台湾和长江流域以南地区。南麂主岛（门屿尾、三盘尾），偶见。

生境：路边、草坡或湿润处。

Description: Herbs annual or short-lived perennial. Culms tufted, erect, terete, base not with bladeless sheaths. Leaves basal, leaf blade long and narrow, subulate linear, papery. Bracts 4-10, leaflike; heads terminal, with numerous spikelets; spikelets subulate-lanceolate, 2-4 flowered. Nutlets broad or narrowly obovate, biconvex, subapically pubescent, distal margin finely ciliate, black-brown when mature, surface finely punctate.

Flower and Fruit: Fl. and fr. May-Nov.

Distribution: Tropical areas of Asia, Africa and Australia. To the south of Yangtze River Basin and Taiwan, China. Main island (Menyuwei, Sanpanwei), occasionally.

Habitat: Road margins, grassy slopes, wet places.

A98. 莎草科
Cyperaceae

毛果珍珠茅
Scleria levis
珍珠茅属 *Scleria*

形态特征：多年生草本，高 70 ~ 90 cm。茎松散丛生或散生，三棱形，粗糙，短柔毛。叶鞘纸质，近半圆形，短；叶片线形。总苞片叶状，小苞片刚毛状，基部耳状。圆锥状花序，具 1 ~ 2 侧枝。小穗棕色，单性，无梗；雄性小穗狭卵形到长圆形卵形；花盘淡黄，3 深裂，裂片边缘反折。小坚果白色，球形到卵形，3 面钝，光滑或具皱纹，先端尖。

花果期：花果期 6 ~ 9 月。

分布：南亚、东南亚至大洋洲。我国长江流域以南地区。南麂主岛（百亩山、国姓岙、美龄居、门屿尾），常见。

生境：干燥的地方，山坡草地或林中灌丛。

Description: Herbs perennial, 70-90 cm tall. Culms laxly tufted or scattered, 3-angled, scabrous, pubescent. Leaf sheath papery, nearly semicircular, short, barbate; leaf blade linear. Involucral bracts leaflike; bractlets setaceous, auriculate at base. Inflorescences paniculate, with 1 or 2 lateral branches. Spikelets 1 or 2 in a cluster, unisexual, sessile; male spikelets narrowly ovoid to oblong-ovoid; disk pale yellow, deeply 3-lobed, lobes lanceolate-triangular, margin reflexed. Nutlets white, spherical to ovoid, bluntly 3-sided, smooth to rugulose, apex tipped.

Flower and Fruit: Fl. and fr. Jun. -Sep.

Distribution: South Asia, Southeast Asia to Oceania. To the south of Yangtze River Basin. Main island (Baimushan, Guoxingao, Meilingju, Menyuwei,), commonly.

Habitat: Dry places, grasslands on slopes, thickets in forests.

A103. 禾本科
Poaceae

看麦娘
Alopecurus aequalis
看麦娘属 ***Alopecurus***

形态特征：一年生草本，松散丛生。秆柔弱，膝曲向上。叶鞘光滑，叶舌膜质，叶片亮绿色，柔软，背面光滑无毛，正面具肋，微糙。圆锥花序圆柱状，灰绿色，小穗椭圆形或卵状长圆形；颖具3脉，基部连合，脊上具细纤毛，外稃膜质，先端钝，芒隐藏或稍外露，花药橙黄色。

花果期：花果期4～8月。

分布：欧亚大陆温带地区。广布我国南北各省区。南麂主岛（三盘尾），偶见。

生境：灌溉沟渠，稻田，潮湿的杂草地。

Description: Herbs annual, loosely tufted.Culms weak, geniculately ascending.Leaf sheath smooth, ligule membranous, leaf blades light green, soft, abaxial surface smooth, glabrous, adaxial surface closely ribbed, scaberulous. Panicle narrowly cylindrical, gray-green, spikelet elliptic or ovate-oblong; glumes 3-veined, keels ciliate-hispid, margins connate at base, lemma membranous, apex obtuse, awn hidden or slightly exposed, anthers orange.

Flower and Fruit: Fl. and fr. Apr. - Aug.

Distribution: Temperate regions of Eurasia. Widely distributed in most areas of south and north China. Main island (Sanpanwei), occasionally.

Habitat: Irrigation ditches, rice fields, damp grasslands, other wet weedy places.

A103. 禾本科
Poaceae

荩草
Arthraxon hispidus
荩草属 *Arthraxon*

形态特征：一年生草本，高 30 ~ 60 cm。秆基部倾斜。叶鞘生短硬疣毛；叶舌膜质，边缘具纤毛；叶片卵状披针形，基部心形，抱茎。总状花序，节间无毛。无柄小穗卵状披针形，灰绿色或带紫；第一颖草质，具 7 ~ 9 脉，生疣基硬毛，第二颖近膜质，第一外稃长圆形，透明膜质；第二外稃近基部伸出一膝曲的芒；花药黄色或带紫色。颖果长圆形，有柄小穗退化为针状刺。

花果期：花果期 9 ~ 11 月。

分布：旧大陆温暖地区。分布几遍我国。南麂主岛各处常见。

生境：溪边、湿地，农田或其他潮湿地。

Description: Herbs annual, 30-60 cm tall. Culm base inclined. Leaf sheaths with short stiff verrucose hairs; ligule membranous, margin ciliate; leaf blade ovate-lanceolate, base cordate, amplexicaul. Racemes, internodes glabrous.Sessile spikelets ovate-lanceolate, gray-green or purplish; lower glame herbaceous, 7-9-veined, verrucose base bristle; upper glume nearly membranous; lower lemma oblong, hyaline and membranous; upper lemma projecting a geniculate awn near base; anthers yellow or purplish. Caryopsis oblong, pedicelled spikelets reduced only to needle-like spines.

Flower and Fruit: Fl. and fr. Sep. -Nov.

Distribution: Warm areas of the Old World.Almost throughout China. Main island of Nanji, commonly.

Habitat: Streamsides, damp meadows, among crops, other moist places.

A103. 禾本科
Poaceae

毛秆野古草（野古草）
Arundinella hirta
野古草属 *Arundinella*

形态特征：多年生草本。秆直立，高 90 ～ 150 cm，质稍硬，被白色疏长柔毛。叶鞘被疣毛，具缘毛；叶舌上缘截平，具长纤毛；叶片先端长渐尖，两面被疣毛。圆锥花序收缩，花序柄、主轴及分枝均被疣毛；孪生小穗柄较粗糙，具疏长柔毛；小穗无毛；第一颖先端渐尖，常 5 脉；第二颖具 5 脉；第一小花雄性外稃具 3 ～ 5 脉。

花果期：花果期 8 ～ 10 月。

分布：日本、朝鲜、俄罗斯东部。我国除新疆、西藏外广泛分布。南麂主岛（百亩山、大山、国姓岙、门屿尾），常见。

生境：草坡、河岸、路旁或田边。

Description: Herbs perennial. Culms erect, 90-150 cm tall, slightly hard, sparsely white villous. Leaf sheaths verrucose, margins ciliate; ligule truncate at upper margin, ciliate; leaf blade apex long acuminate, both surfaces verrucose. Panicle contracted, peduncle, main axis and branches covered with verrucous hairs; twinned spikelet stalks scabrous, sparsely villous; spikelets glabrous; lower glume apex acuminate, often 5-veined; upper glume 5-veined; lower floret male, lemma 3-5-veined.

Flower and Fruit: Fl. and fr. Aug. -Oct.

Distribution: Japan, Korea, East Russia. Almost throughout China except Xinjiang, Xizang. Main island (Baimushan, Dashan, Guoxingao, Menyuwei), commonly.

Habitat: Grassy mountain slopes, river banks, roadsides, field margins.

A103. 禾本科
Poaceae

芦竹
Arundo donax
芦竹属 *Arundo*

形态特征：多年生草本，具发达根状茎。秆粗大直立，高 3 ～ 6 m，具多数节，常生分枝。叶鞘长于节间，无毛或颈部具长柔毛；叶舌截平，先端具短纤毛；叶片扁平，上面与边缘微粗糙，基部白色，抱茎。圆锥花序极大型，分枝稠密，斜升；含 2 ～ 4 小花；外稃中脉延伸成短芒，背面中部以下密生长柔毛。颖果细小黑色。

花果期：花果期 9 ～ 12 月。

分布：亚洲、非洲、大洋洲热带地区。我国华东、华中、华南地区。南麂主岛各处常见。

生境：河岸和潮湿地。

Description: Herbs perennial, with developed rhizomes. Culm thick , erect, 3-6 m tall, with many nodes and often branched. Leaf sheaths longer than internodes, glabrous or neck villous; ligule truncate, apex shortly ciliate; leaf blade flattened, upper and margin slightly scabrous, base white, amplexicaul. Panicle very large, densely branched, obliquely ascending; florets 2-4; lemma midvein extended into a short awn, densely pubescent below the middle of the back. Caryopsis minutely black.

Flower and Fruit: Fl. and fr. Sep. -Dec.

Distribution: Tropical areas of Asia, Africa and Oceania. East, Central and South China. Main island of Nanji, commonly.

Habitat: River banks and other damp places.

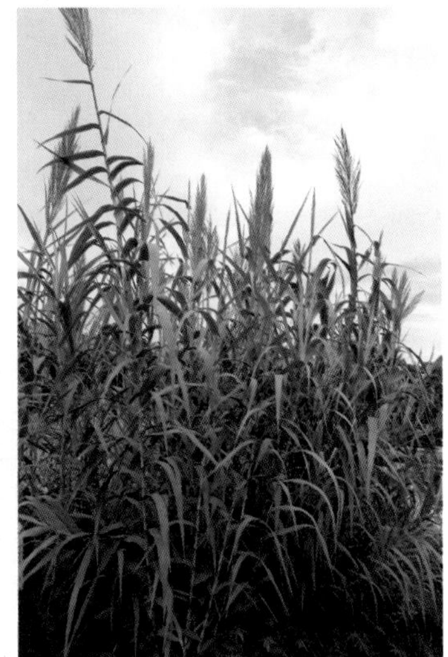

A103. 禾本科 Poaceae

野燕麦
Avena fatua
燕麦属 *Avena*

形态特征：一年生草本。茎直立或在基部弯曲，高 50 ~ 150 cm，不分枝，节 2 ~ 4。叶鞘无毛或基部被毛；叶片正面和边缘具柔毛。圆锥花序呈金字塔形，下垂；分枝粗糙。小穗具小花 2 ~ 3，小花具芒；颖片披针形，草质，9 ~ 11 脉，先端锐尖；外稃革质，先端 2 ~ 4 齿，呈弯曲状。

花果期：花果期 4 ~ 9 月。

分布：亚洲，非洲北部，欧洲。我国大部分地区有分布。南麂主岛各处常见。

生境：耕地、草坡、路旁。

Description: Herbs annual. Culms erect or geniculate at base, 50-150 cm tall, unbranched, 2-4-noded. Leaf sheaths glabrous or basal sheaths puberulous; leaf blades scabrid or adaxial surface and margins pilose. Panicle pyramidal, nodding; branches scabrid. Spikelet with florets 2-3, all florets awned; glumes lanceolate, subequal, herbaceous, 9-11-veined, apex finely acute; lemmas leathery, apex shortly 2-4-toothed, awn geniculate.

Flower and Fruit: Fl. and fr. Apr. -Sep.

Distribution: Asia, North Africa, Europe.Most areas of China. Main island of Nanji, commonly.

Habitat: Cultivated fields, grassy mountain slopes, roadsides.

A103. 禾本科
Poaceae

凤尾竹
Bambusa multiplex* f. *fernleaf
簕竹属 ***Bambusa***

形态特征： 灌木状竹类，茎近直立或顶端稍垂，高 3 ~ 6 m。分枝多数，丛生，枝顶弯曲。茎鞘脱落，不规则四边形，无毛；叶舌有不规则齿。叶片背面浅有白霜，正面翠绿，线形，背面密被短柔毛，正面无毛，排列成羽毛状。花枝从线形到线状披针形，假小穗单生或数个簇生在节上；可育苞片 1 ~ 2，呈卵形或狭卵形，无毛，先端钝或锐尖；小花 5 ~ 13，中间的花能育。

花果期： 成熟颖果未见。

分布： 原产我国，常栽培。我国长江流域以南地区。南麂主岛（美龄居、三盘尾、镇政府），栽培。

生境： 野生或栽培，田野、山地、丘陵和河边。

Description: Shruloby bamboos,culms suberect or apically slightly drooping, 3-6 m. Branches several to many, clustered, the top of branches curved. Culm sheaths deciduous, trapezoid, glabrous; ligule irregularly dentate. Leaf blade abaxially pale glaucous, adaxially bright green, linear, abaxially densely pubescent, adaxially glabrous, arrange in a feather shape. Pseudospikelets solitary or several clustered at nodes of flowering branches, linear to linear-lanceolate; gemmiferous bracts usually 1 or 2, ovate to narrowly ovate, glabrous, apex obtuse or acute; florets 5-13, middle ones fertile.

Flower and Fruit: Not seen mature caryopsis.

Distribution: Native to China,usually cultivated. To the south of the Yangtze River basin. Main island (Meilingju, Sanpanwei, Zhenzhengfu), cultivated.

Habitat: Wild and cultivated, fields, mountains, low hills, riversides.

A103. 禾本科
Poaceae

绿竹
Bambusa oldhamii
箣竹属 *Bambusa*

形态特征：多秆长 6 ～ 12 m，节间略弯曲，初有白粉，无毛，节平。多分枝从秆中向上，多 3 分枝；竿箨脱落，革质，具暗褐色尖毛，不久后脱落；叶耳小，圆形，具缘毛；叶舌近全缘；叶片直立，基部约鞘先端 1/2 宽。叶鞘最初具糙硬毛，叶舌截断；叶耳近圆形，口部刚毛少；叶片长圆状披针形，小穗轴不脱节。颖片 1；外稃类似于颖片，卵形。

花果期：笋期 5 ～ 11 月，花期夏秋季。

分布：我国华东至华南地区。南麂主岛（国姓岙、后隆、门屿尾），栽培。

生境：山坡、山脚、村旁。

Description: Culms 6-12 m, internodes slightly flexuose, initially white powdery, glabrous, nodes flat. Branches many from mid-culm up, 3 dominant; culm sheaths deciduous, leathery, dark brown spinous-hairy, soon glabrescent; auricles small, rounded, ciliate; ligule subentire; blade erect, base ca. 1/2 as wide as sheath apex. Leaf sheaths initially hispid, ligule truncate; auricles suborbicular, oral setae few; blade oblong-lanceolate, rachilla not disarticulating. Glumes 1; lemma similar to glumes, ovate.

Flower and Fruit: New shoots May-Nov. , fl. summer-autumn.

Distribution: East China to South China. Main island (Guoxingao, Houlong, Menyuwei), cultivated.

Habitat: Slopes and foot of moutains, near village.

A103. 禾本科
Poaceae

雀麦
Bromus japonicus
雀麦属 *Bromus*

形态特征：一年生草本。茎直立，高 40 ~ 90 cm。叶鞘具短柔毛；叶片两面具短柔毛。圆锥花序稀疏，有节；分枝 2 ~ 8，纤细，每枝 1 ~ 4 小穗。小穗披针形长圆形，黄绿色，小花 7 ~ 11，紧密重叠；小穗轴节间的短棍棒状；外稃椭圆形，草质，9 脉，通常无毛，先端钝，微 2 齿；内稃短于外稃，刚硬的龙骨具缘毛。

花果期：花果期 5 ~ 7 月。

分布：亚洲，非洲北部，欧洲。我国长江、黄河流域。南麂主岛各处常见。

生境：林缘，路旁，废弃地，河海滩。

Description: Herbs annual. Culms erect, 40-90 cm tall. Leaf sheaths pubescent; leaf blades both surfaces pubescent. Panicle effuse, nodding; branches 2–8, slender, each bearing 1-4 spikelets. Spikelets lanceolate-oblong, yellowish green, florets 7-11, closely overlapping; rachilla internodes shortly clavate; lemmas elliptic, herbaceous, 9-veined, usually glabrous, apex obtuse, minutely 2-toothed; palea shorter than lemma, keels stiffly ciliate.

Flower and Fruit: Fl. and fr. May-Jul.

Distribution: Asia, North Africa, Europe.The Yangtze River and Yellow River Basin. Main island of Nanji, commonly.

Habitat: Forest margins, roadsides, waste ground, river beaches.

A103. 禾本科
Poaceae

拂子茅
Calamagrostis epigeios
拂子茅属 *Calamagrostis*

形态特征：多年生草本，具根状茎，秆直立。叶鞘平滑或稍粗糙，叶舌膜质，先端易破裂，叶片扁平或边缘内卷，上面粗糙，下面较平滑。圆锥花序圆筒形，较紧密，具间断；小穗淡绿色或带淡紫色，颖先端渐尖，第一颖具 1 脉，第二颖具 3 脉，外稃透明膜质，顶端具 2 齿，内稃顶端细齿裂；雄蕊 3，花药黄色。

花果期：花果期 5 ~ 9 月。

分布：亚洲、欧洲。我国各地常见。南麂主岛（关帝岙、三盘尾），常见。

生境：潮湿地和河岸沟渠。

Description: Herbs perennial, rhizomes, culms erect. Leaf sheath smooth or slightly scabrid, ligule membranous, apex easily break, leaf blades flat or margins involute , upper surface rough, lower surface smooth. Panicle cylindric, compact and intermittent; spikelets lightly green or purplish, glume apex acuminate, lower glume 1 vein, upper glume 3-veined, lemma transparent and membranous, apex shortly 2-toothed, top of lemma fine-toothed; Stamens 3, anthers yellow.

Flower and Fruit: Fl. and fr. May-Sep.

Distribution: Asia, Europe. Common in China. Main island (Guandiao, Sanpanwei), commonly.

Habitat: Damp places, especially riversides.

A103. 禾本科
Poaceae

细柄草
Capillipedium parviflorum
细柄草属 *Capillipedium*

形态特征：多年生草本。秆丛生，高 50 ~ 120 cm，不分枝或少分枝。叶舌边缘具短纤毛；叶片线形，顶端长渐尖，基部圆形，两面无毛或被糙毛。圆锥花序长圆形，分枝簇生，枝腋间具细柔毛，小枝为具 3 节的总状花序，节间和花序轴基部具纤毛；无柄小穗第一颖长圆状披针形，背面 2 脉，第二颖上部边缘具纤毛，第二外稃具芒；具柄小穗与无柄小穗等长，常具雄蕊。

花果期：花果期 8 ~ 12 月。

分布：广布旧世界热带与亚热带地区。我国华东、华中至西南地区。南麂主岛（百亩山、门屿尾），偶见。

生境：山坡草地、河边。

Description: Perennial. Culms tufted, 50-120 cm tall, not or little branched. Leaf sheaths glabrous or pilose, ciliate at mouth; leaf blades scaberulous or pubescent, base rounded, apex acuminate. Panicle oblong in outline, branches untidily flexuous, pilose in axils, racemes usually composed of triads, rachis internodes and pedicels shortly ciliate at base; sessile spikelet oblong, lower glume oblong-lanceolate, back 2-veined, upper glume ciliate along upper margins, upper lemma awned; pedicelled spikelet equaling the sessile and often staminate.

Flower and Fruit: Fl. and fr. Aug. -Dec.

Distribution: Widespread in tropical and subtropical areas of the Old World. East China, Central China to Southeast China. Main island (Baimushan, Menyuwei), occasionally.

Habitat: Mountain slopes, streams.

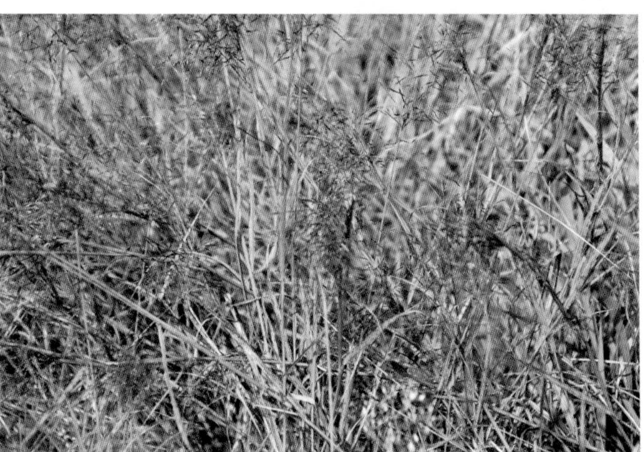

A103. 禾本科
Poaceae

虎尾草
Chloris virgata
虎尾草属 *Chloris*

形态特征：一年生草本，高 15 ～ 100 cm。秆丛生，直立或基部膝曲，稍扁平。基部叶鞘具脊，无毛；叶片平展或折叠，无毛，上面粗糙，先端渐尖。总状花序 5 ～ 12 枚呈指状排列，直立或稍倾斜，淡褐色或略带紫色，花序轴粗糙或具糙硬毛；小穗具 2 或 3 小花，具 2 芒；可育小花外稃倒卵状披针形，脊突起，上部边缘明显具丝状毛，第二小花不育，长圆形，无毛。颖果纺锤形，淡黄色。

花果期：花果期 6 ～ 10 月。

分布：两半球热带至温带地区。分布我国各省区。南麂主岛（大沙岙、马祖岙、美龄居、三盘尾），常见。

生境：石质坡地、河岸沙地、路旁田野，常见于土墙及房顶上。

Description: Herbs annual,15-100 cm tall. Culms tufted, erect or geniculately ascending, slightly flattened. Basal leaf sheaths strongly keeled, glabrous; leaf blades flat or folded, glabrous, adaxial surface scabrous, apex acuminate. Racemes digitate, 5-12, erect or slightly slanting, pale brown or tinged purple, rachis scabrous or hispid; spikelets with 2 or 3 florets, 2-awned, lemma of fertile floret obovate-lanceolate in side view, keel gibbous, conspicuously bearded on upper margins with a spreading tuft of silky hairs, second floret sterile, oblong, glabrous. Caryopsis fusiform, yellowish.

Flower and Fruit: Fl. and fr. Jun. -Oct.

Distribution: Global tropical to temperate regions. Throughout China. Main island (Dashaao, Mazuao, Meilingju, Sanpanwei), commonly.

Habitat: stony slopes, sandy riversides, roadsides, fields, plantations, frequent on walls and roofs.

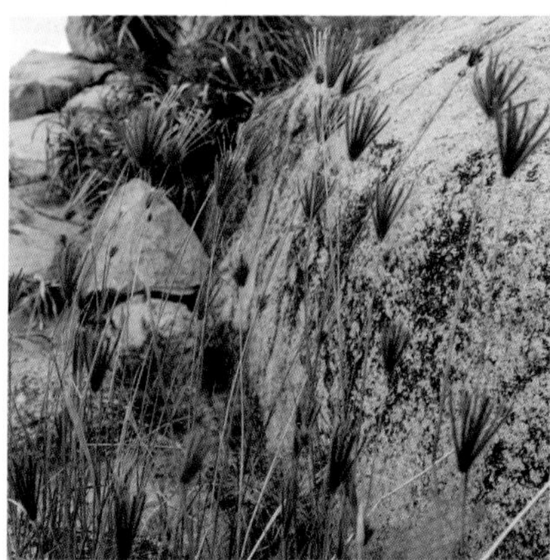

A103. 禾本科
Poaceae

朝阳隐子草（中华隐子草）
Cleistogene shackelii
隐子草属 *Cleistogenes*

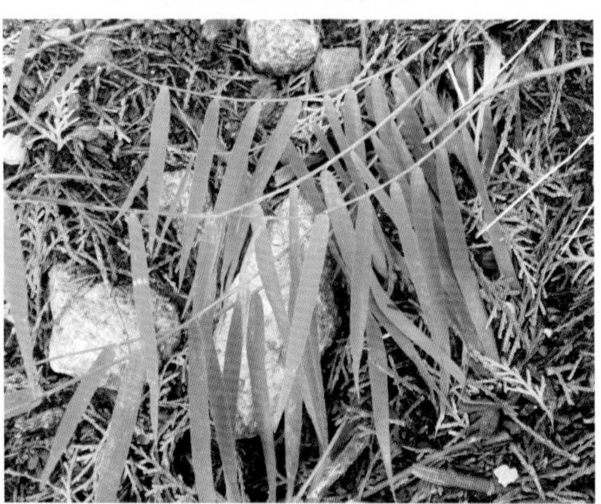

形态特征：多年生草本，高 15 ~ 60 cm。秆丛生，直立，基部密生贴近根头的鳞芽。叶鞘长于节间，鞘口常具柔毛；叶舌短，边缘具纤毛；叶片扁平或内卷。圆锥花序疏展，具 3 ~ 5 分枝，小穗黄绿色或稍带紫色，含 3 ~ 5 小花；颖披针形，先端渐尖；外稃披针形，边缘具长柔毛，具 5 脉，第一外稃先端具芒；内稃与外稃近等长。

花果期：花果期 7 ~ 10 月。

分布：日本、朝鲜。我国长江流域以北地区。南麂主岛（百亩山），偶见。

生境：林缘山坡草地。

Description: Herbs perennial, 15-60 cm tall. Culms tufted, erect, base densely covered with scaly buds close to the root head. Leaf sheath longer than internode, mouth of sheath often pilose; ligule short, margin ciliate; leaf blade flattened or involute. Panicles sparsely expanded, branches 3-5, spikelets yellow-green or slightly purplish, florets 3-5; glumes lanceolate, acuminate at apex; lemma lanceolate, margin villous, 5-veined, lower lemma apex awn; palea subequal to lemma.

Flower and Fruit: Fl. and fr. Jul. - Oct.

Distribution: Japan, Korea .To the north of Yangtze River Basin. Main island (Baimushan), occasionally.

Habitat: Hill slopes in forests, along forest margins.

A103. 禾本科
Poaceae

薏苡
Coix lacryma-jobi
薏苡属 *Coix*

形态特征：一年生草本，秆直立，粗壮，高 1 ~ 3 m，具 10 多节，多分枝。叶茎生，叶鞘短于其节间，无毛；叶片线状披针形，常无毛，中脉粗厚，叶缘粗糙。总状花序腋生，小穗成对，顶端具三分体；雄小穗长圆状卵形，颖片多脉，第一颖龙骨具翅，边缘具纤毛。胞果念珠状，卵球形，骨质，有光泽，无喙。

花果期：花果期 6 ~ 12 月。

分布：热带和亚热带地区。分布于我国南北各地。南麂主岛（门屿尾、兴岙），偶见栽培。

生境：河沟、山谷或溪涧或房前屋后湿润处，常栽培。

Description: Herbs annual. Culms erect, robust, 1-3 m tall, more than 10-noded, branched. Leaves cauline; leaf sheaths shorter than internodes, glabrous; leaf blades linear-lanceolate, usually glabrous, midvein stout, margins scabrous. Raceme axillary, spikelets in pairs with terminal triad; male spikelets oblong-ovate, glumes many-veined, lower glume winged on keels, margin ciliolate. Utricle beadlike, ovoid, bony, glossy, not beaked.

Flower and Fruit: Fl. and fr. Jun. -Dec.

Distribution: Widely in warm regions of the world. Almost throughout China. Main island (Menyuwei, Xingao), occasionally cultivated.

Habitat: Streams, marshy valleys, moist fields, by houses, often cultivated.

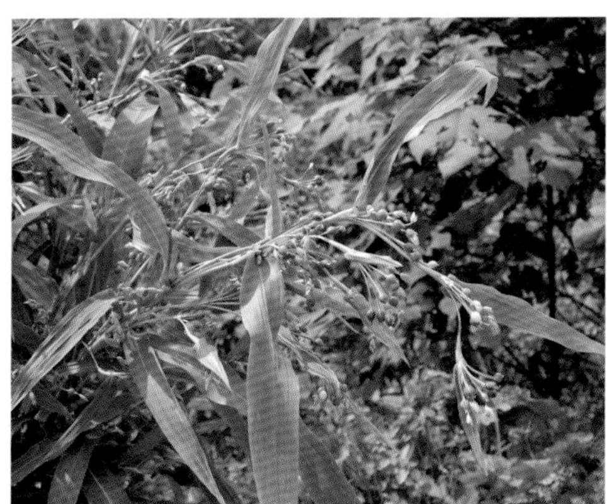

A103. 禾本科
Poaceae

橘草
Cymbopogon goeringii
香茅属 *Cymbopogon*

形态特征：多年生草本。秆直立丛生，高 60 ～ 100 cm，具 3 ～ 5 节，被白粉或微毛。叶鞘无毛，下部者聚集秆基，内面棕红色；叶舌两侧有三角形耳状物并下延为叶鞘边缘的膜质部分；叶片线形，扁平。伪圆锥花序，具 1 ～ 2 回分枝，佛焰苞带紫色，总梗上部生微毛；无柄小穗长圆状披针形；第一颖背部扁平，脊间常具 2 ～ 4 脉；第二外稃 2 裂，芒从先端伸出。

花果期：花果期 7 ～ 10 月。

分布：日本和朝鲜南部。我国华北以南各地。南麂主岛（门屿尾、三盘尾），偶见。

生境：草坡、路旁。

Description: Herbs perennial. Culms erect, tufted, 60-100 cm tall, 3-5-noded, covered with white powder or hairs. Leaf sheath glabrous, lower part gathered at culm base, inner surface brownish red; ligule with triangular ears on both sides and descending into the membranous part of the edge of the leaf sheath, leaf blade linear and flat. False panicle, with 1-2 branches, spathe purplish, pedicels distally microtrichous; sessile spikelet oblong-lanceolate; lower glume flat on the back, 2-4-veind between keels; upper lemma 2-lobed, awn protruding from apex.

Flower and Fruit: Fl. and fr. Jul. -Oct.

Distribution: Japan, south Korea.To the south of North China. Main island (Menyuwei, Sanpanwei), occasionally.

Habitat: Grassy slopes, roadsides.

A103. 禾本科
Poaceae

狗牙根
Cynodon dactylon
狗牙根属 *Cynodon*

形态特征：多年生草本，具匍匐茎。秆纤细，高 10 ~ 40 cm。叶鞘无毛或有疏柔毛，鞘口常具柔毛；叶舌仅为一轮纤毛；叶片线形，短而窄，通常无毛。总状花序 3 ~ 6 枚，指状排列，平展；小穗常有 1/2 ~ 2/3 重叠，颖片线状披针形，略带紫色，1 脉，外稃沿脊具长柔毛，内稃无毛。颖果近圆柱形。

花果期：花果期近全年。

分布：热带与暖温带地区。广布我国黄河以南地区。南麂主岛各处常见。

生境：开阔地，路边、田缘、村庄、河岸或荒地山坡等。

Description: Herbs perennial, stoloniferous. Culms slender, 10–40 cm tall. Leaf sheaths bearded at mouth, otherwise glabrous or thinly pilose; ligule a line of hairs; leaf blades linear, short and narrow, usually glabrous. Inflorescence digitate, racemes 3–6, spreading; spikelets overlapping by 1/2–2/3 their length, glumes linear-lanceolate, often purplish, 1-veined, lemma silky villous along keel, palea glabrous. Caryopsis subterete.

Flower and Fruit: Fl. and fr. nearly all the year.

Distribution: Tropical and warm-temperate regions of the world.Widely distributed to the south of the Yellow River.Main island of Nanji, commonly.

Habitat: Open disturbed situations, roadsides, field margins, village, river bank, hillside of wasteland.

A103. 禾本科
Poaceae

龙爪茅
Dactyloctenium aegyptium
龙爪茅属 *Dactyloctenium*

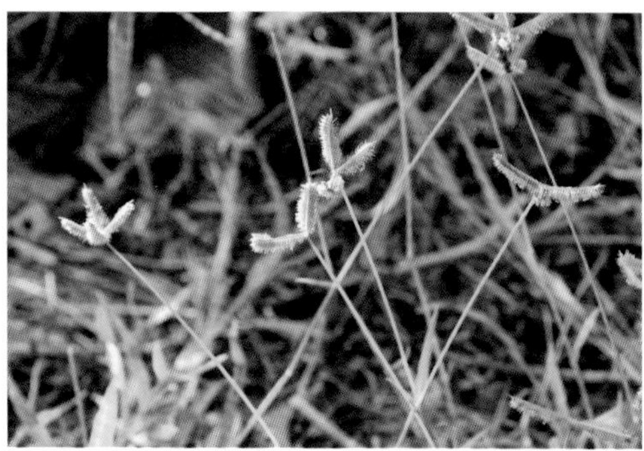

形态特征： 一年生草本，高 15 ～ 60 cm。秆直立，或基部横卧地面。叶鞘松弛，边缘被柔毛；叶舌膜质，顶端具纤毛；叶片扁平，顶端尖或渐尖，两面被疣基毛。穗状花序 2 ～ 7 个指状排列；小穗含 3 小花；第一颖沿脊龙骨状凸起，具短硬毛，第二颖顶端具短芒；外稃中脉成脊，脊上被短硬毛；内稃顶端 2 裂，背缘有翼，翼缘具细纤毛。颖果球状。

花果期： 花果期 5 ～ 10 月。

分布： 旧世界热带及暖温带地区。我国华东、华南至西南等地。南麂主岛（关帝岙、门屿尾、三盘尾），常见。

生境： 受扰杂草地，特别是沙土。

Description: Herbs annual, 15-60 cm tall. Culms erect, or geniculately ascending to shortly stoloniferous. Leaf sheaths loose, margin pilose; ligule membranous, apex ciliate; leaf blade flattened, apex acute or acuminate, both surfaces covered with verrucous basal hairs. Inflorescence digitate, spikes 2-7, spikelets with 3 florets; lower glume keel thick, hispidulous, upper glume extended into a stout scabrid awn; lemma midvein ridged, ridges hispidulous; palea keels winged, wings ciliolate, tip 2-toothed. Caryopsis globose.

Flower and Fruit: Fl. and fr. May-Oct.

Distribution: Tropical and warm-temperate regions of the Old World. East China, South China to Southwest China. Main island (Guandiao, Menyuwei, Sanpanwei), commonly.

Habitat: Disturbed weedy places, especially on sandy soils.

A103. 禾本科
Poaceae

野青茅
Deyeuxia pyramidalis（*Deyeuxia arundinacea*）
野青茅属 *Deyeuxia*

形态特征：多年生草本，高 50 ~ 60 cm。秆直立，节膝曲，丛生。叶鞘疏松裹茎，叶舌膜质，顶端常撕裂；叶片扁平或边缘内卷，无毛，两面粗糙，带灰白色。圆锥花序紧缩似穗状，分枝 3 或数枚簇生；小穗草黄色或带紫色；颖片披针形，先端尖，第一颖具 1 脉，第二颖具 3 脉；外稃稍粗糙，顶端具微齿裂，基盘具柔毛，芒膝曲。

花果期：花果期 6 ~ 9 月。

分布：欧亚大陆的温带地区。分布几遍我国。南麂主岛各处常见。

生境：草坡、开阔林地。

Description: Herbs perennial, 50-60 cm tall. Culms erect, geniculate, tufted. Leaf sheath loosely enclosing stem, ligule membranous, apically often lacerate; leaf blade flattened or margin involute, glabrous, both surfaces scabrid, grayish white.Panicle constricted like spike, branches 3 or several fascicled; spikelets yellow or purplish; glumes lanceolate, apex pointed, lower glume 1-veined, upper glume 3-veined; lemma slightly scabrid, apically microdenticulate, basal disc pilose, awn protruding from lemma near base or below, geniculate, awn column twisted.

Flower and Fruit: Fl. and fr. Jun. -Sep.

Distribution: Temperate regions of Eurasia.Almost throughout China. Main island of Nanji, commonly.

Habitat: Grassy slopes, open woods.

A103. 禾本科
Poaceae

升马唐（纤毛马唐）
Digitaria ciliaris
马唐属 ***Digitaria***

形态特征：一年生草本，高 30 ~ 100 cm。秆基部横卧地面，节处生根和分枝。叶鞘多少具柔毛；叶片线形至线状披针形，上面散生柔毛，边缘加厚，粗糙。总状花序 3 ~ 10 枚，指状或近指状着生；小穗对生，披针形，第一颖很小，三角形，第二颖披针形，3 脉，具柔毛；第一外稃 7 脉，脉间及边缘具柔毛，第二外稃黄绿色，灰色的或淡褐色，椭圆形，先端渐尖。

花果期：花果期 5 ~ 10 月。

分布：热带、亚热带地区。广布我国南北各省区。南麂主岛各处常见。

生境：路旁，杂草荒坡。

Description: Herbs annual, 30-100 cm tall. Culms decumbent at base, branching and rooting at lower nodes. Leaf sheaths ± pilose; leaf blades linear to linear-lanceolate, adaxial surface usually pilose, margins thickened and scabrous. Inflorescence digitate or subdigitate, racemes 3-10; spikelets paired, lanceolate, lower glume very small, triangular; upper glume lanceolate, 3-veined, pilose; lower lemma 7-veined, pubescent to villous, upper lemma yellowish green, gray or pale brown, elliptic, apex acuminate.

Flower and Fruit: Fl. and fr. May-Oct.

Distribution: Tropical and subtropical areas of the world. Throughout the north and south areas of China. Main island of Nanji, commonly.

Habitat: Roadsides, weedy places.

A103. 禾本科
Poaceae

毛马唐
Digitaria ciliaris* var. *chrysoblephara
马唐属 ***Digitaria***

形态特征：一年生草本。与原种升马唐（*Digitaria ciliaris*）的区别在于：第一外稃边缘与侧脉间具柔毛与疣基长刚毛，成熟后开展、变黄，有时刚毛仅存于上部的一对小穗上。

花果期：花果期 6 ~ 10 月。

分布：亚热带和温带地区。我国大部分地区有分布。南麂主岛各处常见。

生境：路旁、田野，杂草荒坡。

Description: Herbs annual. Varietas of *Digitaria ciliaris*, the main difference is the lower lemma pilose and also setose with hard glassy bristles, these spreading and yellowing at maturity; sometimes bristles only present on upper spikelet of a pair.

Flower and Fruit: Fl. and fr. Jun. -Oct.

Distribution: Tropical and warm-temperate regions of the world. Most area of China. Main island of Nanji, commonly.

Habitat: Roadsides, fields, weedy places.

A103. 禾本科
Poaceae

紫马唐
Digitaria violascens
马唐属 *Digitaria*

形态特征：一年生草本，高 20 ～ 60 cm。秆疏丛生。叶鞘无毛或鞘口生柔毛；叶片线状披针形，粗糙，无毛或上面基部具柔毛。总状花序 3 ～ 7 枚近指状着生，斜升；小穗柄稍粗糙，具翅；小穗椭圆状长圆形，三出，具疣基毛，第一颖无，第二颖 3 脉，脉间及边缘生柔毛；第一外稃 5 ～ 7 脉，脉间及边缘生柔毛，第二外稃熟时紫褐色。

花果期：花果期 7 ～ 11 月。

分布：美洲和亚洲的热带地区。分布于我国南北各地。南麂主岛各处常见。

生境：山坡草地、路边、荒野。

Description: Herbs annual, 20-60 cm tall. Culms loosely tufted. Leaf sheaths glabrous or pilose, especially at mouth; leaf blades linear-lanceolate, scabrous, glabrous or adaxial surface pilose at base. Inflorescence subdigitate, racemes 3-7, ascending, rachis ribbonlike, winged; spikelets elliptic-oblong, ternate, hairs verrucose, lower glume absent, upper glume lanceolate, 3-veined, intervein spaces and margins appressed-pubescent; lower lemma 5-7-veined, intervein spaces and margins pubescent, upper lemma dark brown at maturity.

Flower and Fruit: Fl. and fr. Jul. -Nov.

Distribution: Tropical areas of America and Asia. Throughout the north and south areas of China. Main island of Nanji, commonly.

Habitat: Hillsides, grasslands, roadsides and wilderness.

A103. 禾本科
Poaceae

光头稗
Echinochloa colona (Echinochloa colonum)
稗属 *Echinochloa*

形态特征：一年生草本，高 10 ～ 60 cm。秆直立。叶鞘背具脊，无毛；叶舌缺，叶片扁平，无毛，边缘稍粗糙。圆锥花序主轴具棱；小穗卵圆形，具小硬毛，无芒；第一颖三角形，具3脉；第二颖顶端具小尖头，具5 ～ 7脉；第一小花常中性，外稃具7脉，内稃膜质，脊上被短纤毛；第二外稃椭圆形，边缘内卷，包着同质的内稃。

花果期：花果期7 ～ 10月。

分布：世界的温暖地区。我国华东、华中、西南等省区。南麂主岛各处常见。

生境：田野、园圃、路边的湿润地上。

Description: Herbs annual, 10-60 cm tall. Culms erect. Leaf sheaths dorsally crested, glabrous; ligule absent, leaf blade flattened, glabrous, margin slightly scabrid. Panicle main axis ribbed;spikelets ovoid, hispidulous, awnless; lower glume triangular, 3-veined; upper glume with a small tip, 5-7-veined; lower floret usually neutral, lemma 7-veined, palea membranous, crest shortly ciliate; upper lemma elliptic, margin involute, enclosing homogeneous palea.

Flower and Fruit: Fl. and fr. Jul. -Oct.

Distribution: Warm areas of the world. East, Central and Southwest China. Main island of Nanji, commonly.

Habitat: Fields, gardens and wet places along the roadside.

A103. 禾本科
Poaceae

稗
Echinochloa crus-galli (Echinochloa crusgalli)
稗属 *Echinochloa*

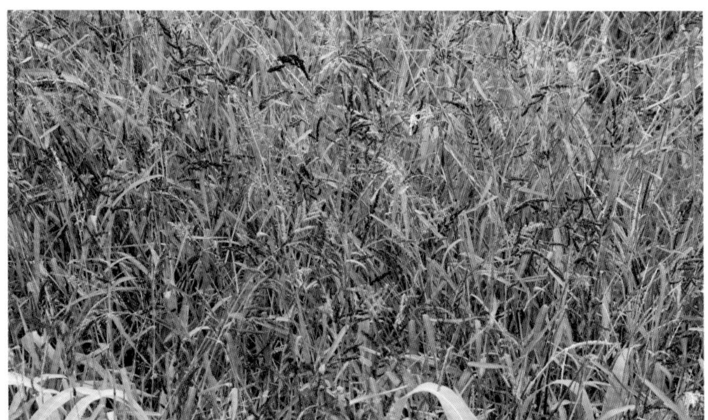

形态特征：一年生草本，秆高 50 ~ 150 cm，无毛。叶鞘疏松裹秆，无毛；叶舌缺；叶片扁平，边缘粗糙。圆锥花序直立；主轴具棱；小穗卵形，脉上密被疣基刺毛，近无柄；第一颖三角形，具 3 ~ 5 脉，脉上具疣基毛；第二颖先端渐尖或具小尖头，具 5 脉；第一小花常中性，外稃草质，7 脉，顶端延伸成芒，内稃薄膜质，具 2 脊；第二外稃椭圆形，顶端具小尖头，包着同质的内稃。

花果期：花果期夏秋季。

分布：全世界温暖地区。分布我国南北各地。南麂主岛各处常见。

生境：沼泽地、沟边及水稻田中。

Description: Herbs annual. Culms 50-150 cm tall, glabrous. Leaf sheath loosely wrapped around culm, glabrous; ligule absent; leaf blade flattened, margin scabrous. Panicle erect; main axis ribbed; spikelets ovate, densely verrucose on veins, subsessile; lower glume triangular, 3-5 veins, with verrucose basal hairs on veins; upper glume apex acuminate or mucronulate, 5-veined; lower floret usually neutral, lemma herbaceous, 7-veined, apically extending into a stout awn, palea membranous, with 2 ridges; upper lemma elliptic, with a small tip, palea homogeneous.

Flower and Fruit: Fl. and fr. summer and autumn.

Distribution: Warm regions of the world. Throughout the north and south areas of China. Main island of Nanji, commonly.

Habitat: Marshes, ditches and rice fields.

A103. 禾本科
Poaceae

牛筋草
Eleusine indica
穇属 *Eleusine*

形态特征：一年生草本，高 10 ～ 90 cm。秆丛生，基部倾斜。叶鞘无毛或疏生疣毛；叶片平展或内折，线形，无毛或上面疏生疣毛。穗状花序 2 ～ 7 枚指状着生，线形，斜升；小穗椭圆形，小花 3 ～ 9，颖披针形，脊粗糙，外稃卵形，脊上具细脉，内稃脊上具狭翼。颖果黑色，卵形，具明显的斜条纹。

花果期：花果期 6 ～ 10 月。

分布：全世界温带和热带地区。分布于我国南北各地。南麂主岛各处常见。

生境：荒芜之地及道路旁。

Description: Herbs annual, 10-90 cm tall. Culms tufted, erect or geniculate at base. Leaf sheaths glabrous or tuberculate-pilose; leaf blades flat or folded, glabrous or adaxial surface tuberculate-pilose. Inflorescence digitate, racemes 2-7, linear, ascending; spikelets elliptic, florets 3-9; glumes lanceolate, scabrid along keel, lemmas ovate, keel with small additional veins, acute; palea keels winged. Caryopsis blackish, ovate, obliquely striate with fine close lines running vertically between the striae.

Flower and Fruit: Fl. and fr. Jun. -Oct.

Distribution: Temperate and tropical areas all over the world. Throughout the north and south areas of China.Main island of Nanji, commonly.

Habitat: Wasteland and roadsides.

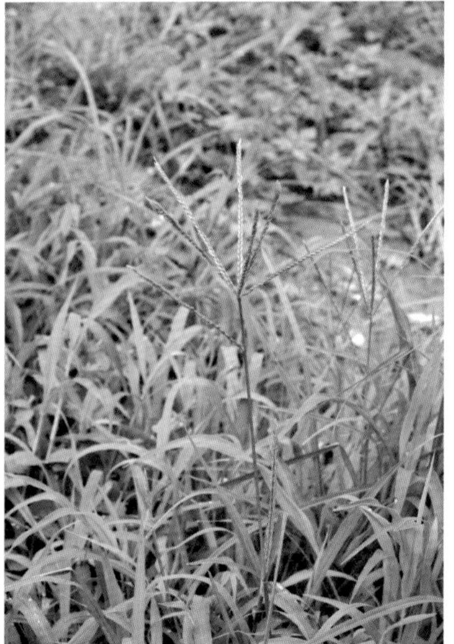

A103. 禾本科
Poaceae

知风草
Eragrostis ferruginea
画眉草属 *Eragrostis*

形态特征：多年生草本。叶鞘两侧压扁，光滑无毛，鞘口与两侧密生柔毛，通常在叶鞘的主脉上生有腺点；叶舌退化为一圈短毛。叶片平展或折叠。圆锥花序大而开展，分枝节密，每节生枝 1 ～ 3 个，向上，枝腋间无毛；小穗长圆形，有 7 ～ 12 小花，多带黑紫色，颖披针形，开展，具 1 脉，先端渐尖；外稃卵状披针形；内稃脊上具有小纤毛，宿存。颖果红棕色。

花果期：花果期 8 ～ 12 月。

分布：东亚、东南亚。分布于我国南北各地。南麂主岛各处常见。

生境：路边、山坡草地。

Description: Herbs perennial.Leaf sheaths laterally compressed, glabrous but along margins and summit densely pilose, sometimes glandular along main vein. Ligules a line of hairs, leaves flat or folded. Panicle large and open, with dense branches, 1-3-branched at each node, upward, glabrous in axils; spikelets oblong, florets 7-12, mostly purplish black, glume lanceolate,open,1-veined, apex acuminate; lemma ovate-lanceolate, palea persi stent, along keels ciliolate. Caryopsis reddish brown.

Flower and Fruit: Fl. andfr. Aug. -Dec.

Distribution: East Asia, Southeast Asia. Throughout the north and south areas of China. Main island of Nanji, commonly.

Habitat: Roadside,hillside grassland.

A103. 禾本科
Poaceae

乱草
Eragrostis japonica
画眉草属 *Eragrostis*

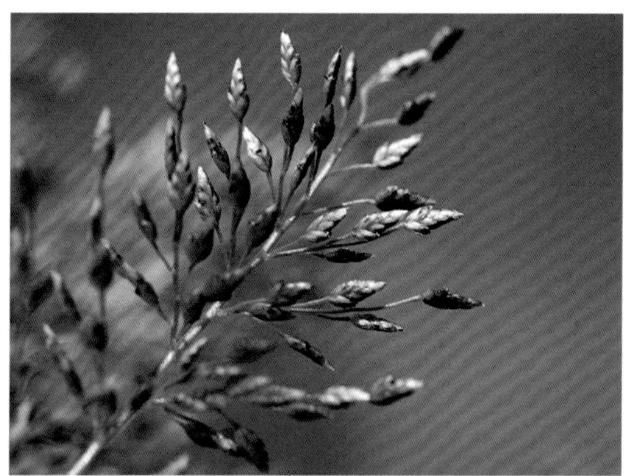

形态特征：一年生草本，高 30 ～ 100 cm。秆直立或膝曲丛生，3 ～ 4 节。叶鞘松裹茎，长于节间，无毛；叶舌干膜质，先端具缘毛；叶片平展，光滑无毛。圆锥花序拉伸，分枝纤细，簇生或轮生，腋间无毛；小穗卵形，熟时略带紫色，小花 4 ～ 8，自小穗轴由上而下逐节断落；颖 1 脉，外稃明显 3 脉，内稃沿 2 脊具短缘毛。颖果红棕色，卵球形。

花果期：花果期 6 ～ 11 月。

分布：东南亚。我国长江流域以南地区。南麂主岛各处常见。

生境：田野，路旁，河边。

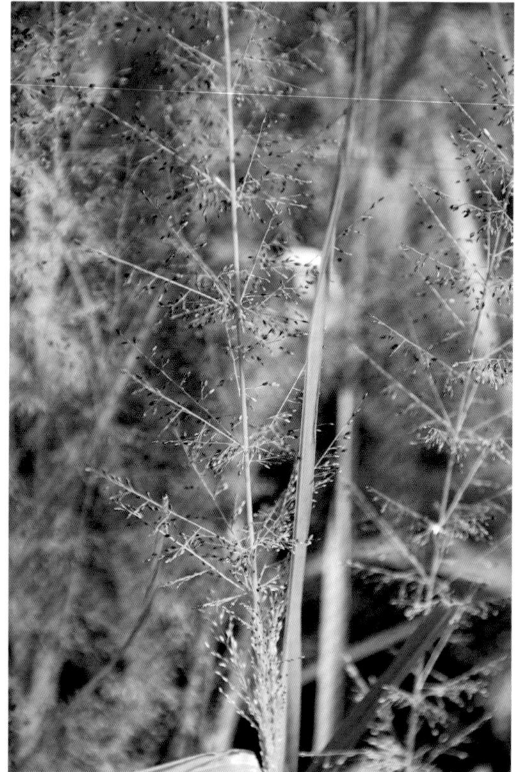

Description: Herbs annual, 30-100 cm tall. Culms erect, or geniculate at base, 3-4-noded. Leaf sheaths usually loose, longer than internodes, glabrous; ligules scarious, fimbriate at apex; leaf blades flat, smooth and glabrous. Panicle elongated, branches slender, clustered or verticillate, glabrous in axils; spikelets ovate, usually purplish at maturity, 4-8-flowered, rachilla distarticulating between florets from top downward at maturity; glumes 1-veined, lemmas distinctly 3-veined, palea along 2 keels ciliolate. Caryopsis red-brown, ovoid.

Flower and Fruit: Fl. and fr. Jun. -Nov.

Distribution: Southeast Asia. To the south of Yangtze River Basin. Main island of Nanji, commonly.

Habitat: Fields, roadsides, stream banks.

A103. 禾本科
Poaceae

野黍
Eriochloa villosa
野黍属 *Eriochloas*

形态特征：一年生草本，高 30 ～ 100cm。秆直立或膝曲上升，基部分枝，节具髭毛。叶片扁平，表面具微毛，背面光滑，边缘微糙，先端锐尖。花序轴长，由 4 ～ 8 枚总状花序组成；总状花序常排列于主轴一侧。小穗卵状椭圆形；第一颖微小；第二颖片和第一外稃 5 ～ 7 脉，被微柔毛；第二外稃微皱，近尖。

花果期：花果期 7 ～ 10 月。

分布：东亚。我国中东部湿润地区。南麂主岛（门屿尾），偶见。

生境：山坡和潮湿地区。

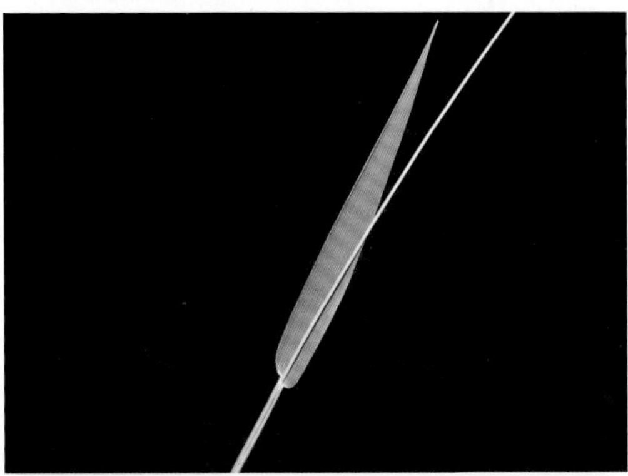

Description: Herbs annual, 30-100 cm tall. Culms erect or geniculately ascending, branching, nodes pubescent. Leaf blades broadly linear, Surface pubescent, abaxially smooth, margins scaberulous, apex acute. Inflorescence axis long, 4-8 racemes; racemes always align one side of main axis. Spikelets ovate-elliptic; lower glume tiny; upper glume and lower lemma 5-7-veined, puberulous; upper lemma weakly rugulose, subacute.

Flower and Fruit: Fl. and fr. Jul. -Oct.

Distribution: East Asia. Humid areas in central and eastern China. Main island (Menyuwei), occasionally.

Habltat: Mountain slopes, moist places.

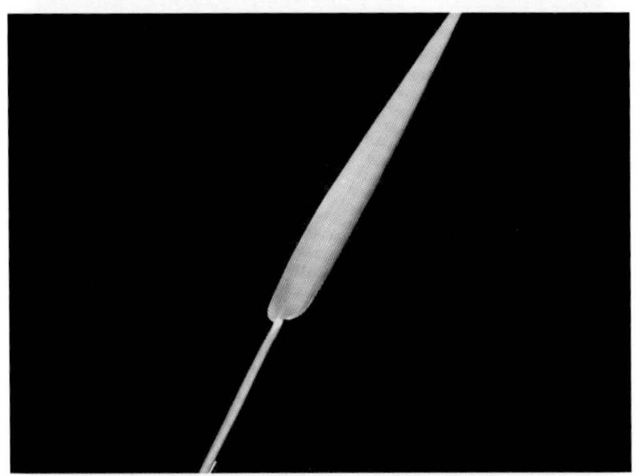

A103. 禾本科
Poaceae

黄茅
Heteropogon contortus
黄茅属 *Heteropogon*

形态特征：多年生草本，高 20 ~ 100 cm。茎纤细，丛生。叶鞘压扁而具脊，叶舌边缘具纤毛；叶片线形，扁平或对折，两面粗糙或表面基部疏生柔毛，顶端钝、渐尖或急尖。总状花序单生于主枝或分枝顶，诸芒常于花序顶扭卷成 1 束，基部 3 ~ 10 对同性小穗；无柄小穗暗褐色，线形，具棕褐色髯毛；有柄小穗，绿色，无毛或被疣基柔毛。

花果期：花果期 4 ~ 12 月。

分布：世界热带亚热带地区，延伸至温暖地区。我国华中、华东、华南、西南地区。南麂主岛（门屿尾、三盘尾），偶见。

生境：干燥的山边、路旁、草地，开阔地或光照充足的阴面。

Description: Herbs perennial, 20-100 cm tall. Culms slender, tufted. Leaf sheaths compressed and keeled, ligule margin ciliate ; leaf blade linear, flat or folded, scabrid or adaxial surface pilose at base, apex obtuse or shortly acute to apiculate.Inflorescence terminal or racemes gathered into a scanty panicle, 3-10 pairs of homogamous spikelets; sessile spikelet dark brown, linear, brown bearded; pedicelled spikelet, greenish, glabrous or tuberculate-hispid.

Flower and Fruit: Fl. and fr. Apr. -Dec.

Distribution: Tropics and subtropics of the world, extending to warm-temperate areas. distributed in Central, East China, South and Southwest China. Main island (Menyuwei, Sanpanwei), occasionally.

Habitat: Dry hillsides, roadsides, grassy places, in the open or light shade.

A103. 禾本科
Poaceae

白茅
Imperata cylindrica
白茅属 *Imperata*

形态特征： 多年生草本，高 25 ~ 120 cm，秆单生或丛生，1 ~ 4 节，节无毛或具髯毛。叶鞘无毛或边缘与鞘口具柔毛；叶片平或卷曲，直立，茎生叶上面被微柔毛，边缘粗糙，先端长渐尖。圆锥花序圆柱形，多毛，有时最下部分枝松散；小穗颖片 5 ~ 9 脉，背面绢毛长约为颖片 3 倍，外稃卵状披针形，具缘毛。颖果椭圆形。

花果期： 花果期 4 ~ 8 月。

分布： 世界热带和温暖地区。我国南北各地有分布。南麂主岛各处常见。

生境： 河岸草地、沙质草甸、荒漠与海滨。

Description: Herbs perennial, 25-120 cm tall. Culms solitary or tufted, 1-4-noded, nodes glabrous or bearded. Leaf sheaths glabrous or pilose at margin and mouth; leaf blades flat or rolled, stiffly erect, culm blades adaxial surface puberulous, margins scabrid, apex long acuminate. Panicle cylindrical, copiously hairy, lowermost branches sometimes loose; spikelets glumes 5-9-veined, back with long silky hairs ca. 3 times glume length, lemma ovate-lanceolate, ciliate. Caryopsis oval.

Flower and Fruit: Fl. and fr. Apr. - Aug.

Distribution: Tropical and temperate areas. Throughout the north and south areas of China. Main island of Nanji, commonly.

Habitat: River and seashore sands, sandy meadow, desert.

A103. 禾本科
Poaceae

毛鸭嘴草
Ischaemum anthephoroides
鸭嘴草属 *Ischaemum*

形态特征：多年生草本。叶鞘疏生疣基毛；叶片先端渐尖，基部楔形，边缘粗糙。总状花序轴节间和小穗均呈三棱形，外侧棱上白色纤毛。无柄小穗披针形；第一颖5～7脉，边缘内折，两侧具脊和翅，第二颖舟形；第一小花雄性；外稃纸质，先端尖，背面微粗糙，具不明显3脉；内稃膜质，具2脊；第二小花两性，芒中部以下膝曲。

花果期：花果期6～9月。

分布：日本、朝鲜。我国浙江、河北、山东。南麂主岛各处常见。

生境：海滩沙地和近海河岸。

Description: Herbs perennial. Leaf sheath sparse wart base hair; leaf blade apex acuminate, base cuneate, margin rough. Internodes and spikelets of raceme axis triangular, with white cilia on the lateral edges. Sessile spikelet lanceolate; lower glume 5-7 veined, edge folded, both sides ridges and wings; upper glume boat-shaped.;lower floret male; lemma papery, apex sharp, back slightly rough, with inconspicuous 3 veins; palea membranous, ridges 2; upper floret bisexual, awn knees below the middle.

Flower and Fruit: Fl. and fr. Jun. -Sep.

Distribution: Japan, Korea. Zhejiang,Hebei, Shandong. Main island of Nanji, commonly.

Habitat: Sand dunes, sandy slopes, near the sea.

A103. 禾本科
Poaceae

千金子
Leptochloa chinensis
千金子属 *Leptochloa*

形态特征：一年生草本，高 30 ~ 90 cm。秆直立，基部膝曲或倾斜，平滑无毛。叶鞘无毛，大多短于节间；叶舌膜质，常撕裂具小纤毛；叶片扁平或多少卷折，先端渐尖，两面微粗糙或下面平滑。圆锥花序分枝及主轴微粗糙；小穗多带紫色，含 3 ~ 7 小花；颖具 1 脉，脊上粗糙，较短而狭窄。颖果长圆球形。

花果期：花果期 8 ~ 10 月。

分布：亚洲东南部。我国东部湿润地区。南麂主岛各处常见。

生境：潮湿的地方。

Description: Herbs annual, 30-90 cm tall. Culms erect, geniculate or decumbent and rooting from nodes, smooth and glabrous. Leaf sheaths glabrous, mostly shorter than internodes; ligule membranous, often torn with small cilia; leaf blades flat or slightly involute, apex acuminate, glabrous, scabrid on both surfaces or abaxial surface smooth. Panicle branches and main axis slightly rough; spikelets mostly purple, florets 3-7；glumes 1-veined, keels rough, short and narrow. Caryopsis oblong and spherical.

Flower and Fruit: Fl. and fr. Aug. -Oct.

Distribution: Southeastern Asia. Moist areas of eastern China. Main island of Nanji, commonly.

Habitat: Moist places.

A103. 禾本科
Poaceae

淡竹叶
Lophatherum gracile
淡竹叶属 *Lophatherum*

形态特征：多年生草本，高 60 ~ 150 cm。秆柔软直立，根结处丛生。叶鞘无毛或具柔毛；叶舌褐色，背有糙毛；叶片披针形，无毛或两面被长毛。总状花序少，具松散毛刺，贴伏小穗初直立，后斜升；小穗狭披针形，近圆柱状，颖卵形，背部圆钝，近革质，具膜质边缘，可育小花外稃长圆形，背部直，无脊，内稃披针形，透明。颖果长椭圆形。

花果期：花果期 6 ~ 10 月。

分布：亚洲、非洲。我国长江流域以南地区。南麂主岛各处常见。

生境：荫蔽的山坡、路边和湿润森林中。

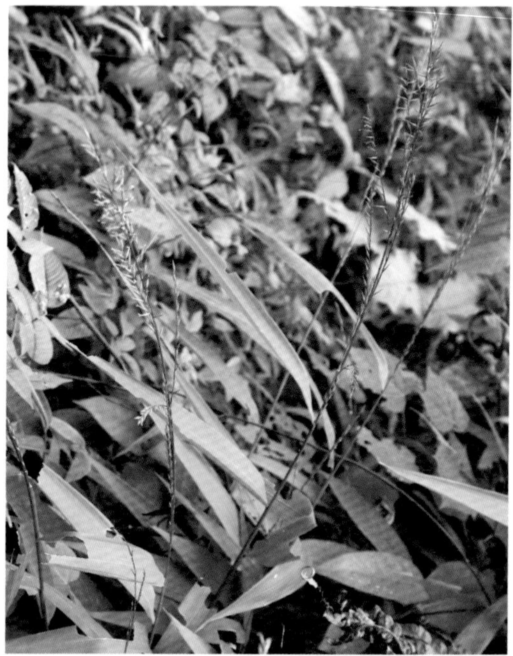

Description: Herbs perennial, 60-150 cm tall. Culms tufted from a knotty rootstock, slender, stiffly erect. Leaf sheaths glabrous or pilose; ligule brown, hispid on backside; leaf blades lanceolate, glabrous or with long hairs on both surfaces. Racemes few, loosely spiculate, erect at first with appressed spikelets, later obliquely spreading; spikelets narrowly lanceolate, subterete, glumes ovate, rounded on back, subleathery with membranous margins; lemma of fertile floret oblong with straight back, not keeled, palea lanceolate, hyaline. Caryopsis oblong.

Flower and Fruit: Fl. and fr. Jun. -Oct.

Distribution: Asia and Africa. To the south of Yangtze River Basin. Main island of Nanji, commonly.

Habitat: Shady slopes, roadsides and in moist forests.

A103. 禾本科
Poaceae

柔枝莠竹（莠竹）
Microstegium vimineum
莠竹属 *Microstegium*

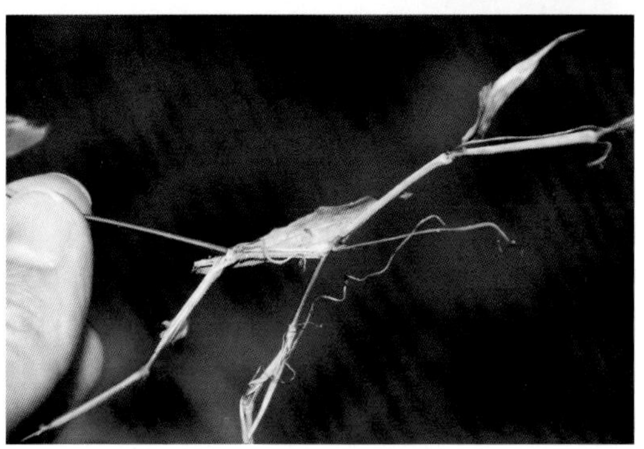

形态特征：一年生草本，高达 1 m。叶鞘短于节间，鞘口具柔毛。叶舌截形，背面无毛。叶片边缘粗糙，顶端渐尖，中脉白色。总状花序 2 ~ 6 枚；第一颖披针形，纸质，背部有凹沟，贴生微毛，先端具网状横脉，沿脊有锯齿状粗糙，内折边缘具丝状毛，顶端尖或有时具二齿；第二颖沿中脉粗糙，顶端渐尖，无芒。颖果长圆形。

花果期：花果期 8 ~ 11 月。

分布：亚洲。我国华北以南地区。南麂主岛各处常见。

生境：林缘与阴湿草地。

Description: Herbs annual, up to 1m tall. Leaf sheath shorter than internode, sheath mouth pilose. Ligule truncated, back glabrous. Leaf blade edge rough, apex acuminate, midvein white. Racemes 2-6, lower glume lanceolate, papery, with concave grooves on the back, attached to the back, reticular transverse veins at the apex, jagged rough along the ridge, filiform hairs at the folded edge, and sharp or sometimes two teeth at the top; upper glume rough along the midvein, apex acuminate ,awanless. Caryopsis oblong.

Flower and Fruit: Fl. and fr. Aug. -Nov.

Distribution: Asia. To the south of North China. Main island of Nanji, commonly.

Habitat: Forest margins, moist grassy places.

A103. 禾本科
Poaceae

五节芒
Miscanthus floridulus
芒属 ***Miscanthus***

形态特征：多年生草本，具发达根状茎。叶茎生，密集；叶鞘无毛，叠生，长于节间；叶片线形，扁平无毛，中脉凸起，边缘粗糙。圆锥花序长圆形或椭圆形，紧密，无毛；分枝较细弱，具多数总状花序，贴伏或斜升；小穗具芒，基盘毛白色，平展；颖片等长，膜质，金褐色；外稃披针形，透明，无脉，具柔毛；芒曲膝状。颖果长圆形。

花果期：花果期 5 ～ 10 月。

分布：东南亚。我国长江流域以南地区。南麂各岛屿常见。

生境：山坡，山谷，长满草的地方。

Description: Herbs perennial, with developed rhizomes. Leaves cauline, congested; leaf sheaths longer than internodes, overlapping, glabrous; leaf blades linear, flat, glabrous, midrib prominent, margins scabrid. Panicle oblong or elliptic, dense,with numerous racemes, appressed or ascending, glabrous; spikelets awned; callus hairs white, spreading; glumes subequal, membranous, golden brown; lower lemma lanceolate, hyaline, veinless, pilose; awn geniculate. Caryopsis oblong.

Flower and Fruit: Fl. and fr. May- Oct.

Distribution: Southeast Asia. To the south of Yangtze River Basin. Throughout Nanji Islands, commonly.

Habitat: Slopes, valleys, grassy places.

A103. 禾本科
Poaceae

山类芦
Neyraudia montana
类芦属 *Neyraudia*

形态特征：多年生草本，秆高约 1 m，密丛型，具 4 ～ 5 节。叶鞘短于其节间，上部者平滑无毛，基部者密生柔毛，枯萎后宿存于秆基；叶舌密生柔毛；叶片内卷，上面有柔毛。圆锥花序分枝微粗糙，斜升；小穗含 3 ～ 6 小花；颖具 1 脉，顶端渐尖或呈锥状；外稃具 3 脉，近边缘生短柔毛，顶端有短芒，基盘具柔毛；内稃稍短于其外稃。

花果期：花果期 8 月

分布：我国华东地区。南麂主岛（百亩山、大山），偶见。

生境：山坡、路旁。

Description: Herbs perennial, culms up to 1 m tall, densely clustered,4-5 noded. Leaf sheath shorter than internodes, upper part smooth, glabrous, base part densely pubescence, persists in culm base after withering; ligule densely pilose; leaf blade involucre, adaxial surface pilose. Panicle branches slightly coarse, obliquely ascending; spikelets contain 3-6 florets; glumes 1-veined, apex acuminate or conical, lemma 3-veined, margin pubescent, apex short awn callus pilose; Palea slightly shorter than lemma.

Flower and Fruit: Fl. and Fr. Aug.

Distribution: East China. Main island (Baimushan, Dashan), occasionally.

Habitat: Mountain slopes, roadsides.

A103. 禾本科
Poaceae

类芦（假芦）
Neyraudia reynaudiana
类芦属 *Neyraudia*

形态特征: 多年生草本，具木质根状茎，须根粗而坚硬。秆直立，通常节具分枝，节间被白粉。叶鞘无毛，仅沿颈部具柔毛；叶舌密生柔毛；叶片扁平或卷折，顶端长渐尖，无毛或上面生柔毛。圆锥花序分枝细长，开展或下垂；小穗含 4 ~ 10 小花，第一外稃不孕，无毛，颖片短小，外稃边脉具白色柔毛，顶端具向外反曲的短芒。

花果期: 花果期 8 ~ 12 月。

分布: 东南亚。我国长江流域以南及西南各地。南麂主岛各处常见。

生境: 溪边、山坡路旁、岩石边或墙角。

Description: Herbs perennial, with woody rhizomes and thick and hard fibrous roots. Culm erect, usually fasciculately branched, internodes somewhat glaucous. Leaf sheath glabrous, pilose only along neck; ligule densely pilose; leaf blade flat or rolled, apex acuminate, glabrous or pilose above. Panicle with slender branches, opening or nodding; spikelets with 4-10 florets, lower lemma infertile, hairless; glumes short, lemmas purplish, lateral veins ciliate with white, soft hairs, awn recurved.

Flower and Fruit: Fl. and fr. Aug. -Dec.

Distribution: Southeast Asia. To the south of Yangtze River Basin and Southeast China. Main island of Nanji, commonly.

Habitat: Streamsides, hill slopes, rocky places, old walls.

A103. 禾本科
Poaceae

求米草
Oplismenus undulatifolius
求米草属 *Oplismenus*

形态特征：多年生草本，高 20 ~ 50 cm。秆纤细，蔓生，基部匍匐上升。叶鞘和叶片常密被疣基毛；叶片披针形至狭卵形，基部近圆形，常歪斜，先端锐尖。花序轴具短硬毛，总状花序 4 ~ 9，退化成簇；小穗 3 ~ 5 对簇生，披针形，具短硬毛，颖草质，具芒，芒粗壮，紫色，有黏性；第一外稃草质，具小尖头，第二外稃近革质，平滑。

花果期：花果期 7 ~ 11 月。

分布：世界温带和亚热带。分布于我国南北各省区。南麂主岛（百亩山、大山、国姓岙、后隆、门屿尾），常见。

生境：林下阴面或潮湿处。

Description: Herbs perennial, 20-50 cm tall. Culms slender, straggling, ascending from a prostrate base. Leaf sheaths and leaf blades usually densely tuberculate-hairy; leaf blades lanceolate to narrowly ovate, base subrounded and usually suboblique, apex acute. Inflorescence axis hispidulous; racemes 4-9, reduced to dense cuneate fascicles; spikelets in 3-5 clustered pairs, lanceolate, hispidulous; glumes herbaceous, awned, the awns stout, purple, viscid; lower lemma herbaceous, apex with a stout mucro, palea absent; upper lemma subcoriaceous, smooth.

Flower and Fruit: Fl. and fr. Jul. -Nov.

Distribution: Temperate and subtropical regions of the world. Throughout the north and south areas of China. Main island (Baimushan, Dashan, Guoxingao, Houlong, Menyuwei), commonly.

Habitat: Light shade in forests, moist places.

A103. 禾本科
Poaceae

铺地黍
Panicum repens
黍属 *Panicum*

形态特征：多年生草本。根茎粗壮发达。秆直立，坚挺，高 50 ～ 100 cm。叶鞘光滑，边缘被纤毛；叶舌顶端被睫毛；叶片质硬，线形，干时常内卷，呈锥形，顶端渐尖，上表皮粗糙或被毛，下表皮光滑。圆锥花序开展，分枝斜上，粗糙，具棱槽；小穗长圆形，无毛，顶端尖；第一颖薄膜质，基部包卷小穗；第二颖具 7 脉；第一小花雄性，第二小花结实，长圆形，顶端尖。

花果期：花果期 6 ～ 11 月。

分布：世界热带和亚热带地区。我国东南各地。南麂主岛各处常见。

生境：海边、溪边以及潮湿之处。

Description: Herbs perennial. Rhizomatous robust. Culms erect and hard, 50-100 cm tall. Leaf sheath smooth, edge ciliated; Ligule apex ciliated; leaf blades hard, linear, often curled in the trunk, tapered, apex acuminate, upper epidermis rough or hairy, lower epidermis smooth. Panicle opening, branches oblique, scabrid, furrowed; spikelets ovate, glabrous, apex acuminate; lower glume membranous, clasping at the base of spikelet; upper glume 7-veined, lower floret male, upper floret female, oblong, apex acuminate.

Flower and Fruit: Fl. and fr. Jun. -Nov.

Distribution: Tropical and subtropical locations worldwide. Southeastern areas of China. Main island of Nanji, commonly.

Habitat: Seaside, stream and moist places.

A103. 禾本科
Poaceae

双穗雀稗
***Paspalum distichum*（*Paspalum paspaloides*）**
雀稗属 *Paspalum*

形态特征：多年生草本，匍匐茎横走、粗壮，节生柔毛。叶鞘龙骨状，无毛，边缘具缘毛；叶片线形，无毛，先端锐尖。总状花序 2 ~ 3 枚对生于短轴上；小穗成 2 列，灰白色，倒卵状长圆形；第一颖退化或微小；第二颖纸质，具柔毛和明显中脉，具 3 ~ 5 脉；第一外稃具 3 ~ 5 脉，通常无毛；第二外稃草质，黄绿色，顶端尖，被毛。

花果期：花果期 5 ~ 9 月。

分布：全世界热带、亚热带地区。我国华东、华南至西南等地。南麂主岛（国姓岙、门屿尾、三盘尾），常见。

生境：田野，路旁，沟边或其他潮湿的肥沃土壤。

Description: Perennial herbs with rhizomes and stolons, Culms nodes usually pubescent. Leaf sheaths keeled, glabrous, margins ciliate; leaf blades linear, glabrous, apex acute. Inflorescence of 2(-3) racemes arising together or separated by a short axis; spikelets single, in 2 rows, pallid, obovate-oblong; lower glume vestigial or small; upper glume papery, 3-5-veined with distinct middle vein, pubescent; lower lemma 3-5-veined, usually glabrous; upper lemma pale green, cartilaginous, apex apiculate and minutely pubescent.

Flower and Fruit: Fl. and fr. May-Sep.

Distribution: Tropical and warm-temperate regions of the world. East, South to Southwest China. Main island (Guoxingao, Menyuwei, Sanpanwei), commonly.

Habitat: Fields, roadsides, ditches and other disturbed places, mostly on moist fertile soils.

A103. 禾本科
Poaceae

圆果雀稗
Paspalum scrobiculatum* var. *orbiculare
雀稗属 *Paspalum*

形态特征：多年生草本，高 30 ~ 90 cm。秆直立，丛生叶鞘较节间长，无毛，鞘口微具长柔毛，基部具白色柔毛。叶长披针形至线形，多数无毛。花序总状，2 ~ 10 枚相互间距排列于主轴上，分枝腋间具长柔毛。小穗椭圆或倒卵形，单生于穗轴一侧，覆瓦状排列为 2 行；第二颖具 3 脉，顶端稍尖；第二外稃等长于小穗，成熟后褐色，革质，有光泽，具细点状粗糙。

花果期：花果期 6 ~ 11 月。

分布：亚洲东南部至大洋洲。分布于我国华东、华南、西南等地区。南麂主岛及大檑山屿各处常见。

生境：山坡、田边、路旁。

Description: Herbs perennial, 30-90 cm tall. Culms erect, clustered; leaf sheath longer than internode, hairless, sheath mouth slightly villous, base white pilose; leaf blade long lanceolate to linear, mostly glabrous. Inflorescences racemose, with 2-10 pieces arranged on the main axis at intervals, villous between the axils of branches; spikelets oval or obovate, solitary on one side of the rachis, in 2 rows, upper glume 3-veined, tip slightly acute; upper lemma as long as spikelets, brown after maturity, leathery and shiny, with pinkish roughness.

Flower and Fruit: Fl. and fr. Jun. -Nov.

Distribution: Southeast Asia to Oceania. East, South and Southwest China. Main island and Daleishan Island, commonly.

Habitat: Hill slopes, roadsides, fields.

A103. 禾本科
Poaceae

显子草
Phaenosperma globosa
显子草属 *Phaenosperma*

形态特征：一年生草本，高 1 ~ 1.5m。秆粗壮，单生或丛生，坚硬，直立或攀援，4 ~ 5 节，不分枝。叶鞘光滑，短于节间，叶片常翻转而使上面向下成深绿色，下面向上成灰白色。圆锥花序成熟时分枝平展；小穗初时狭椭圆状长圆形，熟时开裂，有光泽，第一颖 1 ~ 3 脉，第二颖 3 ~ 5 脉，外稃狭卵形，先端钝。颖果黑褐色，表面具皱纹。

花果期：花果期 5 ~ 9 月。

分布：日本、朝鲜南部、印度东北部。我国华东、华中、西南地区。南麂主岛（百亩山、大山），偶见。

生境：山坡林下、山谷溪旁及路边草丛。

Description: Herbs perennial, 1-1.5 m tall. Culms robust, solitary or tufted, stiff, erect or climbing, 4-5-noded, unbranched. Leaf sheaths smooth, usually shorter than internodes; leaf blades abaxial (upper) surface dark green, adaxial (lower) surface whitish. Panicle branches widely spreading at maturity; spikelets narrowly elliptic-oblong at first, gaping at maturity, glossy, lower glume 1-3-veined, upper glume 3-5-veined, lemma narrowly ovate, apex obtuse. Caryopsis black-brown, rugose.

Flower and Fruit: Fl. andfr. May-Sep.

Distribution: Japan, South Korea, Northeast India. East, Central and Southwest China. Main island (Baimushan, Dashan), occasionally.

Habitat: Grass under forest, beside valley stream and roadside.

A103. 禾本科
Poaceae

芦苇
Phragmites australis
芦苇属 *Phragmites*

形态特征：多年生草本，根状茎发达。秆直立，高 1 ~ 3 m，偶见 8 m，具 20 多节，节下被腊粉。叶鞘淡绿，无毛或覆薄毛；叶舌边缘密生一圈短纤毛；叶披针状线形，无毛，顶端长渐尖。圆锥花序，具多数分枝，着生稠密下垂小穗；小穗具 2 ~ 5 花，颖具锐尖，第一外稃线状披针形，两性外稃狭披针形，顶端长渐尖。

花果期：花果期 7 ~ 11 月。

分布：世界广布。我国各地有分布。南麂主岛各处常见。

生境：沿河岸和湖边的潮湿地区。

Description:Herbs perennial, developed rhizomes. Culm erect, 1-3(8)m tall, 20-knoted, waxed under the knots. Leaf sheath light green, glabrous or thinly hairy; ligule a minute membranous rim, ciliate, leaf blades lanceolate linear, glabrous, apex acuminate. Panicle numerous branched, bearing dense drooping spikelets; spikelets florets 2-5,glumes acute, lowest lemma linear-lanceolate,bisexual lemmas narrowly lanceolate, apex long attenuate.

Flower and Fruit: Fl. and fr. Jul. -Nov.

Distribution: Cosmopolitan. Throughout China. Main island of Nanji, commonly.

Habitat: Moist places along river banks and lake margins.

A103. 禾本科
Poaceae

水竹
Phyllostachys heteroclada
刚竹属 *Phyllostachys*

形态特征：灌木状竹类，竿高可达 6 m，幼竿具白粉并疏生短柔毛。箨鞘背面深绿带紫色，无斑点，被白粉，边缘生纤毛；箨耳小，淡紫色，卵形，缘毛紫色；叶舌短，缘毛白色。叶披针形，每级分枝 2，背面基部有毛。花枝呈紧密的头状，常侧生于老枝上，鳞状苞片 4～6 片；小穗含 3～7 朵小花，外稃披针形，9～13 脉，被短柔毛；鳞被菱状卵形，有 7 条细脉纹，边缘生纤毛；柱头 3，羽毛状。颖果狭卵球形。

花果期：花期 4～8 月。

分布：我国黄河流域及其以南各地。南麂主岛（后隆、门屿尾、三盘尾），大檑山屿，常见。

生境：森林或灌丛山坡上，河岸，山谷。

Description: Shubby bamboos, culms ca. 6 m, initially white powdery, sparsely puberulent. Culm sheaths deep green, tinged with purple, white powdery, margins ciliate; auricles purple, ovate, small, margin ciliate purple; ligule short, white ciliolate. Leaves lanceolate, 2 per ultimate branch, blade abaxially proximally pilose. Flowering branchlets densely capitate, usually lateral on mature leafy branches; scaly bracts 4-6; florets 3-7, lemma lanceolate, 9-13-veined, puberulent; odicules rhomboid-ovate, tenuously 7-veined, margins ciliate; stigmas 3, plumose. Caryopsis narrowly ovoid.

Flower and Fruit: Fl. Apr. -Aug.

Distribution: The Yellow River Basin and to the south of it. Main island (Houlong, Menyuwei, Sanpanwei), Daleishan Island, commonly.

Habitat: Forests or scrub on slopes, river banks, valleys.

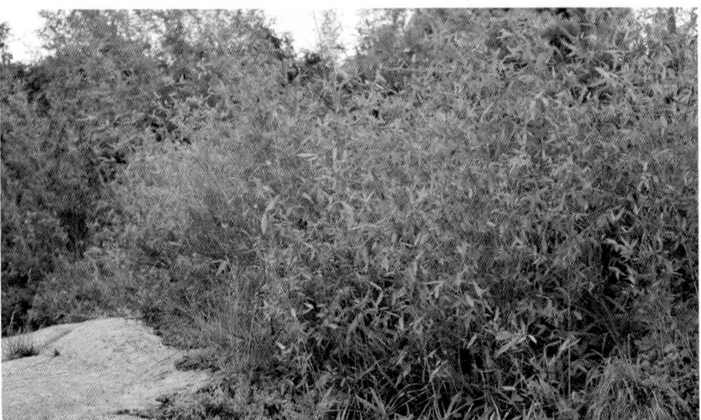

A103. 禾本科
Poaceae

筒轴茅
Rottboellia cochinchinensis
筒轴茅属 ***Rottboellia***

形态特征：一年生草本，须根粗壮。秆直立，高 2 m，无毛；叶鞘具硬刺毛或变无毛；叶舌上缘具纤毛。叶线形，中脉粗壮，无毛或上面疏生短硬毛，边缘粗糙。花序总状，花序轴节间肥厚，易逐节断落。无柄小穗嵌生，第一颖质厚，卵形，先端钝或具 2 ~ 3 微齿，边缘具极窄的翅；第二颖质较薄，舟形；第一小花雄性，第二小花两性。颖果长圆状卵形。

花果期：花果期 9 ~ 11 月。

分布：旧世界热带地区。分布于我国华东、华南、西南等地区。南麂主岛（百亩山），偶见。

生境：光照充足或适阴处，路旁，山坡灌丛，田野，草地。

Description: Herbs annual,with stout fibrous roots. Culms erect, 2 m tall, glabrous; leaf sheaths spiny or become glabrous; ligule upper edge ciliated. Leaves linear, midvein stout, hairless or sparsely short bristles scattered above, rough edges. Inflorescences racemose, internodes of inflorescence axis thick, easy to break off one by one. Sessile spikelet embedded, lower glume thick, ovate, apex blunt or 2-3-toothed, edge very narrow wings; upper glume thin, boat-shaped; lower floret male, upper floret bisexual. Caryopsis oblong-ovate.

Flower and Fruit: Fl. and fr. Sep. -Nov.

Distribution: Throughout the Old World tropics. East, South and Southwest China. Main island (Baimushan), occasionally.

Habitat: Sunny or moderately shady localities, roadsides, hill thickets, fields, grasslands.

A103. 禾本科
Poaceae

甜根子草
Saccharum spontaneum
甘蔗属 *Saccharum*

形态特征：多年生草本，具长根状茎。秆高 1 ～ 4 m，中空，节具短毛，花序以下部分被白色柔毛。叶鞘口和边缘具柔毛，稀为全体被疣基柔毛；叶片线形，无毛，边缘有锯齿。圆锥花序，主轴密生丝状柔毛，总状花序轴疏生丝状长柔毛；第一颖背面无毛，上部边缘具柔毛，先端渐尖，第二颖中脉成脊，边缘具纤毛。

花果期：花果期 7 ～ 9 月。

分布：热带亚热带至暖温带地区。我国南方大部分地区。南麂主岛各处常见。

生境：山坡，砾石河床，低草地。

Description: Herbs perennial, with long rhizomes. Culms 1-4 m tall, often hollow in center, nodes bearded, softly pilose below inflorescence. Leaf sheaths pilose at mouth and margin, sometimes tuberculate-pilose throughout. Leaf blade linear, glabrous, margins serrate. Panicle, axis silky pilose,raceme rachis pilose with long silky hairs; lower glume back glabrous, margins ciliate above, apex acuminate, upper glume midrib ridged, margin ciliated.

Flower and Fruit: Fl. and fr. Jul. -Sep.

Distribution: Tropical and subtropical to warm temperate regions. Most areas in the south of China. Main island of Nanji, commonly.

Habitat: Mountain slopes, gravelly river beds, low grassy places.

A103. 禾本科
Poaceae

囊颖草
Sacciolepis indica
囊颖草属 *Sacciolepis*

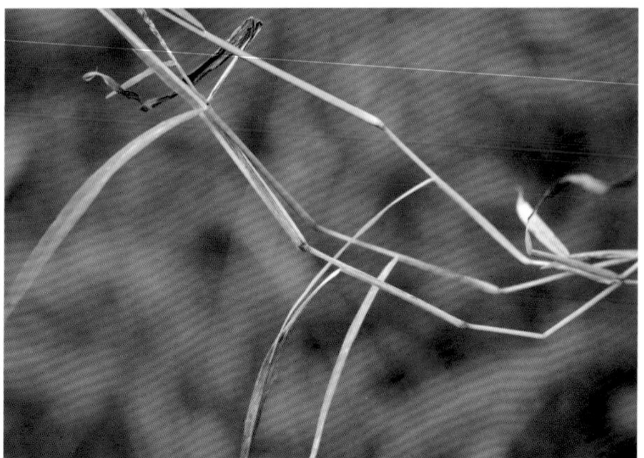

形态特征：一年生草本，秆基膝曲状，高 20 ~ 100 cm。叶鞘具较节间短的棱脊，松弛；叶舌膜质，顶端被短纤毛。叶线形，无毛或被毛。圆锥花序紧缩成圆筒状，主轴无毛，具棱，分枝短；小穗卵状披针形，绿色或带紫色，无毛或具疣基毛；第一颖常具 3 脉，基部包裹小穗，第二颖背部囊状，通常 9 脉，第一外稃膜质，退化或短小，第二外稃边缘包着内稃。颖果椭圆形。

花果期：花果期 7 ~ 11 月。

分布：印度至日本、大洋洲。我国南方大部分地区。南麂主岛（国姓岙、门屿尾、三盘尾），常见。

生境：潮湿地、山谷、溪边。

Description: Herbs annual, culm base knee-curved, 20-100 cm tall. Leaf sheaths shorter ridges than internodes, loose; ligule membranous, apical short fiber hairs; leaves linear, glabrous or coated. Panicle compressed into a cylinder, principal axis glabrous, angular, short-branched; spikelets ovate-lanceolate, green or purplish, glabrous or verrucose; lower glume usually 3-veined, with spikelets wrapped at the base, upper glume saccate at the back, usually 9-veined, lower lemma membranous, degenerated or short, upper lemma wrapped glumelle. Caryopsis oval.

Flower and Fruit: Fl. and fr. Jul. -Nov.

Distribution: India to Japan and Oceania. Most areas in the south of China. Main island (Guoxingao, Menyuwei, Sanpanwei), commonly.

Habitat: Moist places, valleys, streams.

A103. 禾本科
Poaceae

大狗尾草
Setaria faberi
狗尾草属 *Setaria*

形态特征：一年生草本，常具支柱根。秆直立或基部膝曲状，高达 120 cm。叶鞘松弛，边缘具细纤毛，部分基部叶鞘边缘无毛；叶舌密被纤毛；叶线状披针形，边缘具细锯齿。圆锥花序圆柱形，主轴密被长柔毛；小穗椭圆形，下托 1 ~ 3 枚粗糙刚毛；第一颖具 3 脉，第二颖具 5 ~ 7 脉，第一外稃具 5 脉，第二外稃具细横皱纹。颖果椭圆形，顶端尖。

花果期：花果期 7 ~ 10 月。

分布：日本、朝鲜。分布于我国南北各地。南麂主岛各处常见。

生境：山坡、路旁或荒地。

Description: Herbs annual, prop roots. Culms erect or knee-bent at base, up to 120 cm tall. Leaf sheath loose, with fine cilia at the edge, some basal leaf sheath edges glabrous; ligule densely ciliated; leaf blade linear-lanceolate, edge serrulate. Panicle densely cylindrical, axis densely villous; spikelets oval, with 1-3 rough bristles; lower glume 3-veined, upper glume 5-7-veined, lower lemma 5-veined, upper lemma with horizontal wrinkles. Caryopsis elliptic, apex sharp.

Flower and Fruit: Fl. and fr. Jul. -Oct.

Distribution: Japan, Korea. Throughout the north and south areas of China. Main island of Nanji, commonly.

Habitat: Mountain slopes, roadsides, waste places.

A103. 禾本科
Poaceae

金色狗尾草
***Setaria pumila*（*Setaria glauca*）**
狗尾草属 ***Setaria***

形态特征：一年生草本；秆直立或基部膝曲，高 20 ～ 90 cm，光滑无毛。叶鞘无毛，下部具脊，边缘薄膜质；叶舌具纤毛；叶线状披针形，近基部疏生长柔毛。圆锥花序圆柱形，直立，主轴具短细柔毛，刚毛粗糙；通常一簇中仅一个可育小穗；第一颖卵形，具 3 脉，第二颖宽卵形，具 5 ～ 7 脉，第一小花雄性或中性，第一外稃与小穗等长，内稃膜质，第二小花两性，外稃革质，具明显横皱纹。

花果期：花果期 6 ～ 10 月。

分布：原产欧亚大陆的温带和亚热带地区，现广布全世界。分布于我国各地。南麂主岛各处常见。

生境：荒野、山坡、路边和林缘。

Description: Herbs annual; culms erect or geniculate at the base, 20-90cm tall, smooth, glabrous. Leaf sheath glabrous, lower part ridged, margin membranous; ligule ciliated; leaf blades linear-lanceolate, sparsely villous near the base. Panicle densely cylindrical, erect, axis pubescent, bristles coarse; branches reduced to a single mature spikelet; lower glume ovate, 3-veined, upper glume wide ovate 5-7 veined, lower floret male or neutral, lower lemma as long as spikelet, glumelle membranous, upper floret bisexual, leathery, with obvious horizontal wrinkles.

Flower and Fruit: Fl. and fr. Jun. -Oct.

Distribution: Originally from temperate and subtropical Asia and Europe, but now widespread. Throughout China.Main island of Nanji, commonly.

Habitat: Waste places, mountain slopes, roadsides, forest margins.

A103. 禾本科
Poaceae

狗尾草
Setaria viridis
狗尾草属 *Setaria*

形态特征： 一年生草本。秆直立或基部膝曲。叶鞘无毛或具疣状毛，边缘密被长纤毛；叶片线形至线状披针形，平展，无毛或两面具疣状毛，边缘粗糙。圆锥花序紧密，向上疏离，直立或微下垂，主轴被柔毛，刚毛绿色、褐色或紫色；小穗椭圆状长圆形，第一颖 3 脉，长约小穗的 1/3，第二颖 5 ~ 7 脉，几乎与小穗等长，第一外稃与小穗等长，第二外稃苍绿色，长圆形，具点状皱纹，钝圆。颖果灰白色。

花果期： 花果期 5 ~ 10 月。

分布： 旧世界温带、亚热带地区。分布于我国各地。南麂主岛各处常见。

生境： 山坡、路旁或多草的荒地。

Description: Herbs annual. Culms tufted, erect or geniculate. Leaf sheaths glabrous to papillose-pilose, margins densely ciliate; leaf blades linear to linear-lanceolate, flat, glabrous or papillose-pilose on both surfaces, margins scabrous. Panicle dense, usually cylindrical, usually tapering upward, erect or slightly nodding, axis pilose or pubescent, bristles green, brown or purple; spikelets elliptic-oblong, lower glume 3-veined, 1/3 as long as spikelet; upper glume 5-7-veined, as long as spikelet, lower lemma equal to spikelet, upper lemma pale green, oblong, finely punctate-rugose, obtuse. Caropsis grayish white.

Flower and Fruit: Fl. and fr. May-Oct.

Distribution: Temperate and subtropical regions of the Old World. Throughout the north and south areas of China. Main island of Nanji, commonly.

Habitat: Mountain slopes, roadsides, grassy waste places.

A103. 禾本科
Poaceae

光高粱
Sorghum nitidum
高粱属 *Sorghum*

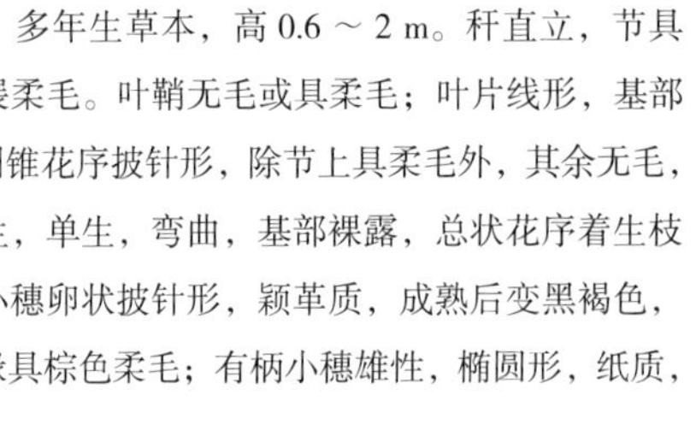

形态特征：多年生草本，高 0.6 ～ 2 m。秆直立，节具灰白色平展柔毛。叶鞘无毛或具柔毛；叶片线形，基部具髯毛。圆锥花序披针形，除节上具柔毛外，其余无毛，初生枝轮生，单生，弯曲，基部裸露，总状花序着生枝顶；无柄小穗卵状披针形，颖革质，成熟后变黑褐色，上部及边缘具棕色柔毛；有柄小穗雄性，椭圆形，纸质，浅棕色。

花果期：花果期夏秋季。

分布：亚洲东部至大洋洲。我国东部湿润地区。南麂主岛（三盘尾），偶见。

生境：草甸，山坡草丛中。

Description: Herbs perennial, 0.6-2 m tall. Culms erect, nodes beared with pale spreading hairs. Leaf sheaths glabrous or pilose; blades linear, beared at base. Panicle lanceolate in outline, glabrous but with soft hairs at the nodes, primary branches whorled, simple, flexuous, lower part bare, racemes borne at branch ends; sessile spikelet ovate-lanceolate; glume leathery, black-brown at maturity, upper part and margins hispid with brown hairs; pedicelled spikelet usually staminate, elliptic, papery, light brown.

Flower and Fruit: Fl. and fr. summer to autumn.

Distribution: East Asia to Oceania. Humid areas in eastern China. Main island (Sanpanwei), occasionally.

Habitat: Meadows, grassy hillsides.

A103. 禾本科
Poaceae

鼠尾粟
Sporobolus fertilis
鼠尾粟属 *Sporobolus*

形态特征：多年生草本，须根较粗壮；秆直立，高 25 ～ 120 cm，平滑无毛。叶鞘疏松裹茎，无毛或边缘具短纤毛；叶舌纤毛状；叶质硬，平滑无毛或仅上面基部具疏柔毛。圆锥花序紧缩成线形，或稠密近穗形；小穗灰绿色，略带紫色；第一颖小，1 脉；外稃先端稍尖。囊果红褐色，长圆状倒卵形或倒卵状椭圆形，顶端截平。

花果期：花果期 3 ～ 12 月。

分布：亚洲东部和东南部。我国南北各地广布。南麂主岛（国姓岙、后隆、门屿尾、三盘尾），常见。

生境：路边、田边、山坡草地及山谷湿处。

Description: Herbs perennial, fibrous roots stout ; culms erect, 25-120 cm tall, smooth, glabrous. Leaf sheath loosely wrapped around the stem, glabrous or with short fiber hairs at the edge; ligule ciliate; leaves hard, smooth, glabrous or only the upper base sparsely pilose. Panicle linear, contracted to spikelike; spikelets grayish green, purplish; lower glume small, 1-veined; lemma with a slightly sharp apex. Grain reddish brown, oblong, obovate or obovate-oval, apex truncated.

Flower and Fruit: Fl. and fr. Mar. -Dec.

Distribution: Eastern and southeastern Asia. Throughout the north and south areas of China. Main island (Guoxingao, Menyuwei, Sanpanwei), commonly.

Habitat: Roadsides, field margins, grassy places on hill slopes, moist ground of mountain valleys.

A103. 禾本科
Poaceae

黄背草（阿拉伯黄背草）
Themeda triandra（*Themeda japonica*）
菅属 *Themeda*

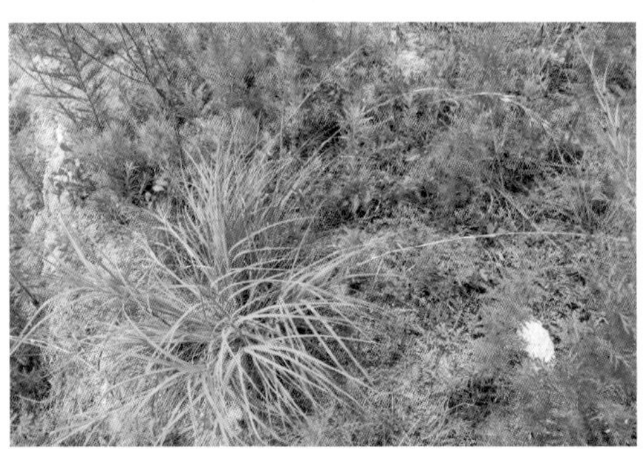

形态特征：多年生草本，高 0.5 ～ 1.5m。秆多毛，淡黄色，常节处被粉。叶鞘常具瘤基硬毛；叶片线形，渐尖，无毛或具柔毛。复合圆锥花序疏松开阔，总状花序由 1 个无柄小穗、2 个有柄小穗和 2 对同性总苞状小穗组成；同性小穗均无柄，雄性，着生于同一平面，无柄小穗两性，基盘被褐色髯毛，具芒，有柄小穗雄性或中性。

花果期：花果期 6 ～ 12 月。

分布：亚洲、非洲、大洋洲。我国大部分地区都有分布。南麂主岛（门屿尾），偶见。

生境：干燥山坡、草地、路旁、林缘。

Description: Herbs perennial, 0.5-1.5 m tall. Culms tussocky, yellowish, usually farinose near nodes. Leaf sheaths usually hispid with tubercle-based hairs; leaf blades linear, finely acuminate, glabrous or pilose. Compound panicle lax, open, raceme composed of a triad of 1 sessile and 2 pedicelled spikelets above the involucre of 2 homogamous pairs; homogamous spikelets all sessile, arising at same level, staminate; sessile spikelet bisexual, callus pungent, brown bearded, awned; pedicelled spikelet male or barren.

Flower and Fruit: Fl. and fr. Jun. -Dec.

Distribution: Asia, Africa, Australia. Most areas of China.Main island (Menyuwei), occasionally.

Habitat: Dry hillsides, grasslands, roadsides, forest margins.

A103. 禾本科
Poaceae

结缕草
Zoysia japonica
结缕草属 *Zoysia*

形态特征：多年生草本，具横走根茎；秆直立，高 15 ~ 20 cm。叶鞘无毛，下部者松弛，上部者紧裹茎；叶舌纤毛状；叶扁平或稍内卷，上面具疏柔毛，下面近无毛。总状花序呈穗状，长 2 ~ 4 cm；小穗柄常弯曲，小穗第一颖退化，第二颖质硬，略有光泽，具 1 脉，顶端钝或渐尖；外稃膜质，长圆形。颖果卵形。

花果期：花果期 5 ~ 8 月。

分布：日本、朝鲜。我国华东、华北、东北等地有分布。南麂主岛（关帝岙、三盘尾），常见。

生境：平原、山坡或海滨草地上。

Description: Herbs perennial, with transverse rhizomes; culms erect, 15-20 cm tall. Leaf sheath glabrous, lower part loose, upper part tightly wraps the stem. Ligule ciliate; leaves flat or slightly curled, adaxially sparsely pubescent and abaxially glabrous; raceme spike-shaped, 2-4 cm long; spikelets often curved, lower glume degenerated, upper glume hard and slightly shiny, 1-vein, apex obtuse or accuminate; lemma membranous, oblong. Caryopsis ovate.

Flower and Fruit: Fl. and fr. May-Aug.

Distribution: Japan, Korea. East, North and Northeast China. Main island (Guandiao, Sanpanwei), commonly.

Habitat: On plain, hillside or seashore grassland.

A103. 禾本科
Poaceae

中华结缕草
Zoysia sinica
结缕草属 *Zoysia*

形态特征：多年生草本，具横走根茎；秆直立，高13～30 cm。叶鞘无毛，鞘口具长柔毛；叶片淡灰绿色，无毛，扁平或边缘内卷。总状花序穗形，伸出叶鞘外；小穗披针形或卵状披针形；颖光滑无毛，侧脉不明显，中脉近顶端与颖分离，延伸成小芒尖，外稃膜质，具1明显的中脉。颖果棕褐色，长椭圆形。

花果期：花果期5～10月。

分布：日本，朝鲜。我国东部沿海地区。南麂主岛（关帝岙、三盘尾），常见。

生境：海边沙滩、延伸至内陆草丛。

Description: Herbs perennial, transverse rhizomes; culms erect, 13-30cm tall. Leaf sheath glabrous, sheath mouth villous; leaf blades pale gray-green, glabrous, flat or margins involute. Racemes spike-shaped, extending out of leaf sheath; spikelets lanceolate or ovate-lanceolate; glumes smooth, glabrous, lateral veins inconspicuous, midvein separated from glumes near the top, and extends into small awn tips; lemma membranous, with 1 obvious midvein. Caryopsis brown, oval.

Flower and Fruit: Fl. and fr. May-Oct.

Distribution: Japan, Korea. Coastal areas of eastern China.Main island (Guandiao, Sanpanwei), commonly.

Habitat: Coastal sands, extending to grazed and trodden places inland.

A106. 罂粟科
Papaveraceae

异果黄堇
Corydalis heterocarpa
紫堇属 *Corydalis*

形态特征：多年生草本，具主根。茎生叶具长柄，二回羽状全裂；一回羽片约5对，具短柄；二回羽片3～5对，近无柄。总状花序生茎和枝顶端，疏具多花和较长的花序轴；花黄色，背部带淡棕色，萼片卵圆形，具短尖；外花瓣顶端圆钝，具短尖，无鸡冠状突起，上花瓣距约占花瓣全长的1/3，末端圆钝，稍下弯，内花瓣基部明显具耳状突起。蒴果长圆形，多少不规则弯曲。

花果期：花果期5月。

分布：日本。我国浙江、山东地区。南麂主岛各处常见。

生境：海岸附近沙石地。

Description: Herbs perennial, with taproot. Cauline leaves with long petiole, bipinnatisect, pinnules ca. 5 pairs, shortly petiolate; bipinnate pinnules 3-5 pairs, nearly sessile. Racemes terminal on stems and branches, densely flowered, inflorescence axis long; flowers yellow, abaxially brownish purple toward apex, sepals ovate-triangular, slightly dentate, outer petals narrowly acute, mucronate, without crest, spur extended through ca. 1/3 of upper petals, apex hooked-curved, inner petals basally auriculate. Capsule oblong, slightly contorted.

Flower and Fruit: Fl. and fr. May.

Distribution: Japan. Zhejiang, Shandong. Main island of Nanji, commonly.

Habitat: Coastal sands.

A109. 防己科
Menispermaceae

木防己
Cocculus orbiculatus
木防己属 *Cocculus*

形态特征：木质藤本。小枝被毛。叶片纸质至近革质，形状变异极大，边全缘至掌状 3 ~ 5 裂不等。聚伞花序具少花，腋生，或具多花组成窄聚伞圆锥花序，顶生或腋生，可至 10 cm 或更长，被柔毛；雄花具 1 或 2 小苞片，被柔毛；萼片 6，外轮卵形或椭圆状卵形，花瓣 6；雄蕊 6，较花瓣短；雌花萼片及花瓣与雄花相同，退化雄蕊 6，微小。核果红或紫红色，近球形。

花果期：花期 5 ~ 8 月，果期 8 ~ 10 月。

分布：亚洲东南部、东部以及夏威夷群岛。分布于我国南北各省区。南麂主岛各处常见。

生境：灌丛、村边、林缘等处。

Description: Woody vines. Young branches striate. Leaf blade papery to subleathery, variable in shape, entire or 3-5-lobed. Inflorescences terminal or axillary, cymose, few flowered, or many flowered arranged in a narrow terminal or axillary thyrse, up to 10 cm or longer, puberulent; male flowers bracteoles 1 or 2, puberulent, sepals 6, outer whorl ovate or elliptic-ovate, petals 6; stamens 6, shorter than petals; female flowers sepals and petals as in male flower; staminodes 6, minute. Drupes rotund, red to reddish purple.

Flower and Fruit: Fl. May-Aug., fr. Aug. -Oct.

Distribution: Southeast and east Asia and the Hawaiian Islands. Throughout the north and south areas of China.Main island of Nanji, commonly.

Habitat: Forests, forest margins, slopes, scrub, open areas, along streams.

A109. 防己科
Menispermaceae

蝙蝠葛
Menispermum dauricum
蝙蝠葛属 *Menispermum*

形态特征：草质藤本。根茎直生，茎自近顶部侧芽生出。叶心状扁圆形，3～9浅裂，基部心形或近平截，掌状脉。圆锥花序单生或双生，具花数朵至20余朵；雄花萼片4～8，倒披针形或倒卵状椭圆形，花瓣6～8，雄蕊常12；雌花具退化雄蕊6～12，雌蕊群具柄。核果紫黑色。

花果期：花期6～7月，果期8～9月。

分布：日本、朝鲜和俄罗斯西伯利亚南部。我国华东、华北和东北地区。南麂主岛（百亩山、国姓岙、后隆），偶见。

生境：路边灌丛或疏林中。

Description: Herbaceous vines. New stems from subapical buds, slender, striate, usually glabrous. Leave blade usually cordate-oblate in outline, shallowly 3-9-lobed; base cordate to subtruncate, margin entire, palmately vein. Inflorescences paniculate, solitary or paired, 20-flowered; male flowers sepals 4-8, oblanceolate to obovate-elliptic,petals 6-8, stamens usually 12; female flowers staminodes 6-12, gynoecium with stalk. Drupes purplish black.

Flower and Fruit: Fl. Jun. -Jul , fr. Aug. -Sep.

Distribution: Japan, Korea and Southern Siberia of Russia. East, North and Northeast China. Main island (Baimushan, Guoxingao, Houlong), occasionally.

Habitat: Shrublands at roadsides, open forests.

A109. 防己科
Menispermaceae

千金藤
Stephania japonica
千金藤属 *Stephania*

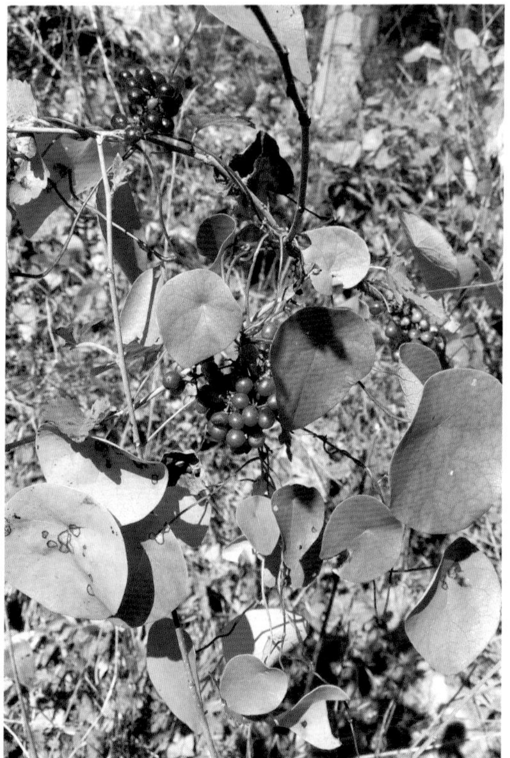

形态特征：稍木质藤本，全株无毛。根褐黄色。小枝纤细。叶盾状着生，三角状圆形或三角状宽卵形，掌状脉 10 ～ 12，叶背粉白色。复伞形聚伞花序腋生，伞梗 4 ～ 8，小聚伞花序近无梗，密集呈头状；雄花萼片 6 或 8，倒卵状椭圆形或匙形，花瓣 3 ～ 4，黄色，稍肉质，聚药雄蕊；雌花萼片及花瓣 3 ～ 4。核果倒卵形或近球形，红色。

花果期：花期春夏季，果期秋冬季。

分布：亚洲东南部至太平洋岛屿。我国长江流域以南各地有分布。南麂主岛各处常见。

生境：村庄，灌丛，林缘以及石灰石山脉等。

Description: Slightly woody vines, glabrous. Root brownish yellow. Stems slender. Leaf blade triangular-rotund or broadly triangular-ovate, conspicuously peltate; palmately 10-12-veined,abaxially glaucous. Inflorescences compound umbelliform cymes, usually axillary, pedicel 4-8, umbellet very condensed, headlike; male flowers sepals 6 or 8, obovate-elliptic to spatulate, petals 3 or 4, slightly fleshy, yellow, synandrium; female flowers sepals and petals 3 or 4. Drupes red, obovate to subglobose.

Flower and Fruit: Fl. spring -summer, fr. autumn -winter.

Distribution: Southeastern Asia to Pacific Islands. To the south of Yangtze River Basin. Main island of Nanji, commonly.

Habitat: Village margins, shrublands, open forests, forest margins, limestone mountains.

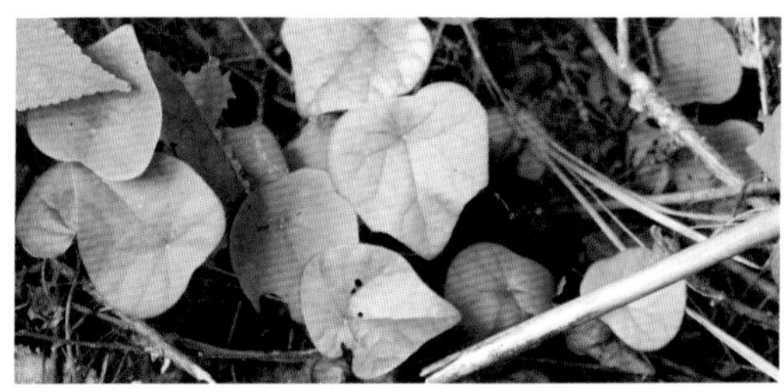

A109. 防己科
Menispermaceae

粪箕笃
Stephania longa
千金藤属 *Stephania*

形态特征：草质藤本，除花序外全株无毛。枝纤细，具纵纹。叶三角状卵形，掌状脉 10 ~ 11；叶柄基部常扭曲。复伞形聚伞花序腋生；雄花萼片 8，楔形或倒卵形，背面被乳头状短毛；花瓣 4 或有时 3，绿黄色；雌花萼片和花瓣均 4 片，偶见 3 片；子房无毛，柱头裂片平叉。核果近球形，红色。

花果期：花期晚春至夏初，果期秋季。

分布：老挝。我国华东、华南地区。南麂主岛（大沙岙、美龄居），偶见。

生境：灌丛或林缘。

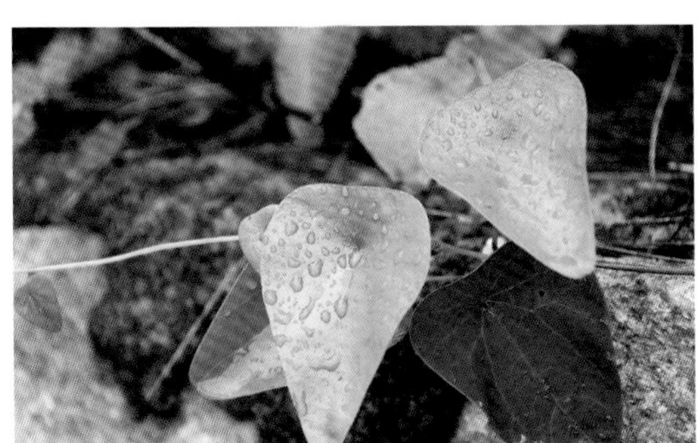

Description: Herbaceous vines, glabrous except for inflorescence. Branches slender, striate. Leaf blade conspicuously peltate, triangular-ovate, palmately 10-11-veined; petiole often twining at base. Inflorescences compound umbelliform cymes, axillary; male flowers sepals 8, cuneate or obovate, abaxially with short papillary hairs petals 4, sometimes 3, greenish yellow; female flowers sepals and petals 4 (or 3); ovary glabrous, stigma lobes divaricate. Drupes red, subglobose.

Flower and Fruit: Fl. late spring-early summer, fr. autumn.

Distribution: Laos. East China, South China. Main island (Dashaao, Meilingju), occasionally.

Habitat: Shrublands, forest margins.

A109. 防已科
Menispermaceae

金线吊乌龟
Stephania cephalantha
千金藤属 *Stephania*

形态特征：草质藤本、落叶、无毛。块根团块状或近圆锥状，有时不规则，褐色，生有许多突起的皮孔。小枝紫红色，纤细。叶纸质，三角状扁圆形至近圆形，边全缘或多少浅波状；掌状脉 7 ~ 9 条；叶柄纤细。雌雄花序同形，均为头状花序，具盘状花托；雄花序常腋生总梗丝状，雌花序单个腋生，总梗粗壮。核果阔倒卵圆形，成熟时红色。

花果期：花期 4 ~ 5 月，果期 6 ~ 7 月。

分布：亚洲和美洲的热带和温带地区。我国长江流域以南地区。南麂主岛（百亩山），偶见。

生境：村边、旷野、林缘。

Description: Herbaceous vines, deciduous, glabrous. Roots tuberous, sometimes irregular, brown, with many projecting lenticels. Branchlets slender, purplish red. Leaf blades membranous or papery, triangular-oblate to rotund, margin entire or subrepand, apex with a finely mucronate acumen; palmately 7-9-veined; petiole slender. Inflorescence homomorphic, capitate, with discoid receptacle; male inflorescences often axillary peduncle filamentous; female inflorescences solitary and axillary, peduncle thicker. Drupes red, broadly rotund.

Flower and Fruit: Fl. Apr. -May, fr. Jun. -Jul.

Distribution: Tropical and temperate regions of Asia and America.To the south of Yangtze River Basin. Main island (Baimushan), occasionally.

Habitat: Village margins, open fields, forest margins.

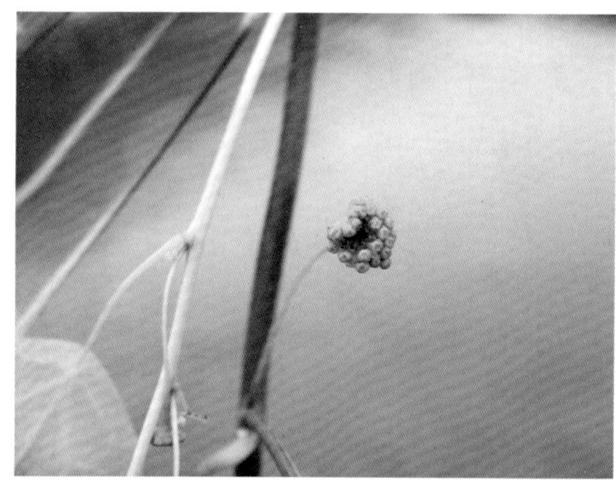

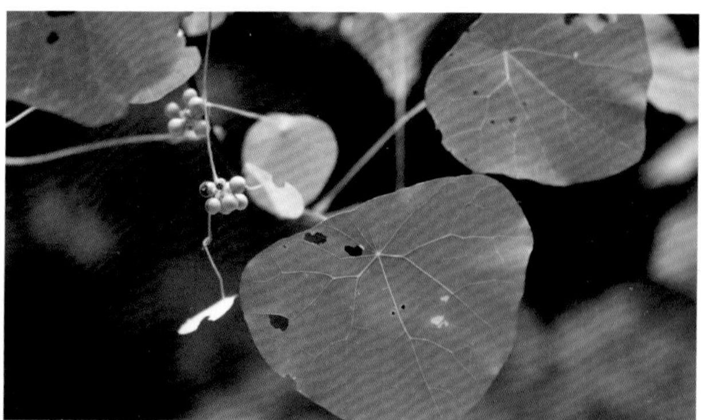

A111. 毛茛科
Ranunculaceae

山木通
Clematis finetiana
铁线莲属 *Clematis*

形态特征：木质藤本。三出复叶；小叶片卵状披针形，狭卵形或卵形，近革质到革质，两面无毛，基部圆形到近心形，边缘全缘，先端锐尖到渐尖。聚伞花序腋生或顶生，苞片三角形到钻形；花药狭长圆形到线形，先端有小尖，子房被短柔毛。瘦果镰刀状狭卵形，有柔毛。

花果期：花期 4 ～ 6 月，果期 7 ～ 11 月。

分布：我国长江流域以南地区。南麂主岛（百亩山、国姓岙、后隆），偶见。

生境：山坡疏林，灌丛及溪边。

Description: Woody vines . Leaves ternately 3-foliolate; leaflet blades ovate-lanceolate, narrowly ovate, or ovate, subleathery to leathery, both surfaces glabrous, base rounded to subcordate, margin entire, apex acute to acuminate. Cymes axillary or terminal, bracts triangular to subulate; anthers narrowly oblong to linear, apex minutely apiculate, ovaries pubescent. Achenes falcate-fusiform, pubsecent.

Flower and Fruit: Fl. Apr. -Jun. , fr. Jul. -Nov.

Distribution: To the south of Yangtze River Basin. Main island (Baimushan, Guoxingao, Houlong), occasionally.

Habitat: Sparse forests, slopes, scrub, along streams.

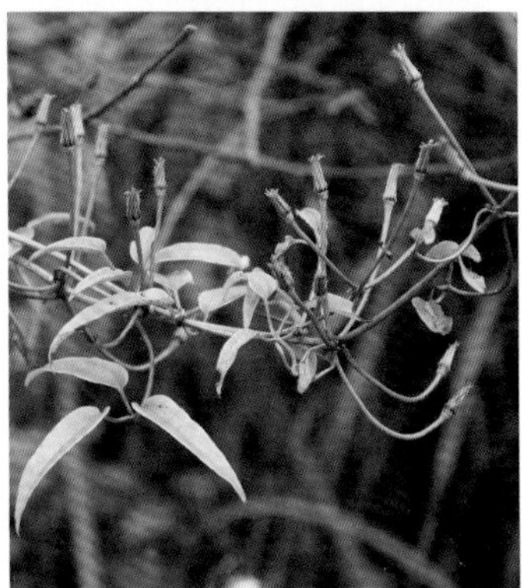

A111. 毛茛科
Ranunculaceae

柱果铁线莲
Clematis uncinata
铁线莲属 *Clematis*

形态特征：木质藤本。分枝具 10 ~ 14 浅纵槽，无毛。一至二回羽状复叶，小叶 5 ~ 15，无毛；小叶薄革质或纸质，卵状椭圆形、卵形、窄卵形或披针形，全缘。聚伞花序腋生并顶生，多花，无毛；苞片钻形或披针形；萼片 4，白色，平展，窄长圆形，边缘被绒毛；雄蕊无毛，花药窄长圆形或线形，顶端具小尖头。瘦果圆柱状钻形。

花果期：花期 6 ~ 7 月，果期 7 ~ 11 月。

分布：日本南部、越南。我国长江流域以南地区。南麂主岛各处常见。

生境：溪边的林缘、灌丛。

Description: Woody vines. Branches shallowly longitudinally 10-14-grooved, glabrous. Leaves pinnate or 2-pinnate, 5-15-foliolate, glabrous; leaflet blades leathery to papery, ovate-elliptic, ovate, narrowly ovate, or lanceolate, margin entire. Cymes axillary or terminal, usually many flowered, glabrous; bracts subulate or lanceolate; sepals 4, white, spreading, narrowly oblong to narrowly obovate-oblong, abaxially glabrous except for velutinous margin; stamens glabrous, anthers narrowly oblong to linear, apex minutely apiculate. Achenes subulate-terete.

Flower and Fruit: Fl. Apr. -Jul. , fr. Jul. -Nov.

Distribution: South Japan, Vietnam. To the south of Yangtze River Basin. Main island of Nanji, commonly.

Habitat: Forest margins, scrub, along streams.

A115. 山龙眼科
Proteaceae

小果山龙眼（越南山龙眼）
Helicia cochinchinensis
山龙眼属 *Helicia*

形态特征：乔木或灌木，高 3 ～ 20m。枝条和叶片均无毛。叶薄革质至纸质，长椭圆形或披针形，顶端短渐尖，基部楔形，全缘或上半部叶缘疏生浅锯齿；侧脉 5 ～ 8 对。总状花序腋生，无毛或有时花序轴和花梗初被白色短毛，后脱落；苞片三角形，小苞片钻形；花被片白色或淡黄色，无毛。果实蓝黑色或黑色，椭球形，果皮厚，革质。

花果期：花期 6 ～ 10 月，果期 11 月至翌年 3 月。

分布：柬埔寨、日本、泰国、越南北部。我国华南地区。南麂主岛（国姓岙），偶见。

生境：常绿阔叶林，混交林，平原，山坡。

Description: Shrubs or trees, 3-20 m tall. Branchlets and leaves glabrous. Leaf blade thin leathery to papery, elliptic or lanceolate, base cuneate, margin entire or upper leaf margin sparsely shallowly serrate; secondary veins 5-8 pairs. Racemes axillary, glabrous, rachis glabrous or sometimes rachis and pedicel whitish pubescent when young, glabrescent; bracts triangular, floral bracts subulate, perianth whitish or yellowish, glabrous. Fruit bluish black to black, ellipsoid, pericarp thick, leathery.

Flower and Fruit: Fl. Jun. -Oct. , fr. Nov. -Mar.

Distribution: Cambodia, Japan, Thailand, North Vietnam, South China. Main island (Guoxingao), occasionally.

Habitat: Broad-leaved evergreen forests, mixed forests, plains, mountain slopes.

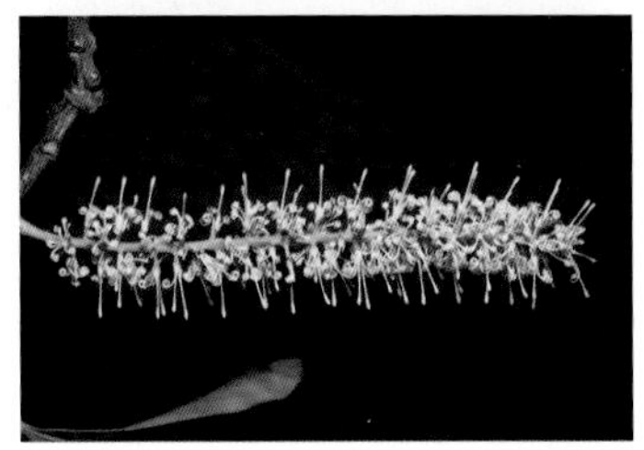

A117. 黄杨科
Buxaceae

雀舌黄杨
Buxus bodinieri
黄杨属 *Buxus*

形态特征：灌木，高 3 ~ 4 m；小枝圆柱形，幼枝四棱形，被短柔毛，后变无毛。叶薄革质，通常匙形，亦有狭卵形或倒卵形，顶端部分最宽，叶面绿色，光亮，基部狭楔形，先端圆或钝，常有浅凹口，中脉凸出，侧脉在两面或仅叶面显著。花序腋生，头状，花密集；苞片卵形，背面无毛或有短柔毛；不育雌蕊有柱状柄，先端膨大。蒴果卵形。

花果期：花期 2 月，果期 5 ~ 8 月。

分布：原产于我国中西部地区，现广为栽培。南麂主岛（美龄居、三盘尾、镇政府），栽培。

生境：温暖湿润和阳光充足环境。

Description: Shrubs, 3-4 m tall; branchlets terete, young branches tetragonous, pubescent or glabrous. Leaf blade thinly leathery, usually spatulate, also narrowly ovate or obovate, widest in apical part, adaxially green and shining, base narrowly cuneate, apex rounded or obtuse, usually with retuse tip, midrib elevated, lateral veins visible on both surfaces or only adaxially. Inflorescence axillary, capitate, flowers dense; bracts ovate, glabrous abaxially or pubescent; sterile pistil with terete gynophore, apex inflated. Capsule ovoid.

Flower and Fruit: Fl. Feb. , fr. May-Aug.

Distribution: Native in central to west China, widely cultivated for ornamental. Main island (Meilingju, Sanpanwei, Zhenzhengfu), cultivated.

Habitat: Warm, humid and sunny environment.

A117. 黄杨科
Buxaceae

黄杨
Buxus sinica
黄杨属 *Buxus*

形态特征：灌木或小乔木，高 1 ~ 6 m；小枝圆柱状，有纵棱，灰白色；幼枝四棱，被短柔毛。叶革质，形状和大小不一，叶面光亮，中脉凸出。花序腋生，头状，被毛，苞片阔卵形，背部多少有毛；不育雌蕊有棒状柄，先端稍膨大；子房较花柱稍长，无毛，花柱粗扁，柱头倒心形，下延达花柱中部。蒴果近球形。

花果期：花期 3 月，果期 5 ~ 6 月。

分布：分布于我国大多数省区，广为栽培。南麂主岛（美龄居、三盘尾、镇政府），栽培。

生境：山谷、山坡林下，灌丛，溪边及石质地区。

Description: Shrubs or small trees, 1-6 m tall;branchlets terete, longitudinally ribbed, grayish white; young branches tetragonous, pubescent. Leaf blade varied in shape and size, leathery, shining adaxially, midrib elevated adaxially. Inflorescences axillary, capitate, pubescent, bracts broadly ovate, ± pubescent abaxially; sterile pistil with clavate gynophore, apex slightly inflated; ovary slightly longer than style, glabrous, style thick and compressed, stigma obcordate, decurrent to middle part of style. Capsule subglobose.

Flower and Fruit: Fl. Mar. , fr. May-Jun.

Distribution: Native in most provinces of China, widely cultivated for ornamental. Main island (Meilingju, Sanpanwei, Zhenzhengfu), cultivated.

Habitat: Forests in mountain valleys and on slopes, thickets, streamsides, stony areas.

A123. 蕈树科
Altingiaceae

枫香树
Liquidambar formosana
枫香树属 *Liquidambar*

形态特征：落叶乔木。小枝被柔毛。叶宽卵形，掌状3裂，基部心形具锯齿；托叶红色，线形，早落。短穗状雄花序多个组成总状；雄蕊多数，花丝不等长；雌花序头状，萼齿4～7，针形，子房被柔毛，花柱卷曲。果序球形，木质，宿存针刺状萼齿及花柱。种子多数，褐色，多角形或具窄翅。

花果期：花期3～6月，果期7～9月。

分布：越南北部，老挝及朝鲜南部。我国黄河以南各地区。南麂主岛偶见栽培。

生境：村落或低山林附近的向阳地。

Description: Trees deciduous. Branchlets puberulent. Leaf blade broadly ovate, palmately 3-lobed, base cordate, margin glandular serrate; stipules red, linear, caducous. Male inflorescence a short spike, several to many together in racemes; stamens many, filaments unequal; female inflorescence capitate, staminode teeth 4-7,needlelike, ovary pilose, style curled.Infructescence globose, woody, with persistent spinescent calyxtooth and style. Seeds many, brown, polygonal or with narrow wings.

Flower and Fruit: Fl. Mar. -Jun. , fr. Jul. -Sep.

Distribution: North Vietnam, Laos and South Korea. To the south of Yellow River. Cultivated on main island of Nanji, occasionally.

Habitat: Sunny places, near villages, montane forests.

A124. 金缕梅科
Hamamelidaceae

檵木
Loropetalum chinense
檵木属 *Loropetalum*

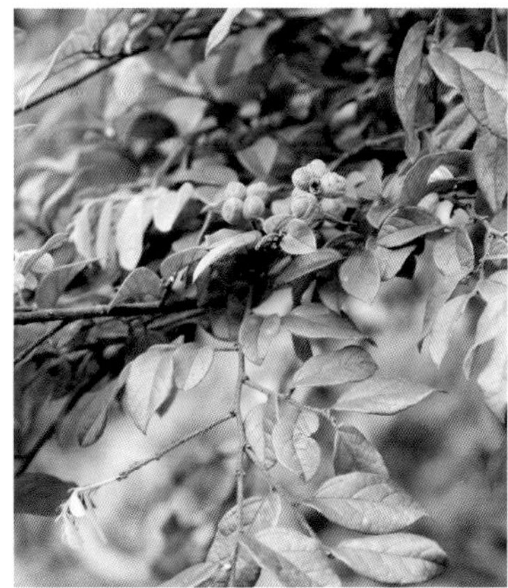

形态特征：灌木或小乔木，多分枝，小枝有星毛。叶革质，卵形，背面密被星状短柔毛，幼时正面疏生短柔毛或星状短柔毛，后脱落，先端尖锐，基部钝，不等侧。花序为短的总状花序或近头状花序，顶生，大部分在短的侧枝上；花有短花梗，先叶开放；花瓣 4，白色或淡黄色，先端圆钝。蒴果卵形或倒卵球形，被星状绒毛。种子卵球形或椭圆形。

花果期：花期 3 ~ 4 月，果期 5 ~ 7 月。

分布：日本、印度。我国长江流域以南地区。南麂主岛各处常见。

生境：向阳的丘陵及山地。

Description: Shrubs or small trees, much branched, branchlets stellately pubsecent. Leaf blade leathery, ovate, abaxially densely stellately pubescent, adaxially sparsely pubescent or stellately pubescent when young, glabrescent, apex acute, base obtuse, asymmetrical. Inflorescence a short raceme or nearly capitate, terminal, mostly on short lateral branches; flowers shortly pedicellate, open before leaves appear; petals 4, white, pale yellow, apex obtuse or rounded. Capsule ovoid or obovoid-globose, stellately tomentose, Seeds ovoid-globose or ellipsoid.

Flower and Fruit: Fl. Mar. -Apr. , fr. May-Jul.

Distribution: Japan, India. To the south of Yangtze River Basin. Main island of Nanji, commonly.

Habitat: Sunny hills and mountains.

A124. 金缕梅科 Hamamelidaceae

红花檵木

Loropetalum chinense var. *rubrum*

檵木属 *Loropetalum*

形态特征：灌木或小乔木。叶革质，卵形，先端尖锐，基部钝。花 3 ～ 8 朵簇生，有短花梗，比新叶先开放，或与嫩叶同时开放，花序柄被毛，花瓣 4，红色或紫红色。蒴果卵圆形或倒卵球形，被褐色星状毛，先端圆。种子卵圆形，黑色发亮。

花果期：花期 3 ～ 4 月。

分布：原产湖南、广西地区，我国南方广泛栽培。南麂主岛（美龄居、三盘尾、镇政府），栽培。

生境：灌丛。

Description: Shrubs or small trees. Leaf blade leathery, ovate, apex acute, base obtuse. Flowers 3-8, cespitose, pedicel short, flowers open before or at the same time as new leaves, inflorescence handle pubescent, sepals 4, red or purple red. Capsule ovoid, ovoid or obovoid-globose, stellately tomentose, hairs brown, apex rounded. Seeds ovoid, black, shiny.

Flower and Fruit: Fl. Mar. -Apr.

Distribution: Native in Hunan, Guangxi. Widely cultivated in southern China. Main island (Meilingju, Sanpanwei, Zhenzhengfu), cultivated.

Habitat: Thickets.

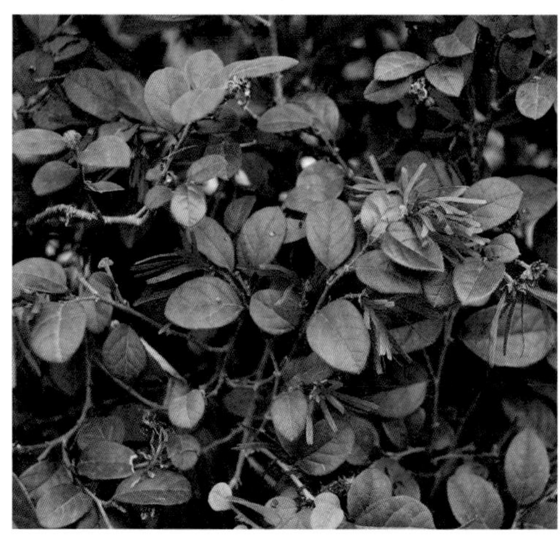

A126. 虎皮楠科 Daphniphyllaceae

琉球虎皮楠
Daphniphylum luzonense
虎皮楠属 *Daphniphyllum*

形态特征：常绿灌木或小乔木，高 1.5 ~ 5 m。小枝粗壮，髓心片状。叶互生，常聚生于枝顶；叶片厚革质，长圆形或长椭圆形，叶缘明显反卷，先端钝尖，基部近圆形，上面具光泽，下面粉绿色，具细小乳头状突起。总状花序腋生；花单性异株；无花被。核果椭球形，成熟时呈紫褐色至紫黑色；果核近黑色，表面有不规则瘤突。

花果期：花期 4 ~ 5 月，果期 10 月至翌年 1 月。

分布：日本。浙江、台湾。南麂主岛（打铁礁、关帝岙、国姓岙、门屿尾、三盘尾），常见。

生境：海岛山坡、灌丛、林中或岩质海岸石缝中。

Description: Evergreen shrubs or small trees, 1.5-5 m tall. Branchlets thick, pith heart flake. Leaves alternate, often clustered on top of branches; leaf blade thick leathery, oblong or oblong, leaf margin distinctly inverted, apex obtuse, base subrounded, upper mask glossy, lower flour green, with fine papillary processes. Racemes axillary; flowers unisexual heterologous; without perianth. Drupe ellipsoid, purplish brown to purplish black when mature; nucleus nearly black, with irregular nodules on surface.

Flower and Fruit: Fl. Apr. -May, fr. Oct. -Jan.

Distribution: Japan. Zhejiang, Taiwan,China. Main island (Datiejiao, Guandiao, Guoxingao, Menyuwei, Sanpanwei), commonly.

Habitat: Island slopes, thickets, under forest, rocky coastal, rock crevices.

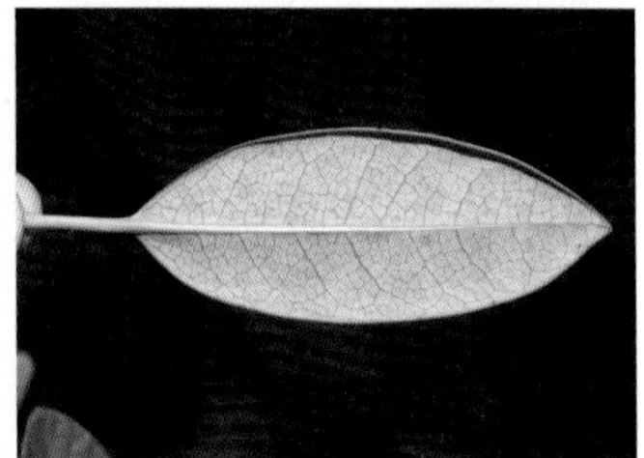

A129. 虎耳草科
Saxifragaceae

虎耳草
Saxifraga stolonifera
虎耳草属 *Saxifraga*

形态特征：多年生草本，高 8 ～ 45 cm。匍匐枝细长，密被卷曲长腺毛，具鳞片状叶。基生叶具长柄，叶片近心形、肾形至扁圆形；茎生叶披针形。聚伞花序圆锥状；花序、花梗被腺毛；花两侧对称，白色，中上部具紫红色斑点，基部具黄色斑点，5 枚。蒴果卵圆形，先端 2 深裂，喙状。

花果期：花果期 4 ～ 11 月。

分布：东亚。我国广布。南鹿主岛（国姓岙、门屿尾），偶见。

生境：林下、灌丛、草甸和阴湿岩隙。

Description: Herbs perennial, 8-45 cm tall. Stolons filiform, densely crisped glandular villous, with scaly leaves. Basal leaves with petiole, leaf blade subcordate or reniform to orbicular; cauline leaves lanceolate. Cymes paniculate; inflorescence and pedicel glandular hair; flower zygomorphy, white, middle and upper with purple-red spots, basally with yellow spots, petals 5. Capsule ovoid, bifid at apexs, coronoid.

Flower and Fruit: Fl. and fr. Apr. -Nov.

Distribution: East Asia. Widely in China. Main island (Guoxingao, Menyuwei), occasionally.

Habitat: Forests, scrub, meadows, shaded rock crevices

A130. 景天科
Crassulaceae

东南景天
Sedum alfredii
景天属 *Sedum*

形态特征：多年生草本，高 10 ~ 20 cm。茎斜上，单生或上部有分枝。叶互生，下部叶常脱落，上部叶常聚生，叶片线状楔形、匙形至匙状倒卵形，有距，全缘。聚伞花序多花，花无梗，花瓣 5 枚，黄色，披针形至披针状长圆形。蓇葖果斜叉开。种子多数，褐色。

花果期：花期 4 ~ 5 月，果期 6 ~ 8 月。

分布：东亚。我国长江流域以南。南麂主岛各处常见。

生境：山坡林下的阴湿石上。

Description:Herbs perennial, 10-20 cm tall. Stems simple or apically branched, ascending. Leaves alternate proximally on stem; usually deciduous, crowded distally on stem, leaf blade linear-cuneate, spatulate, or obovate, base narrowly cuneate and spurred, apex obtuse and sometimes emarginate. Cyme corymbiform, flowers many sessile, unequally 5-merous, petals yellow, lanceolate to lanceolate-oblong. Follicles obliquely divergent. Seeds numerous, brown.

Flower and Fruit: Fl. Apr. -May, fr. Jun. -Aug.

Distribution: East Asia. To the south of Yangtze River Basin. Main island of Nanji, commonly.

Habitat: Shady moist rocks on forested slopes.

A130. 景天科
Crassulaceae

圆叶景天
Sedum makinoi
景天属 ***Sedum***

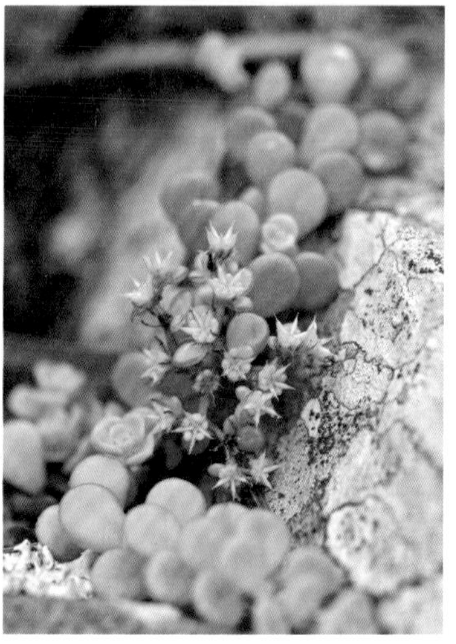

形态特征：多年生草本，高 15 ～ 25 cm。茎下部节上生根，上部直立，无毛。叶对生，倒卵形至倒卵状匙形。聚伞状花序，二歧分枝，花无梗，萼片 5 枚，线状匙形，花瓣 5 枚，黄色，披针形。蓇葖果斜展。种子细小，卵形，有微乳头状突起。

花果期：花期 6 ～ 7 月，果期 8 ～ 9 月。

分布：日本。我国浙江、安徽地区。南麂主岛各处常见。

生境：低山山谷林下阴湿处。

Description: Herbs perennial,15-25 cm tall. Stems basally prostrate and rooting at nodes, apically erect,glabrous. Leaves opposite, obovate to obovate-spatulate. Cyme 2-branched, flowers sessile, unequally 5-merous, sepals linear-spatulate, petals 5, yellow, lanceolate. Follicles obliquely divergent. Seeds ovoid, small, minutely mammillate.

Flower and Fruit: Fl. Jun. -Jul, fr. Aug. -Sep.

Distribution: Japan. Zhejiang, Anhui. Main island of Nanji, commonly.

Habitat: Shady moist forests in low mountain valleys.

A130. 景天科 Crassulaceae

垂盆草
Sedum sarmentosum
景天属 *Sedum*

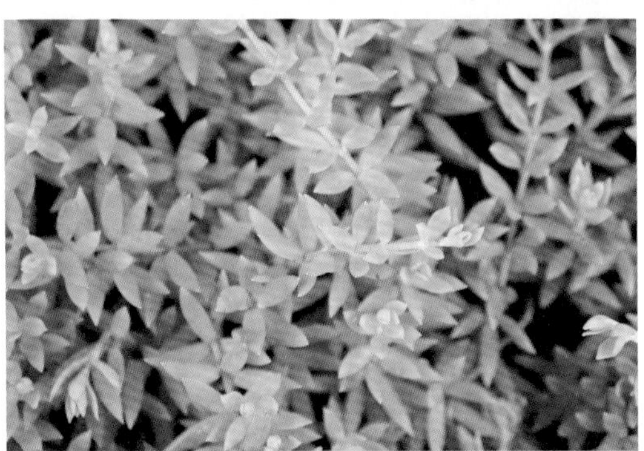

形态特征：多年生草本。不育枝及花茎细，匍匐而节上生根。3 叶轮生，叶倒披针形至长圆形。聚伞花序，有 3～5 分枝，花少，无梗，萼片 5 枚，花瓣 5 枚，黄色，披针形至长圆形。蓇葖果长圆形，略叉开，有长花柱。种子卵形。

花果期：花期 5～7 月，果期 8 月。

分布：日本、朝鲜、泰国北部。我国华南以北各地广泛分布。南麂主岛（大沙岙、镇政府），偶见栽培。

生境：山坡阴处或石上。

Description: Herbs perennial. Sterile and flowering stems creeping and rooting at nodes toward inflorescences, slender. Leaves 3-verticillate, leaf blade oblanceolate to oblong. Cyme 3-5-branched, corymbiform, few flowered, flowers sessile, sepals 5, petals 5, yellow, lanceolate to oblong. Follicles oblong, slightly divaricate, macrostylous. Seeds ovoid.

Flower and Fruit: Fl. May-Jul., fr. Aug.

Distribution: Japan, Korea, North Thailand. Widely except South China. Main island (Dashaao, Zhenzhengfu), occasionally cultivated.

Habitat: Shady places, rocks on slopes.

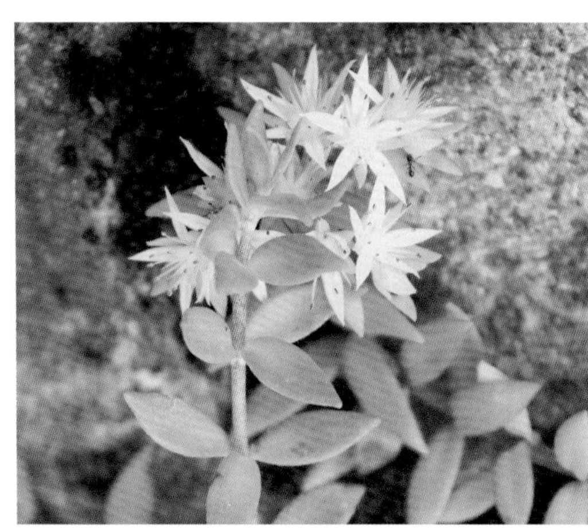

A134. 小二仙草科
Haloragaceae

小二仙草
Gonocarpus micranthus
小二仙草属 *Gonocarpus*

形态特征：多年生草本，高 5 ~ 45 cm，平卧或外倾。茎有时带红色，光滑，无毛。叶交叉互生，卵心形或卵圆形，基部圆形，先端尖或钝，边缘具稀疏锯齿。苞片披针形，膜质，边缘全缘；小苞片棕色，近圆形，膜质，边缘有细锯齿或全缘；柱头红色，头状，流苏状。坚果近球形，无毛，每个果实 1 种子。

花果期：花期 4 ~ 8 月，果期 5 ~ 10 月。

分布：亚洲东部至太平洋岛屿。广布我国东南地区。南麂主岛（关帝岙、门屿尾、三盘尾），偶见。

生境：潮湿沼泽地，或开阔的荒山草丛中。

Description: Herbs perennial, prostrate or decumbent, 5-45 cm tall. Stem sometimes reddish, smooth, glabrous. Leaves decussate, ovate-cordate or elliptic, base rounded, apex acute or obtuse, margin serrulate. Bracts lanceolate, membranous, margin entire; bracteoles brown, suborbicular, membranous, margin serrulate or entire; stigmas red, capitate, fimbriate. Nuts subglobose, glabrous, seed 1 per fruit.

Flower and Fruit: Fl. Apr. -Aug. , fr. May-Oct.

Distribution: Eastern Asia to Pacific islands. Widely in southeastern China. Main island (Guandiao, Menyuwei, Sanpanwei), occasionally.

Habitat: Wet or boggy places, either in open or grassy situations.

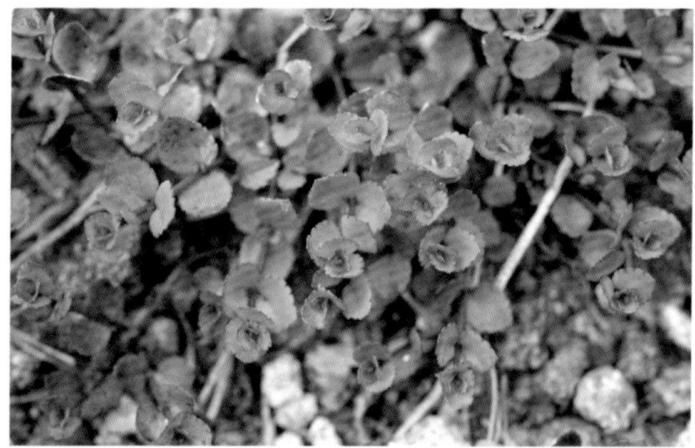

A134. 小二仙草科
Haloragaceae

粉绿狐尾藻
Myriophyllum aquaticum
狐尾藻属 Myriophyllum

形态特征：多年生水生草本，植株长度 50 ～ 80 cm。茎上部直立，下部沉水，具根状茎。叶 5 轮生，圆扇形，一回羽状，两侧有 8 ～ 10 片淡绿色的丝状小羽片。雌雄异株，穗状花序，白色。分果，每果爿种子 1。

花果期：花期 7 ～ 8 月。

分布：原产南美洲，我国各地广为栽培。南麂主岛（镇政府），偶见。

生境：水岸边湿地。

Description: Herbs perennial, aquatic, 50-80 cm long. Stems erect above, submerged below, rhizomatous. Leaves 5-whorled, fan-shaped, unipinnate, 8-10 light green filamentous pinnules on both sides. Dioecious, spica white. Fruit a schizocarp, seed 1 per mericarp.

Flower and Fruit: Fl. Jul. -Aug.

Distribution: Native in South America. Cultivated in most areas. Main island (Zhenzhengfu), occasionally.

Habitat: Waterside wetland.

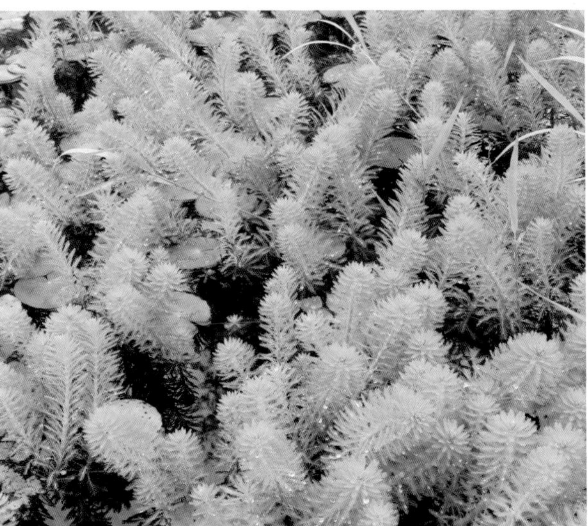

A136. 葡萄科
Vitaceae

蛇葡萄
Ampelopsis glandulosa
蛇葡萄属 *Ampelopsis*

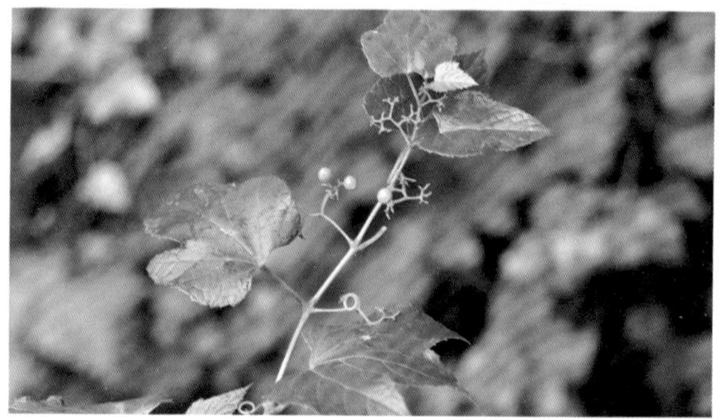

形态特征：木质藤本，小枝圆柱状，有纵棱纹，卷须2或3分枝。单叶，3～5裂，常混生有不分裂者，基部心形，基缺近呈钝角，稀圆形，边缘具锐尖齿。花瓣卵状椭圆形，花药狭椭圆形，子房下部与花盘合生，花柱基部略粗。浆果有籽2～4粒。种子狭椭圆形。

花果期：花期4～8月，果期7～10月。

分布：日本。我国黄河流域以南地区。南麂主岛各处常见。

生境：山谷森林，山坡灌丛，乔木或灌木上。

Description: Woody vines , branchlet terete, with longitudinal ridges, tendrils 2-or 3-branched. Leaves simple, 3-5-cleft, usually mixed with some undivided leaves, base cordate, notches nearly obtuse, rarely rounded, margin with acute teeth. Petals ovate-elliptic, anthers narrowly elliptic, lower part of ovary adnate to disk, style slightly enlarged at base. Berry 2-4-seeded. Seeds narrowly elliptic.

Flower and Fruit: Fl. Apr. -Aug. , fr. Jul. -Oct.

Distribution: Japan. South of the Yellow River Basin. Main island of Nanji, commonly.

Habitat: Forests in valleys, shrublands on hillsides, on trees or shrubs.

A136. 葡萄科
Vitaceae

乌蔹莓
Cayratia japonica
乌蔹莓属 *Cayratia*

形态特征：草质藤本。幼枝绿色，有柔毛，后变无毛，老枝紫绿色。叶为鸟足状5小叶，小叶长椭圆形或狭卵形，先端急尖或渐尖，基部楔形，两面中脉上有短柔毛，中间叶片较大，边缘具8～12枚锯齿；托叶三角形，早落。复二歧聚伞花序腋生；花小，黄绿色，具短柄，外被粉状微毛或近无毛，花萼不明显，花瓣4。浆果卵形，成熟时黑色。

花果期：花期5～6月；果期8～10月。

分布：亚洲东部至南部。我国秦淮以南地区。南麂主岛各处常见。

生境：森林、灌丛、山谷，易受干扰地区、路边。

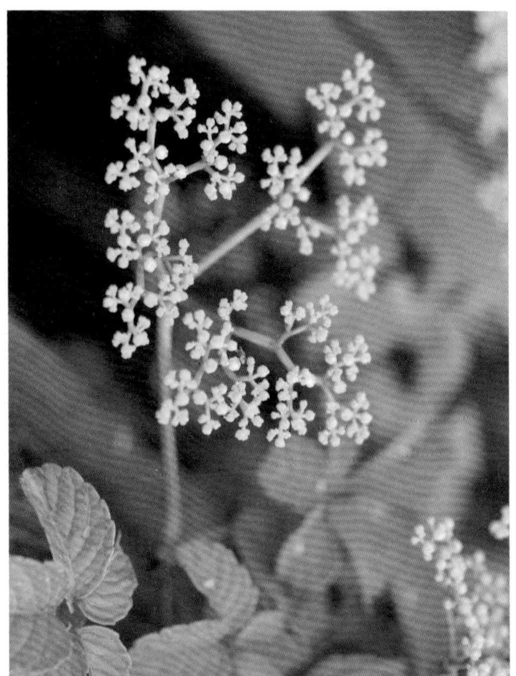

Description: Herbaceous vines . Young branches green, pilose, later glabrous, old branches purplish green. Leaves pedately 5-foliolate, leaflet elliptic or narrowly ovate, apex acute or acuminate, base cuneate, pubescent on both midrib, middle leaf larger, margin 8-12 toothed on each side; stipules triangular, caducous. Compound dichasium axillary; flowers small, yellowish green, stipitate, outside powdery microhairs or subglabrous, calyx indistinct, petals 4. Berry ovate, black when ripe.

Flower and Fruit: Fl. May -Jun. , fr. Aut. -Oct.

Distribution: Eastern to southern Asia.To the south of Qinling and Huaihe. Main island of Nanji, commonly.

Habitat: Forests, shrublands, valleys, disturbed areas, roadsides.

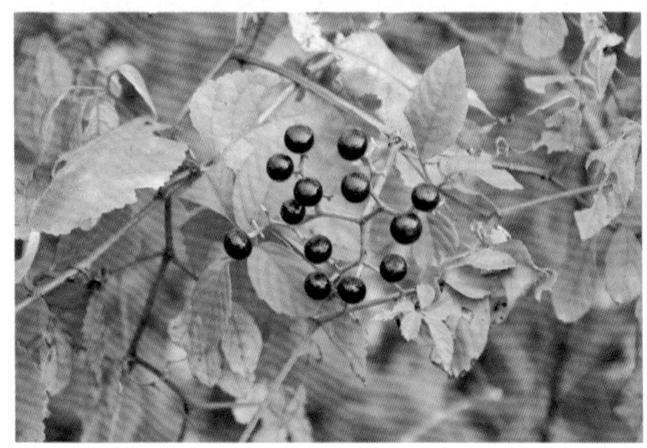

A136. 葡萄科
Vitaceae

地锦（爬山虎）
Parthenocissus tricuspidata
地锦属 ***Parthenocissus***

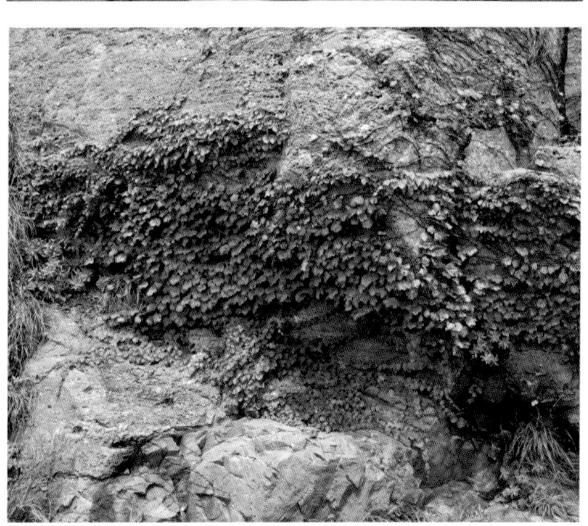

形态特征：落叶木质藤本。枝较粗壮，近无毛或者疏生柔毛；卷须短，多分枝。叶片多为倒卵圆形，3 浅裂，基部心形，边缘有粗锯齿，无毛或中脉上疏生短柔毛，基出脉 5。多歧聚伞花序具短枝，花绿色，5 数；萼片小，全缘；花瓣顶端反折；雄蕊与花瓣对生。浆果，蓝色。

花果期：花期 6 ~ 7 月；果期 9 月。

分布：日本，朝鲜。我国中东部以北。南麂主岛各处常见，常栽培。

生境：灌丛、石壁、岩石山边。

Description: Deciduous woody vines. Branches thicker, nearly glabrous or sparsely pilose; tendril short, much branched, leaves blade obovoid, 3-lobed, base cordate, margin coarsely serrate, glabrous or midvein abaxially sparsely pubescent, basal veins 5. Polychasium with short branches; flowery green, 5; sepals small, entire; petals reflexed at apex; stamens opposite to petals. Berry blue.

Flower and Fruit: Fl. Jun. -Jul, fr. Sep.

Distribution: Japan, Korea.To the north of mid-eastern China. Main island of Nanji, commonly, usually cultivated.

Habitat: Shrublands, cliffs, rocky hillsides.

A136. 葡萄科
Vitaceae

蘡薁
Vitis bryoniifolia
葡萄属 *Vitis*

形态特征：木质藤本。小枝圆柱形，有棱纹，嫩枝、叶柄、叶背密被蛛丝状绒毛或柔毛，以后脱落变稀疏；卷须 2 叉分枝，每隔 2 节间断与叶对生。叶长圆卵形，叶片 3 ～ 5 深裂或浅裂，基出 5 脉；托叶带褐色，椭圆披针形，膜质。圆锥花序与叶对生；花小，无毛；花萼盘形，全缘；花瓣 5，早落。浆果球形，熟时紫红色。

花果期：花期 4 ～ 8 月，果期 6 ～ 10 月。

分布：我国南北各地广布。南麂主岛各处常见。

生境：山谷林中，灌丛、溪边、田野。

Description: Woody vines. Branchlets terete, with longitudinal ridges, becoming sparsely so; tendrils bifurcate, opposite with leaves every 2 knots.Leaf blade broadly ovate or ovate, 3-5 deep or low lobed, basal veins 5, stipules brownish, oval or ovate-lanceolate, membranous. Panicle leaf-opposed; flowers small, glabrous; calyx disc-shaped, entire; petals 5, caducous. Berry globose, purple when ripe.

Flower and Fruit: Fl. Apr. -Aug. , fr. Jun. -Oct.

Distribution: Throughout the north and south areas of China. Main island of Nanji, commonly.

Habitat: Forests in valleys, shrublands, streamsides, fields.

A140. 豆科
Fabaceae

台湾相思
Acacia confusa
金合欢属 *Acacia*

形态特征：常绿乔木，高 6 ～ 15 m，无毛。枝灰色或褐色，无刺，小枝纤细。羽状复叶退化，叶柄变为叶状，革质，披针形。头状花序球形，单生或 2 ～ 3 个簇叶腋；总花梗纤弱；花金黄色，微香，花瓣淡绿色。荚果扁平，干时深褐色，微缢缩，顶端钝而凸，基部楔形。种子 2 ～ 8，椭圆形，扁平。

花果期：花期 3 ～ 10 月；果期 8 ～ 12 月。

分布：原产菲律宾。我国华东、华南至西南等地广泛栽培。南麂各岛屿常见。

生境：山坡、开阔地，极耐干旱和瘠薄

Description: Trees, evergreen, 6-15 m tall, glabrous. Branches gray or brown, unarmed, branchlets slender. Pinnately compound leaves degenerate, petioles become leaflike, leathery, lanceolate. Heads solitary or 2-or 3-fasiculate, axillary, globose; peduncles slender; Flowers golden yellow, fragrant, petals greenish. Legume flat, dark brown when dry, slightly constricted, apex obtuse and convex, base cuneate. Seeds broadly elliptic, compressed, 2-8.

Flower and Fruit: Fl. Mar. -Oct. , fr. Aug. -Dec.

Distribution: Native to Philippines. Widely cultivated in East, South to Southwest China. Throughout Nanji Islands, commonly.

Habitat: Slopes, open areas, hightolerance to drought and barren.

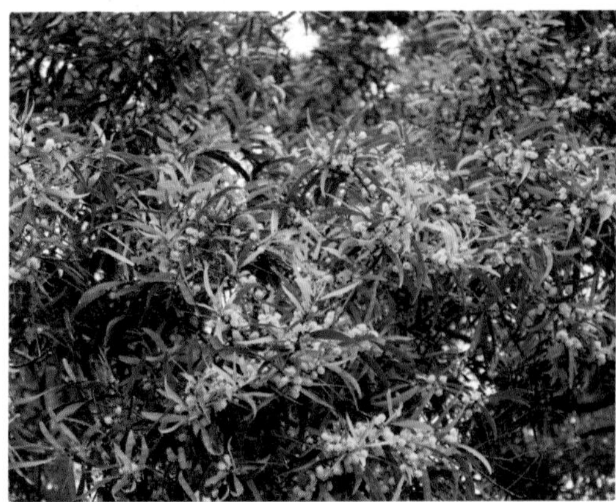

A140. 豆科
Fabaceae

黑荆
Acacia mearnsii
金合欢属 *Acacia*

形态特征：乔木，高 9 ～ 15 m。小枝具棱，被灰白色绒毛。小叶排列紧密，线形，边缘、背面，有时两面均被短柔毛。头状花序圆球状，在叶腋排成总状花序或在枝顶排成圆锥花序，花序轴黄色，密被绒毛。荚果长圆形，扁平，被短柔毛。种子黑色，有光泽，卵圆形。

花果期：花期 6 月，果期 8 月。

分布：原产澳大利亚。我国华东、华南至西南部分地区有引种。南麂主岛（美龄居），栽培。

生境：向阳的山坡、山脚，疏林。

Description: Trees, 9-15 m tall. Branchlets angulate, gray-white tomentose. Leaflets dense, linear, margin, abaxial surface, or sometimes both surfaces pubescent. Heads globose, arranged in axillary racemes or terminal panicles, rachis yellow, densely tomentose. Legume oblong, flat, pubescent. Seeds black, shiny, ovoid.

Flower and Fruit: Fl. Jun. , fr. Aug.

Distribution: Native to Australia. Cultivated in East China, South China and some areas of Southwest China. Main island (Meilingju), cultivated.

Habitat: Sunny slopes, foothills, forests.

A140. 豆科 Fabaceae

山槐（山合欢）
Albizia kalkora
合欢属 *Albizia*

形态特征：落叶小乔木或灌木，高 3 ~ 8 m。枝条暗褐色，被短柔毛，皮孔显著。二回羽状复叶，小叶长圆形，基部偏斜，两面均被短柔毛。头状花序腋生，或于枝顶排成圆锥花序；花初白色，后变黄，小花梗明显；花萼管状，花冠中部以下连合呈管状，裂片披针形，花萼、花冠均密被长柔毛；雄蕊基部连合呈管状。荚果带状，深棕色。种子倒卵形。

花果期：花期 5 ~ 6 月，果期 8 ~ 10 月。

分布：印度，日本，缅甸，越南。我国除西北均有分布。南麂主岛（国姓岙、后隆、门屿尾、三盘尾），常见。

生境：灌木丛，疏林。

Description: Small trees, or shrubs, deciduous, 3-8 m tall. Branchlets dark brown, pubescent, with conspicuous lenticels. Bicyclic pinnate compound leaves; leaflets oblong, base unequal laterally, both surfaces pubescent. Heads in leaf axils, or arranged in panicles on top of branches; flowers primarily white, turning yellow, with conspicuous pedicels; calyx tubular; corolla at middle or lower junctures tubular, lobes lanceolate, calyx and corolla villous; stamens joined at base in tubular form. Pods banded, dark brown. Seeds obovate.

Flower and Fruit: Fl. May-Jun. , fr. Aug. -Oct.

Distribution: India, Japan, Myanmar, Vietnam. Throughout China except Northwest China. Main island (Guoxingao, Houlong, Menyuwei, Sanpanwei), commonly.

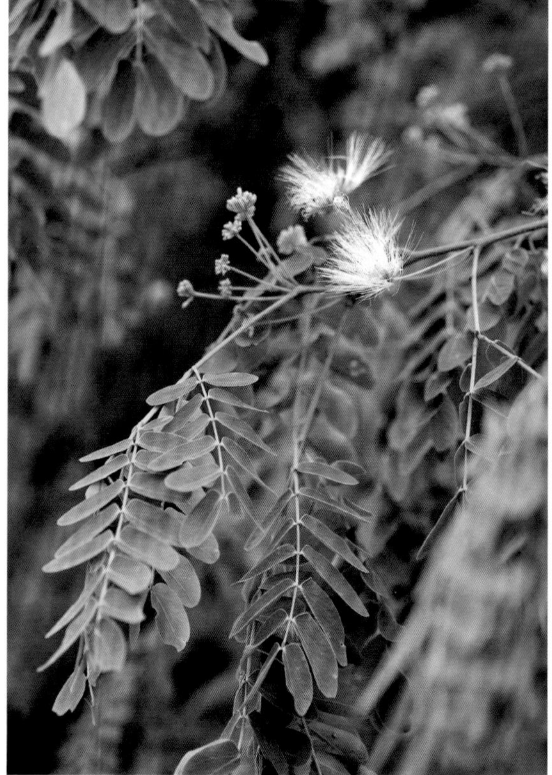

A140. 豆科
Fabaceae

土圞儿
Apios fortunei
土圞儿属 *Apios*

形态特征： 缠绕草本。有球状或卵状块根，茎细长，被稀疏短硬毛。奇数羽状复叶，小叶 3 ~ 7，卵形或菱状卵形，先端急尖，有短尖头，基部宽楔形或圆形。总状花序腋生，苞片和小苞片微小，被短毛，花带黄绿色或浅绿色，花萼稍呈二唇形，龙骨瓣最长，卷成半圆形。荚果浅棕色，无毛，扁平。

花果期： 花期 6 ~ 8 月，果期 9 ~ 10 月。

分布： 日本。我国黄河流域以南地区。南麂主岛（百亩山、大山、国姓岙），偶见。

生境： 山坡灌丛中，缠绕在树上。

Description: Herbs twining. Root tuber spherical or ovoid. Stems slender, sparsely hirsute. Leaves pinnately 3-7-foliolate, ovate or rhomboid-ovate, apex acute, mucronate, base broadly cuneate or rounded. Raceme axillary, bracts and bracteoles minute, ciliate, corolla yellowish green or light green, calyx bowl-like, shallowly 2-lipped, keels much longer than standard, curled into semicircle. Legume light brown, glabrous, compressed.

Flower and Fruit: Fl. Jun. -Aug. , fr. Sep. -Oct.

Distribution: Japan. South of the Yellow River Basin. Main island (Baimushan, Dashan, Guoxingao), occasionally.

Habitat: Hillside thickets, twined in trees.

A140. 豆科
Fabaceae

亮叶猴耳环
Archidendron lucidum（*Pithecellobium lucidum, Abarema lucida*）
猴耳环属 *Archidendron*

形态特征：乔木，高 2 ～ 10 m。小枝无刺，嫩枝、叶柄和花序均被褐色短茸毛。羽片 1 ～ 2 对，小叶斜卵形或长圆形，顶生的一对最大，对生，余互生且较小，先端渐尖而具钝小尖头。头状花序球形，有花 10 ～ 20 朵，排成腋生或顶生的圆锥花序，花冠白色。荚果旋卷成环状，边缘在种子间缢缩。

花果期：花期 4 ～ 6 月；果期 7 ～ 12 月。

分布：印度、越南。我国华东、华南至西南等地。南麂主岛（百亩山、大山、国姓岙、后隆、马祖岙、门屿尾），常见。

生境：疏林或密林中或林缘，灌木丛中。

Description: Trees, 2-10 m tall. Branchlets unarmed, shoots, petiole and inflorescence shortly brown tomentose. Pinnae 1 or 2 pairs, leaflets obliquely ovate or oblong, apical ones larger, opposite, proximal ones alternate and smaller, apex acuminate, mucronate. Heads globose, 10-20-flowered, arranged in panicles, terminal or axillary, corolla white. Legume twisted into a circle, margin between seeds constricted.

Flower and Fruit: Fl. Apr. -Jun. , fr. Jul. -Dec.

Distribution: India, Vietnam. East, South to Southwest China. Main island (Baimushan, Dashan, Guoxingao, Houlong, Mazuao, Menyuwei), commonly.

Habitat: Thin or thick forests, forests margin, thickets.

A140. 豆科
Fabaceae

杭子梢
Campylotropis macrocarpa
杭子梢属 *Campylotropis*

形态特征：灌木，高 1 ~ 2 m，偶见 3 m。小枝贴生或近贴生短或长柔毛，嫩枝毛密，少有具绒毛，老枝常无毛。羽状复叶具 3 小叶，椭圆形或宽椭圆形，有时过渡为长圆形。总状花序单一（稀二）腋生并顶生，花冠紫红色或近粉红色。荚果长圆形、近长圆形或椭圆形，先端具短喙尖。

花果期：花果期 6 ~ 10 月。

分布：朝鲜。我国南北各地广布。南麂主岛（国姓岙、门屿尾），偶见。

生境：山坡、灌丛、林缘、山谷沟边及林中。

Description: Shrubs, 1-2(3) m tall. Branchlets appressed or ascending hairy, young branches densely hairy, rarely villi; old branches often glabrous. Pinnately compound leaves with 3 leaflets, elliptic or broadly elliptic, sometimes oblong. Raceme simple (rarely diploid), axillary, terminal; corolla purple or pink. Legume oblong, nearly oblong, or elliptic, apex mucronate.

Flower and Fruit: Fl. and fr. Jun. -Oct.

Distribution: Korea. Throughout the north and south areas of China. Main island (Guoxingao, Menyuwei), occasionally.

Habitat: Mountain slopes, thickets, forest margins, valleys and forests.

A140. 豆科
Fabaceae

狭刀豆（海刀豆）
Canavalia lineata
刀豆属 *Canavalia*

形态特征： 多年生缠绕草本。小叶卵形或倒卵形，两面薄被短柔毛，基部截平或楔形，先端圆形或骤尖。总状花序腋生；花萼被短柔毛，上唇2裂，裂齿顶端具小尖头，下唇具3齿；花冠紫红色，旗瓣宽卵形，顶微凹，基部具2痂状附属体及2耳，翼瓣线状长圆形，稍呈镰状，龙骨瓣倒卵状长圆形。荚果长椭圆形，扁平。种子2～3颗，卵形，棕色，有斑点。

花果期： 花果期6～10月。

分布： 东亚、东南亚。我国东南沿海地区。南麂主岛（东方岙、门屿尾），栽培。

生境： 砂质海滩。

Description: Herbs perennial, twining. Leaflets ovate or obovate, sparsely pubescent on both surfaces, base truncate or cuneate, apex rounded or cuspidate. Racemes axillary; calyx pubescent, upper lip broadly 2-lobed and lobes apiculate at apex, lower lip 3-lobed; corolla purplish red, standard broadly ovate, emarginate, base with 2 thickenings, wings linear-oblong, slightly falcate, keel obovate-oblong. Legumes oblong, compressed. Seeds 2 or 3, ovate, brown, blotched.

Flower and Fruit: Fl. and fr. Jun. -Oct.

Distribution: East Asia, Southeast Asia. Coastal areas of southeastern China. Main island (Dongfangao, Menyuwei), cultivated.

Habitat: Sandy beaches.

A140. 豆科
Fabaceae

紫花野百合（农吉利、野百合）
Crotalaria sessiliflora （***Crotalaria brevipes***）
猪屎豆属 ***Crotalaria***

形态特征：一年生或短暂多年生草本，直立，高 30 ～ 100 cm。茎通常从下部分枝，圆柱状，密被粗硬毛。托叶线形，宿存或早落；单叶，叶片形状变异较大，通常为线形或线状披针形，背面密被丝状短柔毛，基部渐狭，先端渐尖。总状花序顶生、腋生或密生枝顶形似头状，亦有叶腋生出单花，花 1 或多数。荚果短圆柱形，种子 10 ～ 15，无毛。

花果期：花果期 5 月至翌年 2 月。

分布：东亚至东南亚。我国长江流域以南地区。南麂主岛（三盘尾），偶见。

生境：荒地路旁及山谷草地。

Description: Herbs annual or short-lived perennial, erecrt, 30-100 cm tall. Stems often branching from lower parts, terete, densely coarsely hirsute. Stipules linear, persistent or caducous; leaves simple, leaf blade variable in shape, usually linear to linear-lanceolate, abaxially densely silky pubescent, base attenuate, apex acuminate. Racemes terminal or leaf-opposed or densely congested and headlike on branch apices, few to many flowered or flowers solitary in axils. Legume cylindric, 10-15-seeded, glabrous.

Flower and Fruit: Fl. and fr. May-Feb.

Distribution: East to Southeast Asia. To the south of Yangtze River Basin. Main island (Sanpanwei), occasionally.

Habitat: Wasteland roadside and valley meadows.

A140. 豆科
Fabaceae

藤黄檀
Dalbergia hancei
黄檀属 *Dalbergia*

形态特征：藤本。枝纤细，有时钩状或扭曲。羽状复叶，小叶 7 ~ 13，狭长圆或倒卵状长圆形，先端钝或圆，微缺。总状花序远较复叶短，幼时包藏于苞片内，数个总状花序常再集成腋生短圆锥花序；花萼阔钟状，具缘毛；花冠绿白色，芳香，各瓣均具长柄，旗瓣椭圆形，具耳，翼瓣与龙骨瓣长圆形。荚果扁平，长圆形或带状，无毛，种子 1。种子肾形，扁平。

花果期：花期 3 ~ 5 月，果期 6 ~ 11 月。

分布：我国长江流域以南地区。南麂主岛（百亩山、大山、国姓岙、后隆、马祖岙、门屿尾），常见。

生境：山坡灌丛，山谷溪旁。

Description: Woody climbers. Branchlets slender, sometimes hooked or twisted. Pinnate, leaflets 7-13, narrowly oblong or obovate-oblong, apex obtuse or rounded, emarginate. Racemes much shorter than compound leaves, often several compressed axillary short panicles; calyx broadly campanulate, ciliate; corolla greenish white, fragrant, rather long clawed, standard elliptic, auriculate, wings and keel oblong. Legume flattened, oblong or banded, glabrous, 1-seeded. Seeds reniform, compressed.

Flower and Fruit: Fl. Mar. -May, fr. Jun. -Nov.

Distribution: South of the Yangtze River Basin in China. Main island (Baimushan, Dashan, Guoxingao, Houlong, Mazuao, Menyuwei), commonly.

Habitat: Bushes on mountain slopes, by streams along valleys.

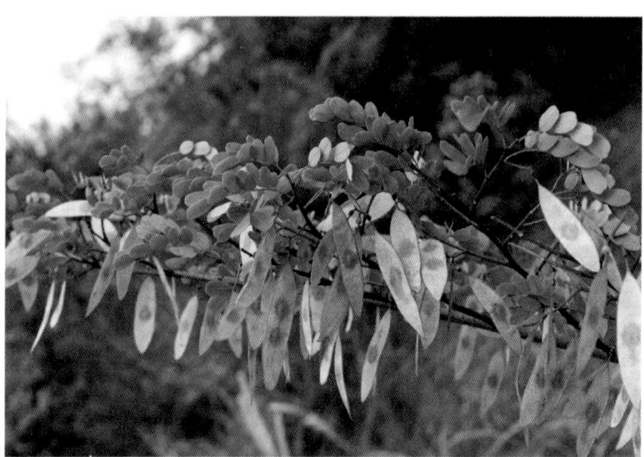

A140. 豆科
Fabaceae

黄檀
Dalbergia hupeana
黄檀属 *Dalbergia*

形态特征：落叶乔木，高 10 ～ 20 m。树皮暗灰色；嫩枝淡绿色，无毛。羽状复叶，小叶 7 ～ 11，椭圆形到长圆状椭圆形，近革质，两面无毛。圆锥花序顶生或生于最上部叶腋，疏被锈色短柔毛；花萼钟状，5 齿，花冠白色或浅紫色，各瓣具柄，旗瓣圆形，先端微缺，翼瓣倒卵形，龙骨瓣半月形，与翼瓣内侧均具耳。荚果长圆形或阔舌状，薄革质。种子肾形。

花果期：花期 5 ～ 7 月。

分布：我国华南，华中，华东和西南地区。南麂主岛（百亩山、大山、国姓岙、后隆、马祖岙、门屿尾），常见。

生境：林中、山坡、峡谷、溪旁和山地林中的灌丛中。

Description: Deciduous trees, 10-20 m tall. Bark dull gray; young shoots pale green, glabrous. Pinnate, leaflets 7-11, elliptic to oblong-elliptic, subleathery, both surfaces glabrous. Panicles terminal or extending into axils of uppermost leaves, sparsely rusty puberulent; calyx campanulate, 5-toothed, corolla white or light purple; petals clawed; standard orbicular, emarginate; wings obovate and half-moon-shaped; keel auriculate on upper side below. Legume oblong or broadly ligulate, thinly leathery. Seeds reniform.

Flower and Fruit: Fl. May-Jul.

Distribution: South, Central, East and Southwest China. Main island (Baimushan, Dashan, Guoxingao, Houlong, Mazuao, Menyuwei), commonly.

Habitat: Forests, among bushes on mountain slopes, ravines, by streams, woodland slopes.

A140. 豆科
Fabaceae

假地豆
Desmodium heterocarpon
山蚂蝗属 *Desmodium*

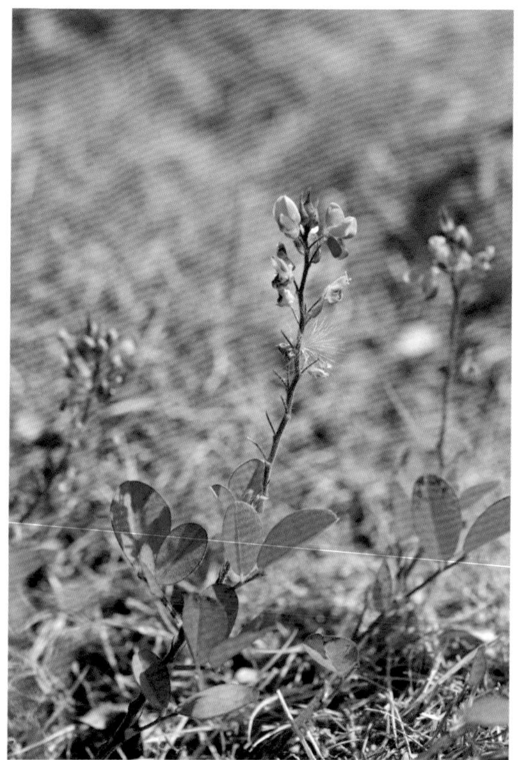

形态特征：小灌木，高 30 ~ 150 cm，基部多分枝。叶具 3 小叶，顶生叶椭圆形或宽倒卵形，侧生叶较小，先端圆钝，微凹。总状花序顶生或腋生，总花梗密被淡黄色钩毛，花密集；花萼 4 裂，花冠紫红色，紫色或白色，旗瓣倒卵状长圆形，具短柄，翼瓣倒卵形，具耳和瓣柄，龙骨瓣极弯曲，先端钝。荚果狭长圆形，腹背两缝线被钩状毛，荚节 4 ~ 7，近方形。

花果期：花期 7 ~ 10 月，果期 10 ~ 12 月。

分布：东南亚、大洋洲。我国长江以南各地。南麂主岛各处常见。

生境：山坡草地、水旁、灌丛或林中。

Description: Shrubs or subshrubs, 30-150 cm tall, stem much branched from base, appressed rough hairy. Leaves 3-foliolate, terminal leaflet blade elliptic or broadly obovate, lateral leaflet blade small, apex rounded or obtuse, emarginate. Racemes terminal or axillary, peduncles densely covered with yellowish spreading hooked hairs, densely flowered; calyx 4-lobed, corolla purple-red, purple, or white, standard obovate-oblong, shortly clawed, wings obovate, auriculate, clawed, keel extremely curved, apex obtuse. Legume narrowly oblong, both sutures hooked hairy, 4-7-jointed, articles quadrate.

Flower and Fruit: Fl. Jul. -Oct. , fr. Oct. -Dec.

Distribution: Southeast Asia, Oceania. To the south of Yangtze River. Main island of Nanji, commonly.

Habitat: Grassy slopes, watersides, thickets, forests.

A140. 豆科
Fabaceae

小叶三点金
Desmodium microphyllum
山蚂蝗属 *Desmodium*

形态特征：多年生草本，平卧或直立，高 10 ~ 50 cm。茎多分枝，纤细，通常红褐色。羽状三出复叶，托叶披针形，具条纹；小叶薄纸质，倒卵状长椭圆形或长椭圆形，两端圆钝。总状花序顶生或腋生，被黄褐色开展柔毛，有花 6 ~ 10 朵；花冠粉红色，旗瓣倒卵形或倒卵状圆形。荚果腹背两缝线浅齿状，荚节 3 ~ 4，近圆形，扁平。

花果期：花期 5 ~ 9 月，果期 9 ~ 11 月。

分布：东南亚。我国长江以南各地。南麂主岛各处常见。

生境：荒地、草丛或灌木林。

Description: Herbs perennial, erect or prostrate, 10-50 cm tall. Stems much branched, slender, usually reddish brown. Leaves 3-foliolate, stipules lanceolate, striate; leaflet papery, narrowly obovate-elliptic or narrowly elliptic, both ends rounded. Racemes terminal or axillary, yellow-brown spreading pubescent, 6-10-flowered; corolla pink, standard obovate or obovate-orbicular. Legume both surfaces shallowly dentiform, 3-or 4 -jointed, articles nearly orbicular, flat.

Flower and Fruit: Fl. May-Sep. , fr. Sep. -Nov.

Distribution: Southeast Asia. To the south of Yangtze River.Main island of Nanji, commonly.

Habitat: Wastelands, grasslands, thickets.

A140. 豆科
Fabaceae

河北木蓝（马棘）
Indigofera bungeana（Indigofera pseudotinctoria）
木蓝属 *Indigofera*

形态特征：直立灌木，高40～100 cm。茎褐色，圆柱形，有皮孔，枝银灰色，被灰白色丁字毛。羽状复叶，小叶2～4对；对生，椭圆形，稍倒阔卵形，先端钝圆。总状花序腋生，花冠紫色或紫红色。荚果褐色，线状圆柱形，被白色丁字毛。

花果期：花期5～7月，果期8～10月。

分布：日本、朝鲜。分布几遍我国。南麂主岛各处常见。

生境：山坡、草地或河滩地。

Description: Shrubs, erect, 40-100 cm tall. Stems brown, terete, with lenticel; branches whitish gray, with appressed white medifixed sym-metrically 2-branched trichomes. Pinnately compound leaves, leaflet blades opposite, 2-4 pairs, elliptic, slightly broad obovate, apex obtuse. Racemes axillary, corolla purple to purplish red. Legume brown, cylindric, stright, with appressed white medifixed trichomes.

Flower and Fruit: Fl. May-Jul. , fr. Aug. -Oct.

Distribution: Japan, Korea. Almost throughout China. Main island of Nanji, commonly.

Habitat: Slopes, grasslands, river beaches.

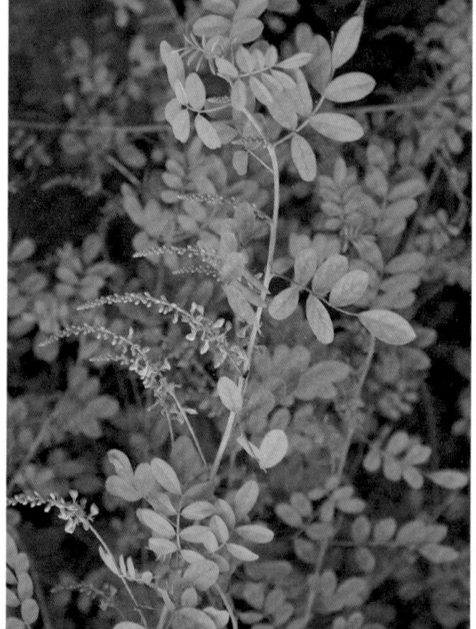

A140. 豆科
Fabaceae

长萼鸡眼草
Kummerowia stipulacea
鸡眼草属 *Kummerowia*

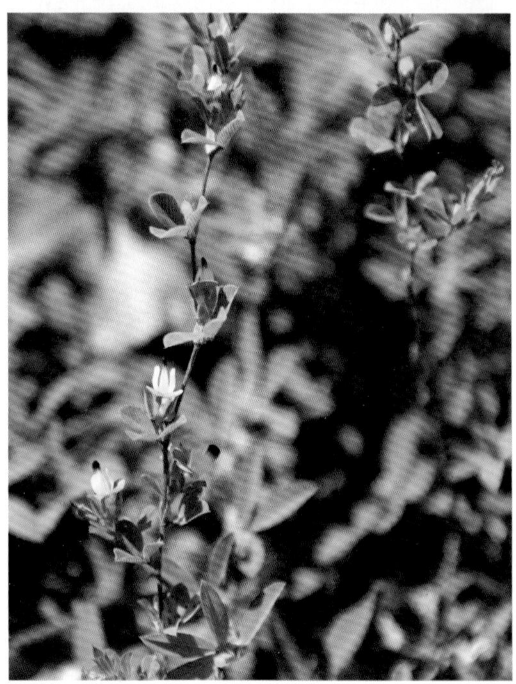

形态特征：一年生草本，高 7 ～ 15 cm。茎和枝上被疏生向上的白毛，有时仅节处有毛。羽状 3 小叶，纸质，倒卵形、宽倒卵形或倒卵状楔形，全端先端微凹。花常 1 ～ 2 朵腋生，花梗有毛，花冠上部暗紫色。荚果椭圆形或卵形，稍侧偏。

花果期：花期 7 ～ 8 月，果期 8 ～ 10 月。

分布：东亚。分布几遍我国。南麂主岛（关帝岙、三盘尾），偶见。

生境：路旁、草地、山坡、固定或半固定沙丘。

Description: Herbs annual, 7-15 cm tall. Stem and branches with sparse upward-pointing white hairs, sometimes only nodes. Leaves 3-foliolate, leaflets papery, obovate, broadly obovate or obovate cuneate, entire apex emarginate. Flowers 1 or 2, axillary; pedicel hairy, upper corolla dark purple. Legume ovoid or elliptic, slightly compressed.

Flower and Fruit: Fl. Jul. -Aug. , fr. Aug. -Oct.

Distribution: East Asia. Almost throughout China. Main island (Guandiao, Sanpanwei), occasionally.

Habitat: Roadsides, grasslands, mountain slopes, stable or semistable sand dunes.

A140. 豆科
Fabaceae

鸡眼草
Kummerowia striata
鸡眼草属 *Kummerowia*

形态特征：一年生草本，高 10 ～ 45 cm。茎和枝上被倒生的白色细毛。羽状 3 小叶，小叶纸质，倒卵形、长倒卵形或长圆形，全缘先端常圆形；两面沿中脉及边缘有白色粗毛，侧脉多而密。花小，1 ～ 3 朵簇生叶腋，花梗无毛，花冠粉红色或紫色。荚果圆形或倒卵形，稍扁，先端短尖，被小柔毛。

花果期：花期 7 ～ 9 月，果期 8 ～ 10 月。

分布：东亚。分布几遍我国。南麂主岛（关帝岙、三盘尾），偶见。

生境：路旁、田边、溪旁、砂质地或缓山坡草地。

Description: Herbs annual, 10-45 cm tall. Stem and branch with downward-pointing white hairs. Leaves 3-foliolate, leaflets papery, obovate, narrowly obovate, or oblong, entire apex usually rounded; both sides with white coarse hairs along the midrib and margin, lateral veins dense. Flowers 1-3 in upper axils of leaves, pedicel glabrons, corolla pink or purple. Legume orbicular or obovoid, slightly compressed, apex mucronate, pubescent.

Flower and Fruit: Fl. Jul. -Sep. , fr. Aug.-Oct.

Distribution: East Asia. Almost throughout China. Main island (Guandiao, Sanpanwei), occasionally.

Habitat: Roadsides, field, streamsides, sandy soils, grassy slopes.

A140. 豆科
Fabaceae

胡枝子
Lespedeza bicolor
胡枝子属 *Lespedeza*

形态特征：直立灌木，高 1 ～ 3 m，多分枝。小枝被稀疏短柔毛。羽状 3 小叶，小叶卵形、倒卵形或卵状长圆形，先端钝圆或微缺，具短尖。总状花序腋生，长于叶，常构成大型疏散圆锥花序；花冠紫红色，旗瓣倒卵形，先端微凹，翼瓣较短，近长圆形，基部具耳和瓣柄。荚果斜倒卵形，稍扁，具网纹，密被短柔毛。

花果期：花期 7 ～ 9 月，果期 9 ～ 10 月。

分布：东亚。分布几遍全国。南麂主岛（美龄居、门屿尾、三盘尾），常见。

生境：山坡、林缘、路旁、灌丛及杂木林间。

Description: Shrubs, erect, 1-3 m tall, much branched. Branchlets sparsely pubescent. Leaves 3-foliolate, ovate, obovate, or ovate-oblong, apex obtuse-rounded or emarginate, mucronate. Racemes axillary, longer than leaves, often branched in large lax panicles; corolla reddish purple, standard obovate, apex emarginate, wings suboblong, short, base auriculate, clawed. Legume obliquely obovoid, slightly compressed, reticulate veined, densely pubescent.

Flower and Fruit: Fl. Jul.-Sep. , fr. Sep. -Oct.

Distribution: East Asia. Almost throughout China. Main island (Meilingju, Menyuwei, Sanpanwei), commonly.

Habitat: Mountain slopes, forest margins, roadsides, thickets, forests.

A140. 豆科
Fabaceae

中华胡枝子
Lespedeza chinensis
胡枝子属 *Lespedeza*

形态特征：亚灌木，高达 1 m，全株被白色伏毛。茎直立或疏散。羽状 3 小叶，小叶倒卵状长圆形、长圆形、卵形或倒卵形，边缘稍反卷，先端平截或微凹，具小刺尖。总状花序腋生，花少，花冠白色或黄色，旗瓣椭圆形，翼瓣狭长圆形，具爪。荚果卵圆形，先端具喙，基部稍偏斜，表面有网纹，密被白色伏毛。

花果期：花期 8 ~ 9 月，果期 10 ~ 11 月。

分布：我国华东至华中等地。南麂主岛（百亩山、国姓岙、后隆、美龄居、门屿尾、三盘尾），常见。

生境：灌木丛中、林缘、路旁、山坡、林下草丛等处。

Description: Subshrubs, to 1 m tall, adpressed white hairy throughout. Stems erect or diffuse. Leaves 3-foliolate, leaflets obovate-oblong, oblong, or ovate-obovate, margin slightly involute, apex truncate, subtruncate, emarginate, or obtuse, mucronate. Racemes axillary, few flowered, corolla white or yellow, standard elliptic, wings narrowly oblong, clawed. Legume ovoid, apex rostrate, base slightly oblique, reticulate veined, densely adpressed white hairy.

Flower and Fruit: Fl. Aug. -Sep. , fr. Oct. -Nov.

Distribution: East China to Central China. Main island (Baimushan, Guoxingao, Houlong, Meilingju, Menyuwei, Sanpanwei), commonly.

Habitat: Thickets, forest margins, roadsides, mountain slopes, grasslands, forests.

A140. 豆科
Fabaceae

截叶铁扫帚
Lespedeza cuneata
胡枝子属 *Lespedeza*

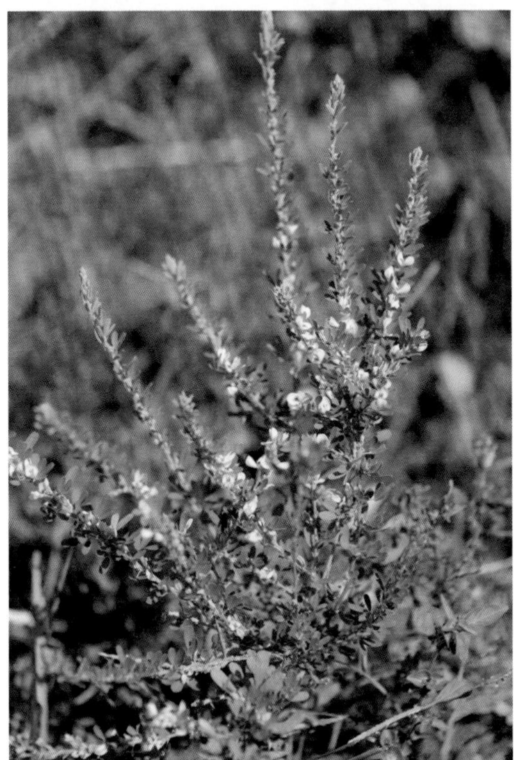

形态特征：亚灌木或多年生草本，高达 1 m。茎直立或斜升，被毛。叶密集，羽状 3 小叶，楔形或线状楔形，先端截形成近截形，具小刺尖，上面近无毛，下面密被伏毛。总状花序腋生，具 2 ～ 4 朵花，花萼狭钟状，5 裂，花冠淡黄色或白色，旗瓣基部具紫色斑点。荚果宽卵形或近球形，被伏毛。

花果期：花期 7 ～ 8 月，果期 9 ～ 10 月。

分布：亚洲东部。我国秦淮以南各地常见。南麂主岛各处常见。

生境：山坡、路旁。

Description: Subshrubs or perennial herbs, to 1 m tall. Stems erect or ascending, hairy. Leaves crowded, 3-foliolate, leaflets cuneate or linear-cuneate, apex truncate or subtruncate, mucronate, adaxially subglabrous, abaxially densely adpressed hairy. Racemes axillary, 2-4-flowered, calyx narrowly campanulate, 5-lobed; corolla yellowish or white, standard with purple spots at base. Legume broadly ovoid or subglobose, adpressed hairy.

Flower and Fruit: Fl. Jul. -Aug. , fr. Sep. -Oct.

Distribution: Eastern Asia. Commonly to the south of Qinling and Huaihe. Main island of Nanji, commonly.

Habitat: Mountain slopes, roadsides.

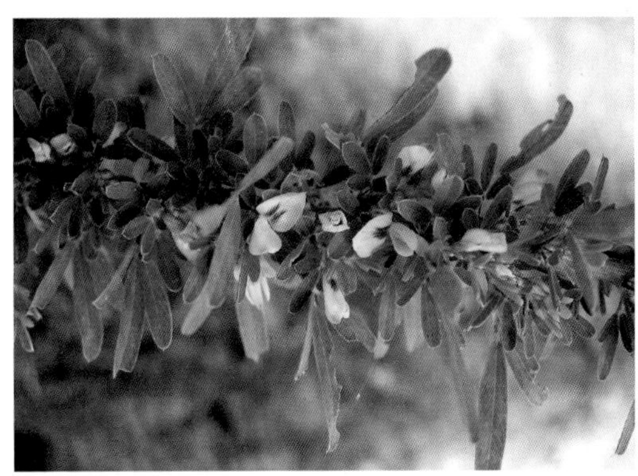

A140. 豆科
Fabaceae

铁马鞭
Lespedeza pilosa
胡枝子属 *Lespedeza*

形态特征：多年生草本，密被长柔毛。茎平卧，细长，长 60 ～ 100 cm。羽状 3 小叶，小叶宽倒卵形或倒卵形，先端圆形、近截形或微凹，有小尖刺。总状花序腋生，比叶短，花冠黄白色或白色，旗瓣椭圆形，具瓣柄，翼瓣比旗瓣与龙骨瓣短实。荚果广卵形，两面密被毛，先端具尖喙。

花果期：花期 7 ～ 9 月，果期 9 ～ 10 月。

分布：日本、朝鲜。我国长江以南各地。南麂主岛各处常见。

生境：荒坡、草地。

Description: Herbs perennial, densely villous throughout. Stems procumbent, 60-100 cm, slender. Leaves 3-foliolate, leaflets broadly obovate or obovate, apex rounded, subrounded, or emarginate, mucronate. Racemes axillary, shorter than leaves, corolla yellowish white or white, standard elliptic, clawed; wings shorter than standard and keel. Legume broadly ovoid, both surfaces densely villous, apex acute-rostrate.

Flower and Fruit: Fl. Jul. -Sep. , fr. Sep. -Oct.

Distribution: Japan, Korea. To the south of Yangtze River. Main island of Nanji, commonly.

Habitat: Waste slopes, grasslands.

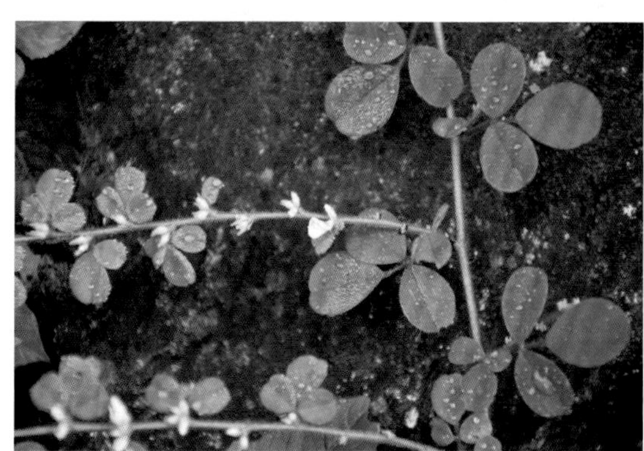

A140. 豆科
Fabaceae

银合欢
Leucaena leucocephala
银合欢属 *Leucaena*

形态特征：灌木或小乔木，高 2 ~ 6 m。小枝被短柔毛，老时无毛，具棕色皮孔。托叶早落，三角状，极小；叶轴被柔毛，在最下一对羽片着生处有黑色腺体 1 枚；小叶线形长圆形，中脉偏向小叶上缘，基部楔形，边缘具缘毛，先端锐尖。花白色，花瓣狭倒披针形，外面被短柔毛，子房具短柄，疏生短柔毛。荚果带状，扁平。

花果期：花期 4 ~ 7 月，果期 8 ~ 10 月。

分布：原产热带美洲，热带亚热带地区广布。我国南方部分地区有栽培和归化。南麂主岛（国姓岙、美龄居、门屿尾、镇政府），偶见，常栽培。

生境：荒地或疏林中。

Description: Shrubs or small trees, 2-6 m tall. Branchlets pubescent, glabrous when old, with brown lenticels. Stipules caducous, deltoid, very small; rachis pubescent with black glands at location of lowest pinnae; leaflets linear-oblong, main vein close to upper margin, base cuneate, margin ciliate, apex acute. Flowers white, petals narrowly oblanceolate, outside pubescent, ovary shortly stipitate, sparsely pubescent. Legume strap-shaped,flat.

Flower and Fruit: Fl. Apr. -Jul. , fr. Aug. -Oct.

Distribution: Originally from tropical America, widely distributed in tropical and subtropical regions. Cultivated and naturalized in some places of southern China. Main island (Guoxingao, Houlong, Meilingju, Menyuwei, Sanpanwei, Zhenzhengfu), occasionally, usually cultivated.

Habitat: Waste land or open forest.

A140. 豆科
Fabaceae

天蓝苜蓿
Medicago lupulina
苜蓿属 *Medicago*

形态特征：一年生或短暂多年生草本，高 15 ~ 60 cm。茎平卧或上升，多分枝。托叶卵状披针形，基部圆或戟状，先端渐尖；小叶椭圆形，卵形或倒卵形，纸质，被短柔毛，基部楔形，边缘在上部具不明显锯齿，先端截形或微凹，具细尖。荚果肾形，表面具同心弧形脉纹，被稀疏毛，熟时变黑。种子 1，棕色，卵形，光滑。

花果期：花期 7 ~ 9 月，果期 8 ~ 10 月。

分布：欧亚大陆广布。分布于我国各地。南麂主岛各处常见。

生境：常见于河岸、路边、荒地及林缘。

Description: Herbs annual or short-lived perennial, 15-60 cm tall. Stems prostrate or ascending, much branched. Stipules ovate-lanceolate, base rounded or hastate, apex acuminate; leaflets elliptic, ovate, or obovate, papery, pubescent, base cuneate, margin distally obscurely serrate, apex truncate or retuse, apiculate. Legume reniform, sculptured with concentric arcuate veins, sparsely hairy, black when ripe. Seed 1, brown, ovoid, smooth.

Flower and Fruit: Fl. Jul. -Sep. , fr. Aug. -Oct.

Distribution: Eurasia. Throughout China. Main island of Nanji, commonly.

Habitat: Stream banks, roadsides, waste fields, woodland margins.

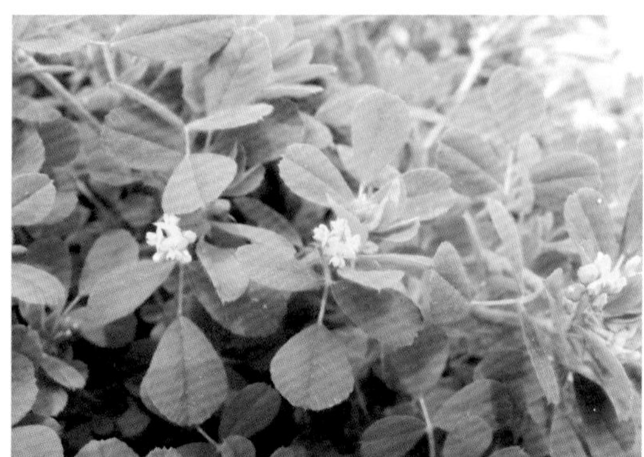

A140. 豆科 Fabaceae

紫苜蓿
Medicago sativa
苜蓿属 *Medicago*

形态特征：多年生草本，高 30 ～ 100 cm。茎直立、丛生以至平卧，四棱形，无毛或微被柔毛。羽状 3 小叶，小叶长卵形、倒长卵形至线状卵形，纸质，先端钝圆，具由中脉伸出的长齿尖，边缘 1/3 以上具锯齿。花序总状或头状；花冠各色：淡黄、深蓝至暗紫色。荚果螺旋状紧卷 2 ～ 4 圈，偶见 6 圈，被柔毛或渐脱落。

花果期：花期 5 ～ 7 月，果期 6 ～ 10 月。

分布：原产北亚和西南亚，现世界广布。我国各地有栽培，常逸生。南麂主岛各处常见。

生境：路旁、旷野、草地、河岸。

Description: Herbs perennial , 30-100 cm tall. Stems erect, ascending, rarely prostrate, quadrangular, glabrous or puberulent, much branched. Leaves 3-foliolate, leaflets long ovate, obovate, to linear-ovate, papery, apex rounded, mucronate with a denticle from midrib, margin serrulate in upper 1/3. Heads or racemes; corolla variable in color, yellowish, deep blue, to dark purple. Legume tightly coiled in 2-4(-6) spirals, pubescent or gradually off.

Flower and Fruit: Fl. May-Jul. , fr. Jun. -Oct.

Distribution: Native to North and Southwest Asia, now cosmopolitan. Cultivated throughout China, often escaped. Main island of Nanji, commonly.

Habitat: Roadsides, fields, grasslands, and stream banks.

A140. 豆科
Fabaceae

草木樨（黄香草木樨）
Melilotus officinalis
草木樨属 *Melilotus*

形态特征：二年生草本，高 40 ～ 100 cm。茎直立，具纵棱。托叶镰状线形，全缘或基部有 1 尖齿；叶柄细长；小叶倒卵形，宽卵形，倒披针形至线形，边缘具浅齿。总状花序初时稠密，花开后渐疏松。荚果卵形，先端具宿存花柱。种子 1 ～ 2 粒，黄褐色，卵球形，光滑。

花果期：花期 5 ～ 9 月，果期 6 ～ 10 月。

分布：欧亚大陆广布。分布我国各地。南麂主岛各处常见。

生境：沙质草原，山坡，沟壑海岸，混合林地的边缘。

Description: Herbs biennial, 40-100 cm. Stems erect, longitudinally ridged. Stipules linear-falcate, entire or with 1 tooth at base; petiole slender; leaflets obovate, broadly ovate, oblanceolate, to linear, margins shallowly serrate. Racemes dense at first, becoming lax in anthesis. Legume ovoid, apex with persistent style. Seeds 1 or 2, yellowish brown, ovoid, smooth.

Flower and Fruit: Fl. May-Sep. , fr. Jun. -Oct.

Distribution: Eurasia. Throughout China. Main island of Nanji, commonly.

Habitat: Sandy grasslands, hillsides, ravine shores, margins of mixed woodlands.

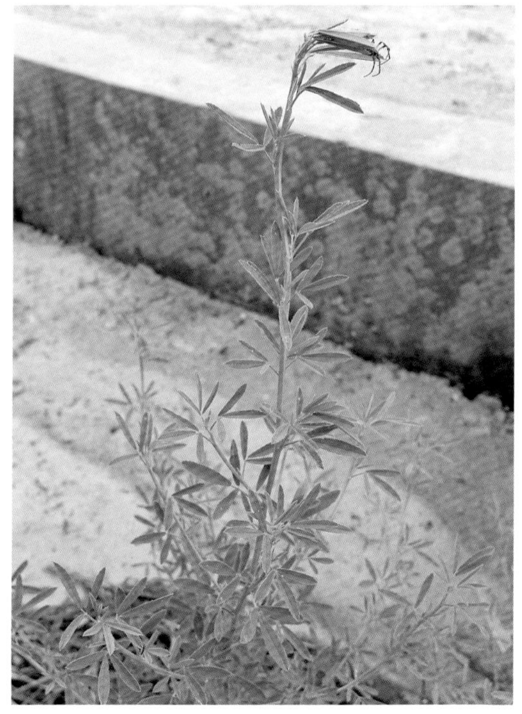

A140. 豆科
Fabaceae

葛（葛藤、野葛）
Pueraria montana（*Pueraria lobata*）
葛属 *Pueraria*

形态特征：粗壮藤本，长可达 8 m，全体被黄色长硬毛，茎基部木质，有粗厚的块状根。托叶背着，卵状长圆形，具线条；小叶三裂，偶尔全缘，顶生小叶宽卵形或斜卵形。总状花序长 15 ~ 30 cm，中部以上有颇密集的花；花冠紫色。荚果长椭圆形，扁平，被褐色长硬毛。

花果期：花期 7 ~ 10 月，果期 10 ~ 12 月。

分布：东南亚，澳大利亚。除新疆、青海及西藏外，分布几遍我国。南麂主岛各处常见。

生境：山地林中、灌丛、开阔地。

Description: Robust vines, with tuberous roots. Stems to 8 m, woody at base, hirsute with yellowish hairs in all parts. Stipules dorsi-fixed, ovate-oblong, striate; leaflets 3-lobed, rarely entire, terminal one broadly ovate or obliquely ovate. Racemes 15-30 cm, dense flowers above the middle; corolla purple. Legumes long elliptic, flattened, brown hirsute.

Flower and Fruit: Fl. Jul. -Oct. , fr. Oct. -Dec.

Distribution: Southeast Asia to Australia. Throughout China except Xinjiang, Qinghai, and Xizang. Main island of Nanji, commonly.

Habitat: Mountain forests, thickets, open places.

A140. 豆科
Fabaceae

三裂叶野葛
Pueraria phaseoloides
葛属 *Pueraria*

形态特征：草质藤本，茎纤细，长 2 ~ 4 m，具开展褐硬毛。托叶基生，卵状披针形；小叶宽卵形，菱形或卵状菱形，顶生小叶较宽，侧生的较小，偏斜，全缘或 3 裂。总状花序单生；苞片和小苞片线状披针形，多毛；花具短梗，聚生在稍疏离的节上；子房线形，略被毛。荚果近圆柱状，初时稍被紧贴的长硬毛，后近无毛。

花果期：花期 8 ~ 9 月，果期 10 ~ 11 月。

分布：东南亚，热带地区广泛种植。我国华东、华南地区。南麂主岛（门屿尾），偶见。

生境：山地和丘陵灌丛。

Description: Herbaceous vines, stem slender, 2-4 m, brownish hirsute. Stipules basifixed, ovate-lanceolate; leaflets broadly ovate, rhomboid, or ovate-rhomboid, terminal one broader, lateral ones smaller, oblique, entire or 3-lobed. Racemes solitary; bracts and bracteoles linear-lanceolate, hirsute; flowers with short pedicels, clustered at slightly distant nodes; ovary linear, thinly hairy. Legumes subcylindric, first adpressed hir-sute, later subglabrous.

Flower and Fruit: Fl. Aug. -Sep. , fr. Oct. -Nov.

Distribution: Southeast Asia, widely cultivated elsewhere in the tropics. East China, South China. Main island (Menyuwei), occasionally.

Habitat: Thickets of mountainous and hilly areas.

A140. 豆科
Fabaceae

鹿藿
Rhynchosia volubilis
鹿藿属 ***Rhynchosia***

形态特征：缠绕藤本，全株各部多少被灰色至淡黄色柔毛，茎略具棱。叶为羽状或有时近指状 3 小叶；小叶纸质，顶生小叶菱形或倒卵状菱形。总状花序，花冠黄色。荚果长圆形，红紫色，极扁平，在种子间略收缩。种子通常 2 颗，椭圆形或近肾形，黑色，光亮。

花果期：花期 5 ~ 8 月，果期 9 ~ 12 月。

分布：日本、朝鲜、越南。我国长江以南各地。南麂主岛各处常见。

生境：山坡路旁草丛。

Description: Herbs densely gray to light yellow villous, ribbed. Leaves pinnately or sometimes almost digitately 3-foliolate, leaflets papery, terminal leaflet rhomboid or obovate-rhomboid. Racemes, corolla yellow. Legume oblong, reddish purple, extremely compressed, slightly constricted between seeds. Seeds usually 2, black, lustrous, elliptic or subreniform.

Flower and Fruit: Fl. May-Aug. , fr. Sep. -Dec.

Distribution: Japan, Korea, Vietnam. To the south of Yangtze River. Main island of Nanji, commonly.

Habitat: Grass along the hillside.

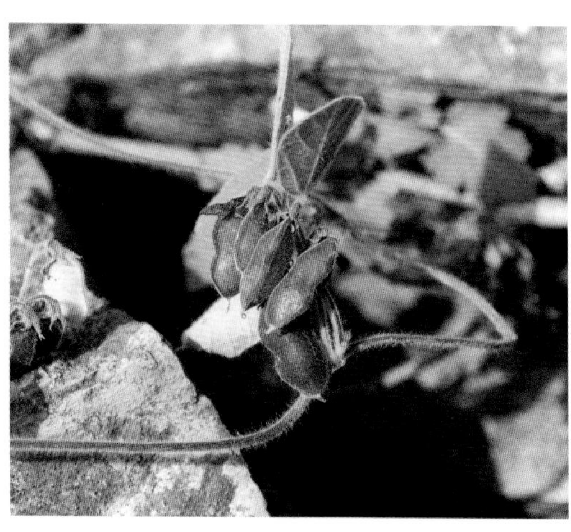

A140. 豆科
Fabaceae

双荚决明
Senna bicapsularis
决明属 *Senna*

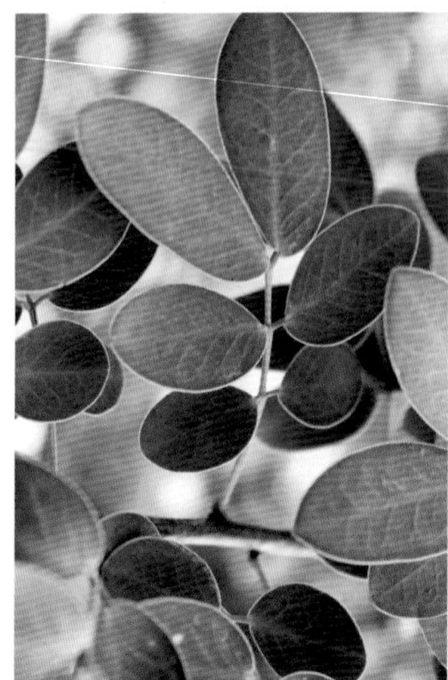

形态特征：直立灌木，多分枝，无毛，达 3 m 高。有小叶 3 ~ 4 对，小叶倒卵状或倒卵状长圆形，膜质，先端钝圆，基部渐狭，偏斜，在最下方的一对小叶间有黑褐色线形而钝头的腺体 1 枚。总状花序疏松，枝条顶端的叶腋间；子房具无毛的柄和花柱。荚果棕色，圆柱状，直或微曲。种子卵形，扁平。

花果期：花期 10 ~ 11 月，果期 11 月至翌年 3 月。

分布：原产美洲热带，现热带地区广泛栽培。我国华南地区常见栽培。南麂主岛（美龄居、镇政府），栽培。

生境：池边、路旁、公园和草地边缘。

Description: Shrubs erect, much branched, glabrous, to 3 m tall. Leaves with 3 or 4 pairs of leaflets, obovate or obovate-oblong, membranous, apex obtusely rounded, base obliquely cuneate, with a blackish brown, clavate to ovoid gland between lowest pair of leaflets. Racemes lax, in axils of apical leaves; ovary with glabrous stalk and style. Legume brown, terete, straight or slightly curved. Seeds ovoid, flat.

Flower and Fruit: Fl. Oct. -Nov. , fr. Nov. -Mar.

Distribution: Native to tropical America; widely cultivated in the tropics. Commonly cultivated in South China. Main island (Meilingju, Zhenzhengfu), cultivated.

Habitat: Poolsides, roadsides, parks and grasslands.

A140. 豆科
Fabaceae

蔓茎葫芦茶
Tadehagi pseudotriquetrum
葫芦草属 *Tadehagi*

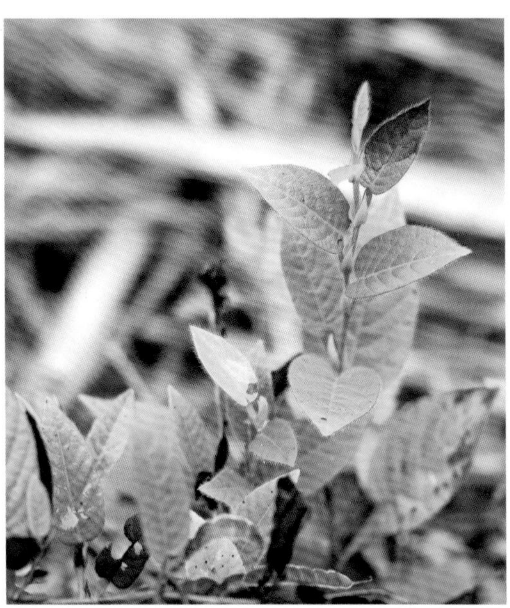

形态特征：亚灌木，茎蔓生，长 30 ~ 60 cm。叶仅具单小叶，小叶卵形，有时卵圆形，先端急尖，基部心形，正面无毛，背面沿脉疏被短柔毛，近叶缘处弧曲联结。花冠紫红色，旗瓣近圆形，先端凹入，翼瓣倒卵形，基部具耳，龙骨瓣镰刀状，无耳，有瓣柄。荚果 5 ~ 8 节，缝线密被白色短柔毛。

花果期：花期 8 月，果期 10 ~ 11 月。

分布：印度、尼泊尔、菲律宾。我国华东至华南地区。南麂主岛（关帝岙、三盘尾），偶见。

生境：向阳山坡、疏林灌丛。

Description: Subshrubs, stem ascending, 30-60 cm. Leaves 1-foliolate, blade ovate, sometimes ovate-orbicular, apex acute, base cordate, adaxially glabrous, abaxially sparsely pubescent on veins, not reaching margin but arching and joining together. Corolla purplish red, standard nearly orbicular, emarginate at apex, wings obovate, base auriculate, keel falcate, not auriculate, clawed. Legume 5-8-jointed, densely white pubescent on both sutures.

Flower and Fruit: Fl. Aug. , fr. Oct. -Nov.

Distribution: India, Nepal, Philippines. East China to South China. Main island (Guandiao, Sanpanwei), occasionally.

Habitat: Sunny slopes, sparse forest thickets.

A140. 豆科
Fabaceae

小巢菜
Vicia hirsuta
野豌豆属 *Vicia*

形态特征：一年生草本，高 15 ~ 90 cm。茎攀援，细长，后脱落。偶数羽状复叶，卷须分枝；小叶线形或狭长圆形，无毛。总状花序明显短于叶；花萼钟形，花冠白色或浅紫色，很少粉红色，旗瓣椭圆形，与翼瓣近等长，比龙骨瓣长，子房无柄，密被硬毛。荚果长圆菱形，多毛。种子 2，扁圆形。

花果期：花期 2 ~ 6 月，果期 2 ~ 8 月。

分布：北半球温暖地区。我国南北各地广布。南麂主岛（大沙岙、关帝岙、门屿尾、三盘尾），常见。

生境：山沟、河滩、草坡、田边或路旁草丛。

Description: Herbs annual, 15-90 cm. Stem climbing, slender, glabrescent. Leaves paripinnate, tendril branched; leaflets linear or narrowly oblong, glabrous. Raceme obviously shorter than leaf; calyx campanulate, corolla white to light purple, rarely pink, standard elliptic, subequaling wings and longer than keel, ovary sessile, densely rigidly hairy. Legume oblong-rhomboid, hirsute. Seeds 2, oblate-spheroid.

Flower and Fruit: Fl. Feb. -Jun. , fr. Feb. -Aug.

Distribution: Warm areas of the Northern Hemisphere. Throughout the north and south areas of China. Main island (Dashaao, Guandiao, Menyuwei, Sanpanwei), commonly.

Habitat: Valleys, creek banks, grassy slopes, grasslands of roadsides, fields and field margins.

A143. 蔷薇科
Rosaceae

龙芽草
Agrimonia pilosa
龙芽草属 *Agrimonia*

形态特征：多年生草本，高 30 ～ 120 cm。茎疏生柔毛和短柔毛，或下部密被硬毛。奇数羽状复叶，小叶 3 ～ 4 对，上部叶退化为 3 小叶，小叶片倒卵形，边缘有尖锐锯齿或裂片。花序穗状总状顶生，花冠黄色。果实倒卵圆锥形，外面有 10 条肋，具柔毛，顶端有数层钩刺。

花果期：花果期 5 ～ 12 月。

分布：东亚、中欧。分布于我国各地。南麂主岛（百亩山、大山、国姓岙、门屿尾），偶见。

生境：疏林、林缘、灌丛、草地、溪边、路旁。

Description: Herbs perennial, 30-120 cm tall. Stems sparsely pilose and pubescent, or densely rigidly hairy in lower part. Leaf blade interrupted imparipinnate with (2 or) 3 or 4 pairs of leaflets, reduced to 3 leaflets on upper leaves, leaflets obovate, margin acutely to obtusely serrate. Inflorescence terminal, spicate-racemose, corolla yellow. Fruit obovoid-conic, abaxially 10-ribbed, pilose, with a multiseriate crown of prickles.

Flower and Fruit: Fl. and fr. May-Dec.

Distribution: East Asia, Central Europe. Throughout China. Main island (Baimushan, Dashan, Guoxingao, Menyuwei), occasionally.

Habitat: Thinned forests, forest margins, thickets, meadows, stream banks, roadsides.

A143. 蔷薇科
Rosaceae

毛柱郁李
Prunus pogonostyla（*Cerasus pogonostyla*）
李属 *Prunus*

形态特征：灌木或小乔木，高 0.5 ~ 1.5 m。小枝灰色，嫩枝绿色，无毛或微被柔毛。叶片倒卵状椭圆形至椭圆状披针形，边缘圆钝，稀为尖重锯齿；叶柄被稀疏短柔毛；托叶线形，边有腺齿。花单生或 2 朵一束，花叶同放，花瓣粉红色。核果椭圆形或近球形，核表面光滑。

花果期：花期 3 月，果期 4 ~ 5 月。

分布：我国浙江、福建、广东、湖南、江西、台湾。南麂主岛（百亩山、国姓岙、后隆、门屿尾），偶见。

生境：山坡林下、峡谷向阳处。

Description: Shrubs or small trees, 0.5-1.5 m tall. Branchlets gray, young branchlets green, glabrous or sparsely pilose. Leaf blade obovate-elliptic to elliptic-lanceolate, margin obtusely rarely acutely biserrate; petiole sparsely pubescent; stipules linear, margin glandular serrate. Flowers solitary or 2 in a fascicle, opening at same time as leaves, petals pink. Drupe ellipsoid to subglobose, endocarp smooth.

Flower and Fruit: Fl. Mar. , fr. Apr. -May.

Distribution: Zhejiang, Fujian, Guangdong, Hunan, Jiangxi, Taiwan, China. Main island (Baimushan, Guoxingao, Houlong, Menyuwei), occasionally.

Habitat: Forests on mountain slopes, sunny places in ravines.

A143. 蔷薇科
Rosaceae

野山楂
Crataegus cuneata
山楂属 *Crataegus*

形态特征：落叶灌木，高达 15 m。分枝密，常具细刺。叶宽倒卵形至倒卵状长圆形，边缘有不规则重锯齿，顶端 3 浅裂，少有 5 裂或不裂；托叶草质，镰刀状，边缘有齿。伞房花序，花 5 ~ 7 朵，花梗被柔毛；花瓣白色，近圆形或倒卵形。梨果近球形，红色或黄色，萼片常宿存反折。

花果期：花期 5 ~ 6 月，果期 9 ~ 11 月。

分布：日本。我国秦淮地区以南。南麂主岛（百亩山、大山、打铁礁、关帝岙、国姓岙、后隆、门屿尾、三盘尾），常见。

生境：山谷、灌丛。

Description: Shrubs deciduous, to 15 m tall. Branched densely, usually with slender thorns. Leaf blade broadly obovate to obovate-oblong or obovate-elliptic, margin irregularly doubly serrate or serrate, 3-lobed, rarely 5-lobed in apical part or not lobed; stipules falcate, herbaceous, margin serrate. Corymb, 5-7-flowered, peduncle pubescent; petals white, suborbicular or obovate. Pome red or yellow, subglobose, sepals often persistent, reflexed.

Flower and Fruit: Fl. May-Jun. , fr. Sep. -Nov.

Distribution: Japan. To the south of Qinling and Huaihe. Main island (Baimushan, Dashan, Datiejiao, Guandiao, Guoxingao, Houlong, Menyuwei, Sanpanwei), commonly.

Habitat: Valleys, thickets.

A143. 蔷薇科
Rosaceae

翻白草
Potentilla discolor
委陵菜属 *Potentilla*

形态特征：多年生草本，高 15 ~ 50 cm。花茎直立，密被白色绵毛。基生小叶 2 ~ 4 对，叶柄密被白色绵毛，小叶片长圆形，边缘具圆钝锯齿。聚伞花序有花数朵至多朵，疏散；萼片三角状卵形，副萼片披针形，外被白色绵毛，花瓣黄色，倒卵形，顶端微凹或圆钝。瘦果近肾形，光滑。

花果期：花果期 5 ~ 9 月。

分布：日本、朝鲜。分布于我国南北各地。南麂主岛（关帝岙、三盘尾），偶见。

生境：荒地、山谷、沟边、山坡草地、草甸及疏林下。

Description: Herbs perennial,15-50 cm. Flowering stems erect, densely white lanate. Radical leaves with 2-4 pairs of leaflets, petioles densely white lanate, small leaves broadly cuneate, or obliquely rounded, margin obtusely serrate. Inflorescence cymose, laxly several to many flowered; sepals triangular-ovate, epicalyx segments lanceolate, abaxially white lanate,petals yellow, obovate, apex rounded or emarginate. Achenes subreniform, smooth.

Flower and Fruit: Fl. and fr. May-Sep.

Distribution: Japan, Korea. Throughout the north and south areas of China. Main island (Guandiao, Sanpanwei), occasionally.

Habitat: Wastelands, valleys, ravines, hillside grasslands, meadows, sparse forests.

A143. 蔷薇科
Rosaceae

厚叶石斑木
Rhaphiolepis umbellata
石斑木属 *Rhaphiolepis*

形态特征：常绿灌木或小乔木，高 2 ～ 4 m。枝粗壮极叉开，枝和叶在幼时有褐色柔毛，后脱落。叶厚革质，长椭圆形、卵形或倒卵形，全缘或有疏生钝锯齿。圆锥花序顶生，直立，密生褐色柔毛；萼筒倒圆锥状，萼片三角形；花瓣白色，倒卵形。梨果球形，黑紫色带白霜，顶端有萼片脱落残痕，种子 1 粒。

花果期：花期 4 ～ 6 月，果期 9 ～ 11 月。

分布：日本。我国浙江东部、台湾。南麂主岛（打铁礁、关帝岙、国姓岙、门屿尾、三盘尾），常见。

生境：沿海岛屿的山坡、路旁岩石上。

Description: Evergreen shrubs or small trees, 2-4 m tall. Branchlets stout, divaricate, initially brown pubescent, glabrescent. Leaf blade , thickly leathery, narrowly elliptic, ovate, or obovate, margin entire or remotely crenate. Panicle terminal, erect, peduncle densely brown pubescent; hypanthium obconical, sepals lanceolate or triangular-lanceolate; petals white, obovate. Pome blackish purple, globose, glaucescent, sepals caducous, leaving an annular ring, seed 1.

Flower and Fruit: Fl. Apr. -Jun. , fr. Sep. -Nov.

Distribution: Japan. East Zhejiang, and Taiwan, China. Main island (Datiejiao, Guandiao, Guoxingao, Menyuwei, Sanpanwei), commonly.

Habitat: Slopes and rocks of coastal islands.

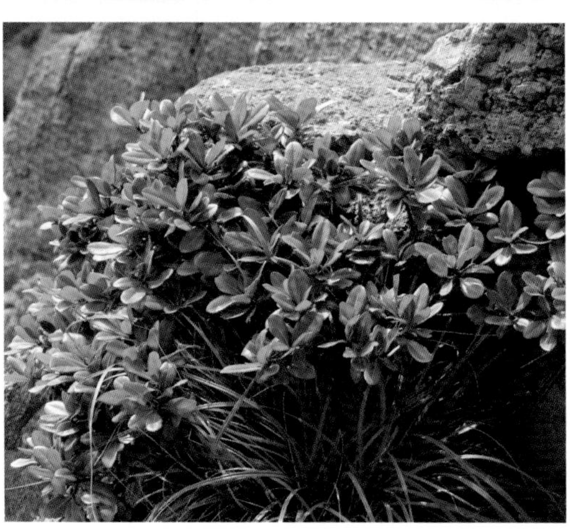

A143. 蔷薇科
Rosaceae

硕苞蔷薇
Rosa bracteata
蔷薇属 *Rosa*

形态特征：铺散常绿灌木，有长匍枝。小枝粗壮，密被黄褐色柔毛，混生针刺和腺毛。小叶 5 ~ 9，革质，椭圆形、倒卵形；托叶大部分离生，呈篦齿状深裂。花单生或 2 ~ 3 朵集生；苞片大，宽卵形，边缘有不规则缺刻状锯齿，花瓣白色或黄白色，倒卵形，先端微凹。蔷薇果球形，密被黄褐色柔毛，萼片宿存、反折。

花果期：花期 5 ~ 7 月，果期 8 ~ 11 月。

分布：日本南部。我国华东至西南地区。南麂各岛屿常见。

生境：混交林、灌丛、沙丘、溪边、海岸或路边。

Description: Shrubs evergreen, diffuse, with long repent branches. Branchlets robust, densely tawny puberulent, tomentose, smaller prickles and glandular bristles often present and dense to scattered. Leaflets 5-9, leathery, elliptic or obovate; stipules mostly free, deeply divided into often pectinate lobes. Flowers solitary or 2 or 3 and fasciculate; bracts several, large, broadly ovate, margin irregularly incised serrate, petals white or yellowish white, obovate, apex emarginate. Hip globose, densely tawny pubescent, with persistent, reflexed sepals.

Flower and Fruit: Fl. May-Jul. , fr. Aug. -Nov.

Distribution: South Japan. East China to Southwest China. Throughout Nanji Islands, commonly.

Habitat: Mixed forests, scrub, sandy hills, stream sides, seashores, roadsides.

A143. 蔷薇科
Rosaceae

小果蔷薇
Rosa cymosa
蔷薇属 *Rosa*

形态特征：攀援灌木，攀援成斜升；小枝圆柱形，有钩状皮刺。小叶 3 ~ 5，偶见 7，叶卵状披针形或椭圆形，边缘有紧贴或尖锐细锯齿，两面无毛，上面亮绿色，下面颜色较淡。复伞房花序，花多朵，萼片卵形，花瓣白色或黄色，倒卵形，先端凹，基部楔形。蔷薇果球形，红色至黑褐色，萼片脱落。

花果期：花期 5 ~ 6 月，果期 7 ~ 11 月。

分布：老挝、越南。我国长江流域以南。南麂各岛屿常见。

生境：向阳山坡、路旁、溪边或丘陵地。

Description: Shrubs evergreen, climbing or scandent; branchlets terete, prickles scattered, hooked. Leaflets 3-5, rarely 7, ovate-lanceolate or elliptic, margin acutely serrulate, both surfaces glabrous, adaxially bright green, abaxially lighter. Flowers numerous, in compound corymbs, sepals ovate, petals white or yellow, obovate, base cuneate, apex emarginate. Hip red to black-brown, globose, sepals caducous.

Flower and Fruit: Fl. May-Jun. , fr. Jul. -Nov.

Distribution: Laos, Vietnam. To the south of Yangtze River Basin. Throughout Nanji Islands, commonly.

Habitat: Open slopes, roadsides, stream sides, hills.

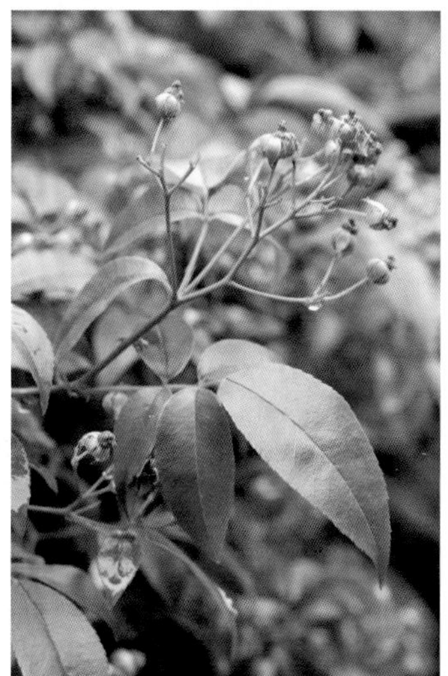

A143. 蔷薇科
Rosaceae

金樱子
Rosa laevigata
蔷薇属 ***Rosa***

形态特征： 常绿攀援灌木，高可达 5 m。小枝粗壮，散生扁弯皮刺，无毛。小叶 3 枚，革质，椭圆状卵形、倒卵形或披针状卵形，边缘有锐锯齿，上面亮绿色，无毛，下面黄绿色。花单生叶腋；花梗和萼筒密被腺毛，随果实成长变为针刺；花瓣白色，宽倒卵形，先端微凹。蔷薇果梨形、倒卵形，密被刺毛，萼片宿存、直立。

花果期： 花期 4 ～ 6 月，果期 7 ～ 11 月。

分布： 越南。我国华东、华中、华南地区。南麂主岛（百亩山、大山、门屿尾、三盘尾），常见。

生境： 灌丛，开阔的山地、田野。

Description: Shrubs evergreen, climbing, to 5 m. Branchlets robust, prickles scattered, curved, flat, glabrous. Leaflets 3, leathery, elliptic-ovate, obovate, or lanceolate-ovate, margin acutely serrate,glabrous, bright green above, yellow green below. Flower solitary, axillary; pedicel and hypanthium densely glandular bristly; petals white, broadly obovate,apex emarginate. Hip purple-brown, pyriform or obovoid, rarely subglobose, densely glandular bristly, with persistent, erect sepals.

Flower and Fruit: Fl. Apr. -Jun. , fr. Jul. -Nov.

Distribution: Vietnam. East China, Central China and South China. Main island (Baimushan, Dashan, Menyuwei, Sanpanwei), commonly.

Habitat: Thickets, open montane areas, open fields.

A143. 蔷薇科
Rosaceae

光叶蔷薇
Rosa luciae
蔷薇属 *Rosa*

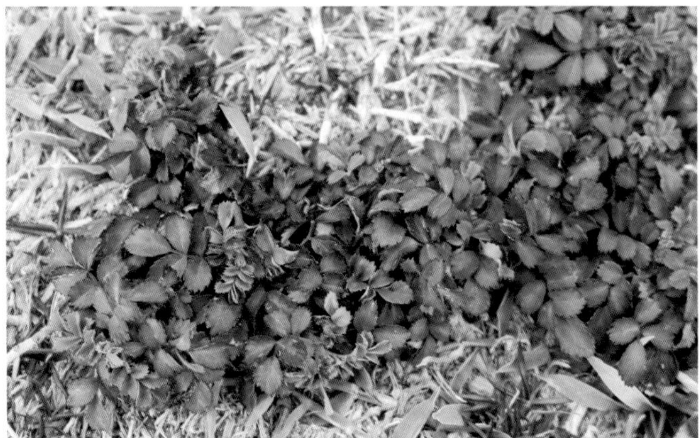

形态特征：匍匐灌木，蔓生或平卧，3～5 m。小枝红棕色，圆柱状，幼时具毛。小叶椭圆形，卵形，或倒卵形，无毛，中脉背面突出。花多朵成伞房花序或单生；苞片早落，萼片 5，披针形，边缘全缘；花瓣 5，白色的或粉红色，芳香，倒卵形，先端圆钝；花柱合生成束，外露。蔷薇果紫黑褐色，球形，疏生腺毛。

花果期：花期 4～7 月，果期 10～11 月。

分布：日本，朝鲜，菲律宾。我国华东和华南地区。南麂主岛各处常见。

生境：灌丛，海边悬崖，海岸，石灰岩上。

Description: Shrubs prostrate, sprawling, or procumbent, 3-5 m. Branchlets red-brown, terete, pubescent when young. Leaf blades elliptic, ovate, or obovate, glabrous, abaxially with prominent midvein. Flowers numerous in corymb or solitary; bracts caducous, sepals 5, lanceolate, margin entire; petals 5, white or pink, fragrant, obovate, apex rounded-obtuse; styles connate into column, exserted. Hip purple-black-brown, globose, sparsely glandular-pubescent.

Flower and Fruit: Fl. Apr. -Jul. , fr. Oct. -Nov.

Distribution: Japan, Korea, Philippines. East and South China. Main island of Nanji, commonly.

Habitat: Thickets, sea cliffs, coasts, on limestone.

A143. 蔷薇科
Rosaceae

野蔷薇（多花蔷薇）
Rosa multiflora
蔷薇属 ***Rosa***

形态特征：攀援灌木。小枝圆柱状，通常无毛。叶下有成对皮刺，粗壮，弯曲，平展；托叶栉状，多贴生叶柄，叶轴具短刺；小叶 5 ~ 9，倒卵形，长圆形，边缘有尖锐单锯齿。伞房花序多花，苞片生于花梗基部，小；萼片 5，脱落，披针形；花瓣 5，单瓣，白色，芳香，倒卵形，先端微凹；花柱合生成束，外露。蔷薇果红褐色或紫褐色，近球形，无毛，发亮。

花果期：花期 5 ~ 6 月，果期 7 ~ 10 月。

分布：日本，朝鲜。我国南北多数地区有分布。南麂主岛各处常见。

生境：灌丛、山坡、河边。

Description: Shrubs climbing. Branchlets terete, usually glabrous. Prickles paired below leaves, curved, stout, flat; stipules pectinate, mostly adnate to petiole, rachis shortly prickly; leaflets 5-9, obovate, oblong, margin simply serrate. Flowers numerous in corymb, bracts at base of pedicel, small; sepals 5, deciduous, lanceolate; petals 5, white, obovate, apex emarginate; styles connate into column, exserted. Hip red-brown or purple-brown, subglobose, glabrous, shiny.

Flower and Fruit: Fl. May-Jun. , fr. Jul. -Oct.

Distribution: Japan, Korea. Most areas in northern and southern China. Main island of Nanji, commonly.

Habitat: Thickets, slopes, river sides.

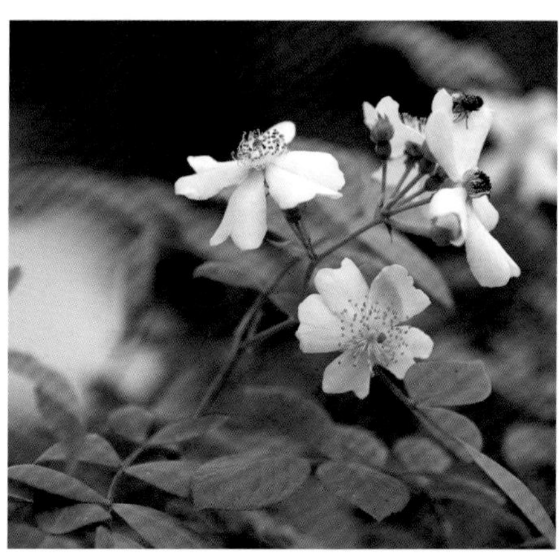

A143. 蔷薇科
Rosaceae

茅莓
Rubus parvifolius
悬钩子属 ***Rubus***

形态特征：灌木，高 1 ~ 2 m，枝呈弓形弯曲，被柔毛和稀疏钩状皮刺。奇数羽状复叶，小叶 3 ~ 5，菱状圆卵形或倒卵形，边缘有不整齐粗锯齿或缺刻状粗重锯齿。伞房花序顶生或腋生，花瓣粉红或紫红色。聚合果球形，红色肉质，可食，酸甜多汁。

花果期：花期 5 ~ 6 月，果期 7 ~ 8 月。

分布：日本、朝鲜、越南。我国东部湿润半湿润地区。南麂各岛屿常见。

生境：山坡杂木林下、向阳山谷、路旁或荒野。

Description: Shrubs 1-2 m tall, with arching branches, branchlets with soft hairs and sparse, curved prickles. Leaves imparipinnate, 3-5-foliolate, rhombic-orbicular or obovate, margin unevenly coarsely serrate or coarsely incised-doubly serrate. Inflorescences terminal, corymbose, rarely short racemes, axillary inflorescences corymbose, petals pink to purplish red. Aggregate fruit red, ovoid-globose, esculent, sweet and juicy.

Flower and Fruit: Fl. May-Jun. , fr. Jul. -Aug.

Distribution: Japan, Korea, Vietnam. Humid and subhumid areas of eastern China. Throughout Nanji Islands, commonly.

Habitat: Forests, thickets, clearings, slopes, sunny valleys, roadsides, waste places.

A143. 蔷薇科
Rosaceae

锈毛莓
Rubus reflexus
悬钩子属 ***Rubus***

形态特征：攀援灌木，高达 2 m。枝被锈色绒毛状毛，有稀疏小皮刺。单叶，心状长卵形，上面有明显皱纹，下面密被锈色绒毛，边缘 3 ～ 5 裂，偶见 7 裂，有不整齐的粗锯齿或重锯齿。花数朵生于叶腋或成顶生短总状花序；总花梗和花梗密被锈色长柔毛，花瓣长圆形至近圆形，白色。果实近球形，深红色；核有皱纹。

花果期：花期 6 ～ 7 月，果期 8 ～ 9 月。

分布：我国华东至西南地区。南麂主岛各处常见。

生境：山区的山坡、疏林、灌丛、山谷、溪边，潮湿的地方，近水边。

Description: Shrubs scandent, to 2 m tall. Branchlets rust colored tomentose, with sparse, minute prickles. Leaves simple, blade narrowly or broadly ovate to suborbicular, adaxially prominently rugose, abaxially rust colored tomentose, margin 3-5(-7)-lobed or -parted, margin unevenly coarsely serrate to doubly serrate. Inflorescences terminal, short subracemes, or flowers several in clusters in leaf axils; rachis and pedicels rusty villous, petals white, oblong to suborbicular. Aggregate fruit dark red, subglobose; pyrenes rugose.

Flower and Fruit: Fl. Jun. -Jul. , fr. Aug. -Sep.

Distribution: East China to Southwest China. Main island of Nanji, commonly.

Habitat: Mountainous regions, slopes, sparse forests, thickets, valleys, stream sides, moist places, near water courses.

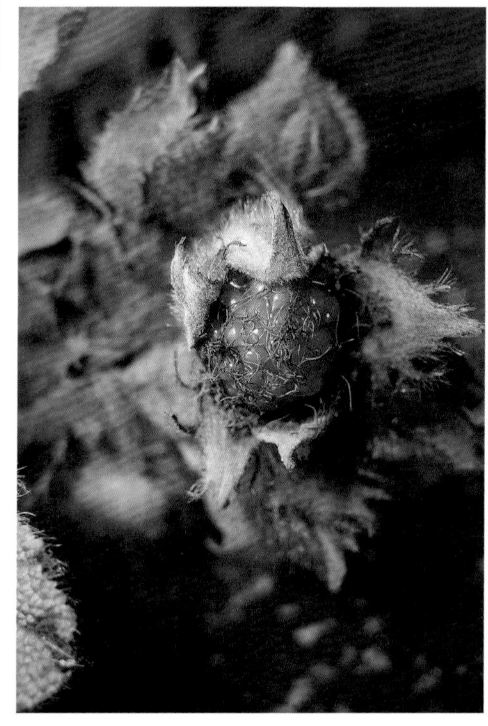

A143. 蔷薇科 Rosaceae

中华绣线菊
Spiraea chinensis
线菊属 *Spiraea*

形态特征：灌木，高 1.5 ～ 3m。小枝呈拱形弯曲，红褐色，幼时被黄色绒毛。叶片菱状卵形至倒卵形，边缘有缺刻状粗锯齿，上面暗绿色，被短柔毛，叶脉下陷，下面密被黄色绒毛，叶脉凸起；叶柄被短绒毛。伞形花序，花梗具短绒毛；花萼筒钟状，萼片卵状披针形，花瓣近圆形，白色。蓇葖果开张，被短柔毛。

花果期：花期 3 ～ 6 月，果期 7 ～ 10 月。

分布：我国黄河流域以南。南麂主岛各处常见。

生境：灌丛，山坡，开阔地带，路旁。

Description: Shrubs, 1.5-3 m tall. Branchlets arched, red-brown and yellow tomentose initially. Leaf blade rhombic-ovate to obovate, margin with laceration coarsely serrate, above dark green, pubescent, veins sunken, below densely covered with yellow villi, veins protruding; petiole shortly tomentose. Umbel inflorescence; pedicel shortly tomentose; calyx tube campanulate, sepals ovate-lanceolate, petals subcircular, white. Follicles open, all pubescent.

Flower and Fruit: Fl. Mar. -Jun. , fr. Jul. -Oct.

Distribution: South of the Yellow River Basin. Main island of Nanji, commonly.

Habitat: Thickets, slopes, open places, roadsides.

A146. 胡颓子科
Elaeagnaceae

蔓胡颓子
Elaeagnus glabra
胡颓子属 *Elaeagnus*

形态特征：常绿蔓生或攀援灌木，高达 5 m。幼枝密被锈色鳞片，老枝鳞片脱落，灰棕色。叶纸质，卵形或卵状椭圆形，边缘全缘，微反卷。花淡白色，下垂，密被银白色和散生少数褐色鳞片，常 3 ~ 7 花密生于叶腋短小枝上成伞形总状花序。果实圆矩形，稍有汁，被锈色鳞片，成熟时红色。

花果期：花期 9 ~ 11 月，果期 4 ~ 5 月。

分布：日本、朝鲜南部。我国华东至华南地区。南麂主岛各处常见。

生境：灌丛。

Description: Shrubs, evergreen, rampant or scandent, to 5 m tall. Young branches with dense silvery brown scales, scales fall off old branches, gray. Leaf blade papery, ovate or ovate elliptic, margin entire, slightly recurved. Flowers pale white, nutant, densely silver scaly and few scattered brown scales. 3-7 flowers densely axillary on short branches into umbels. Drupe broadly elliptic, slightly succulent, rust-colored scales, red when ripe.

Flower and Fruit: Fl. Sep. -Nov. , fr. Apr. -May.

Distribution: Japan, South Korea. East China to South China. Main island of Nanji, commonly.

Habitat: Thickets.

A146. 胡颓子科
Elaeagnaceae

大叶胡颓子
Elaeagnus macrophylla
胡颓子属 *Elaeagnus*

形态特征：常绿直立灌木，高 2 ~ 3 m。幼枝扁棱形，灰褐色，密被淡黄白色鳞片，老枝鳞片脱落，黑色。叶厚纸质或薄革质，卵形至宽卵形或阔椭圆形至近圆形，顶端钝形或锐尖，幼时被银白色鳞片，成熟后脱落。花白色，被鳞片，略开展，常 1 ~ 8 花生于叶腋短小枝上。果实长椭圆形，被银白色鳞片。

花果期：花期 9 ~ 10 月，果期翌年 3 ~ 4 月。

分布：日本、朝鲜南部。我国东部沿海岛屿。南麂主岛各处常见。

生境：海岸灌丛。

Description: Shrubs, evergreen, erect, 2-3 m tall. Young branchlets flat prisms, gray-brown, densely covered with yellowish white scales, old branch scales flabrescent, black. Leaf blade thick papery or thin leathery, orbicular-ovate to broadly elliptic-ovate, apex rounded or obtuse, upper densely white scaly when young, fall off when ripe. Flowers white, scaly, slightly spreading, 1-8 flowers axillary on short shoot. Drupe oblong, white scaly.

Flower and Fruit: Fl. Sep. -Oct. , fr. Mar. -Apr. of 2nd year.

Distribution: Japan, South Korea. Coastal islands of eastern China. Main island of Nanji, commonly.

Habitat: Thickets near seashores.

A146. 胡颓子科
Elaeagnaceae

胡颓子
Elaeagnus pungens
胡颓子属 *Elaeagnus*

形态特征：常绿直立灌木，高 3 ~ 4 m。叶革质，椭圆形或宽椭圆形，上面幼时被银白色和少数褐色鳞片，下面密被鳞片。花白色，下垂，密被鳞片，1 ~ 3 花腋生锈色短枝上。核果椭圆形，幼时被褐色鳞片，熟时红色。

花果期：花期 9 ~ 12 月，果期翌年 4 ~ 6 月。

分布：日本。我国华东至华南地区。南麂主岛各处常见。

生境：开阔的山坡、路旁或灌丛，常靠海边。

Description: Shrubs evergreen, erect, 3-4 m tall. Leaf blade leathery, oblong to narrowly so, upper with dense whitish and few brown scales when young, densely scaly underside. Flowers white, nutant, densely scaly, 1-3 flowers axillary on rusty short shoot. Drupe oblong, brown scaly when young, red when ripe.

Flower and Fruit: Fl. Sep. -Dec. , fr. Apr. -Jun of 2nd year.

Distribution: Japan. East China to South China. Main island of Nanji, commonly.

Habitat: Open slopes, roadsides or thickets, often near the sea.

A147. 鼠李科
Rhamnaceae

雀梅藤
Sagereta thea
雀梅藤属 *Sageretia*

形态特征：攀缘灌木或直立，小枝具刺。叶互生或近对生，长圆形或卵状椭圆形，边缘具细锯齿，侧脉在背面明显凸起。花无梗，黄色，数个簇生排成顶生或腋生疏散穗状或圆锥状穗状花序；花萼被疏柔毛，花瓣短于萼片，匙形，顶端 2 浅裂。核果近圆球形，成熟时紫黑色，1 ~ 3 分核。

花果期：花期 7 ~ 9 月，果期翌年 3 ~ 5 月。

分布：亚洲东部、印度。我国长江流域以南地区。南麂各岛屿常见。

生境：丘陵、山地林下或灌丛。

Description: Shrubs scandent or erect, armed. Leaves alternate or subopposite, usually oblong, or ovate-elliptic, margin serrated, leaf surface under lateral veins distinctly convex. Flowers sessile, yellow, few fascicled in terminal or axillary lax spikes or paniculate spikes; calyx tube sparsely pubescent; petals shorter than sepals, spatulate, apex 2-fid. Drupe subglobose, black or purple-black at maturity, with 1-3 pyrenes.

Flower and Fruit: Fl. Jul. -Sep. , fr. Mar. -May of 2nd year.

Distribution: Eastern Asia, India. To the south of Yangtze River Basin. Throughout Nanji Islands, commonly.

Habitat: Hills, under forests, thickets.

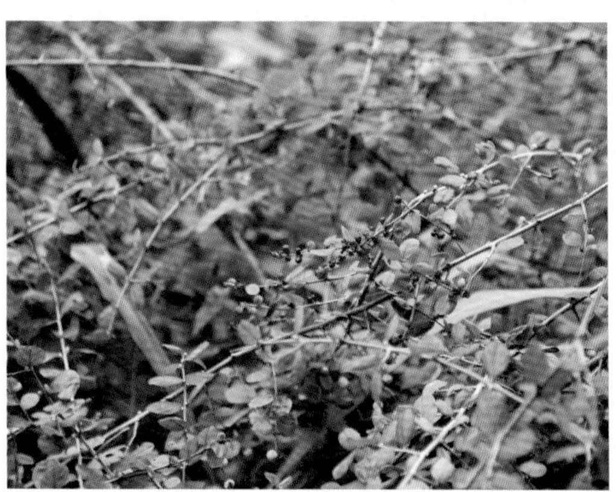

A149. 大麻科
Cannabaceae

朴树
Celtis sinensis
朴属 *Celtis*

形态特征：落叶乔木，高达 20 m。树皮平滑，灰色。一年枝被密毛，老枝无毛。叶革质，互生，宽卵形至狭卵形，基部圆楔，对称或稍偏斜，中部以上边缘有浅锯齿，三出脉。花杂性同株，1 ~ 3 枚着生于当年枝的叶腋，花被片 4，被毛，雄蕊 4，柱头 2。核果近球形，红褐色；果梗较叶柄粗 1 ~ 1.5 倍，疏被柔毛。

花果期：花期 3 ~ 4 月，果期 9 ~ 10 月。

分布：日本。我国长江流域以南地区。南麂各岛常见，常栽培。

生境：路旁、山坡。

Description: Trees deciduous, to 20 m tall. Bark smooth, gray. One year branches hairy, old branches glabrous. Leaf blade leathery, alternate, ovate to ovate-elliptic, base rounded, obtuse, or obliquely truncate, ± symmetric to moderately oblique, margin subentire to crenate on apical half, trinervious. Polygamous homophytic, flowers polygamic, fascicled in leaf axils, 1(-3) per leaf axil, tepal 4, pilose, stamens 4, stigma 2. Drupe ± globose, rufous, fruiting pedicel 1-1.5 × as long as subtending petiole, sparsely pilose.

Flower and Fruit：Fl. Mar. -Apr. , fr. Sep. -Oct.

Distribution：Japan. To the south of Yangtze River Basin. Throughout main island of Nanji, usually cultivated.

Habitat：Roadsides, slopes.

A149. 大麻科
Cannabaceae

葎草
Humulus scandens
葎草属 *Humulus*

形态特征： 一年生缠绕藤本。叶片掌状(3或)5～9浅裂，纸质，背面叶脉具刚硬具刺的毛，正面短柔毛但不密，基部心形；裂片卵形三角形，边缘有锯齿。雄花黄绿色，雌花苞片卵球形，纸质，具刺，先端渐尖。瘦果成熟时露出苞片。

花果期： 花期春夏季，果期秋季。

分布： 日本，朝鲜、越南。我国大部分区域广泛分布。南麂各岛屿常见。

生境： 林缘，荒地，溪边

Description: Herbs annual, twining. Leaf blade palmately (3 or)5-9-lobed, papery, abaxially with rigid spinulose hairs on veins, adaxially pubescent but not densely so, base cordate; lobes ovate-triangular, margin serrate. Male flowers yellowish green; female flower bracts ovoid, papery, spinulose, apex acuminate. Achenes exerted from bracts when mature.

Flower and Fruit: Fl. spring to summer, fr. autumn

Distribution: Japan, Korea, Vietnam. Most areas of China. Throughout Nanji Islands, commonly.

Habitat: Forest margins, wastelands, along streams.

A150. 桑科
Moraceae

雅榕
Ficus concinna
榕属 *Ficus*

形态特征：常绿乔木，高15～25m。树皮深灰色，具皮孔，分枝气生根少，小枝粗壮，无毛。叶干后灰绿色，狭椭圆形，全缘，光滑无毛；基生侧脉短，4～8对，两面明显。榕果成对腋生或3～4个簇生于无叶小枝叶腋，球形，无柄或近无柄；雄花、瘿花、雌花同生于一榕果内。

花果期：花果期3～6月。

分布：热带亚洲。我国华东至西南地区。南麂主岛（国姓岙），栽培。

生境：密林，村边。

Description: Evergreen trees, 15-25 m tall. Bark dark gray, lenticellate. Branches producing few aerial roots; branchlets thick, glabrous. Leaf blade grayish green when dry, narrowly elliptic, margin entire, glabrous; basal lateral veins short, secondary veins 4-8 on each side of midvein, conspicuous on both surfaces. Figs axillary on leafy branchlets, paired, or in clusters of 3 or 4 on leafless older branchlets, globose, sessile or subsessile; involucral bracts caducous; male, gall, and female flowers within same fig.

Flower and Fruit: Fl. and fr. Mar. -Jun.

Distribution: Tropical Asia. East China to South China. Main island (Guoxingao), cultivated.

Habitat: Dense forests, near villages.

A150. 桑科
Moraceae

天仙果（矮小天仙果）
Ficus erecta（*Ficus erecta* var. *beecheyana*）
榕属 *Ficus*

形态特征：落叶灌木，高 3 ～ 4 m。树皮灰褐色，小枝密生硬毛。叶厚纸质，互生，倒卵形至狭倒卵形，基部圆形至浅心形，全缘或上部偶有疏齿，被柔毛或硬毛；托叶三角状披针形，膜质，早落。隐头花序单生或成对腋生，雄花和瘿花同生于一隐头花序中，雌花生于另一隐头花序中。榕果单生叶腋，球形，无毛，成熟时红色；果柄较短。

花果期：花果期 5 ～ 6 月。

分布：日本、朝鲜、越南。我国华东至西南地区。南麂各岛屿常见。

生境：山坡林下阴湿处，山谷、溪边灌木丛。

Description: Shrubs deciduous, 3-4 m tall. Bark grayish brown, young branchlets densely hispid. Leaves thickly papery, alternate, obovate to narrowly obovate, apex shortly acuminate, base rounded to shallowly cordate, margin entire or upper part occasionally remotely serrate, pilose or hirsute; stipules triangular-lanceolate, membranous, caducous; Hypanthodium single or axillary in pairs, male and gall flowers in one hypanthodium, female in another. Figs axillary, solitary globose, glabrous, red at maturity; petiole short.

Flower and Fruit: Fl. and fr. May-Jun.

Distribution: Japan, Korea, Vietnam. East China to Southwest China. Throughout Nanji Islands, commonly.

Habitat: Shaded moist places under slope forests, or thickets along valleys and streams.

A150. 桑科
Moraceae

薜荔
Ficus pumila
榕属 *Ficus*

形态特征：常绿灌木，攀援或匍匐。生根小枝不育。叶二型，营养枝的叶卵状心形，顶端渐尖，叶柄很短；生殖枝的叶革质，卵状椭圆形，全缘，网脉明显，呈蜂窝状，叶柄短粗。隐头花序具短柄；雄花和瘿花同生于一花序中；雌花生另一花序中。榕果单生叶腋，大梨形或近球形，果柄粗短。瘦果近球形，有粘液。

花果期：花果期 5 ~ 8 月。

分布：日本、越南。我国长江流域以南地区。南麂主岛各处常见。

生境：村寨附近或墙壁上。

Description: Evergreen Shrubs, climbers or scandent. Rooting branchlets sterile. Leaves distichous, leaf blade on fertile branchlets ovate-cordate, apex acuminate, petiole short; leaf blade on sterile branchlets ovate-elliptic, veins conspicuous, honeycomblike, margin entire, petiole short and thick. Hypanthium with short stalks; Male and gall flowers within same fig, female flowers within anther fig. Figs axillary on normal leafy branches, pear-shaped to ± globose or cylindric; peduncle short and thick. Achenes ± globose, with adherent liquid.

Flower and Fruit: Fl. and fr. May-Aug.

Distribution: Japan, Vietnam. To the south of Yangtze River Basin. Main island of Nanji, commonly.

Habitat: Near village or on the wall.

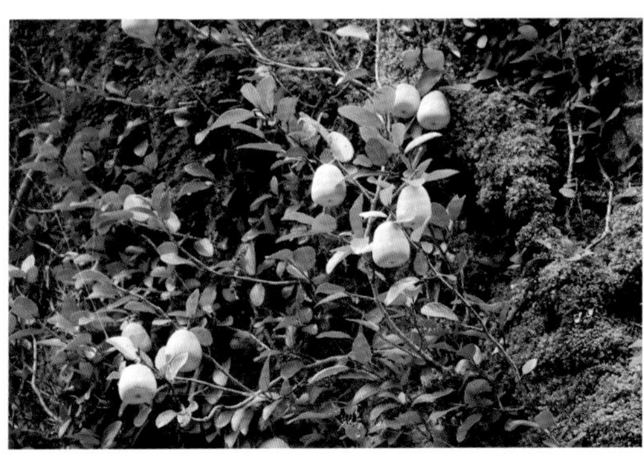

A150. 桑科
Moraceae

爱玉子
Ficus pumila var. *awkeotsang*
榕属 *Ficus*

形态特征： 薜荔（*Ficus pumila*）变种，叶长椭圆状卵形，长 7 ～ 12 cm，宽 3 ～ 5 cm，背面密被锈色柔毛。榕果长圆形，表面被毛，顶部渐尖；脐部凸起；总梗短，密被粗毛。

花果期： 花果期 5 ～ 8 月。

分布： 我国华东地区。南麂主岛各处常见。

生境： 岩石或墙上。

Description: Variety of *Ficus pumila*. Leaf blade oblong-ovate, 7-12 × 3-5 cm, abaxially densely covered with rust-colored pubescence. Figs cylindric, piliferous, apical pore acuminate; umbilical region raised; peduncle short, densely covered with thick hairs.

Flower and Fruit: Fl. and fr. May-Aug.

Distribution: East China. Main island of Nanji, commonly.

Habitat: Usually climbing on rocks or walls.

A150. 桑科
Moraceae

笔管榕
Ficus subpisocarpa
榕属 *Ficus*

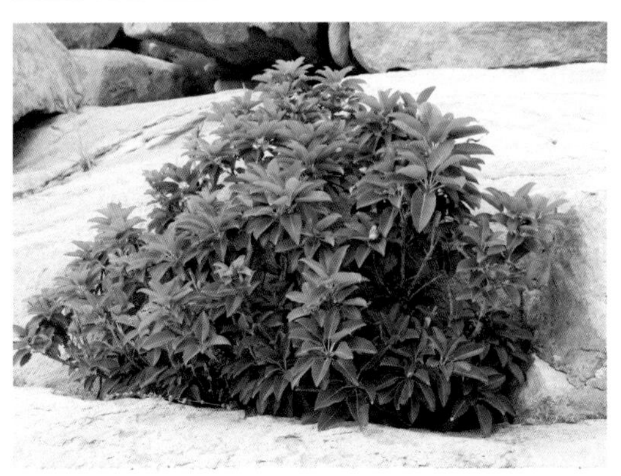

形态特征：落叶乔木。树皮黑棕色。分枝气生根少；小枝淡红，无毛。叶互生或簇生；叶片椭圆形至长圆形，纸质，无毛，全缘或稍波状；侧脉 7 ～ 9 对。榕果腋生于叶状小枝，单生或对生，总苞片宽卵形，基部革质；雄花、瘿花和雌花在同一榕果内；雄花少，瘿花多，具长柄；雌花无梗或有梗。

花果期：花期 4 ～ 6 月。

分布：日本、东南亚。我国华东至华南地区。南麂主岛和大檑山屿，常见野生或栽培。

生境：靠近海岸的平原或村庄。

Description: Trees deciduous. Bark blackish brown. Branches with few aerial roots; branchlets pale red, glabrous. Leaves alternate or fasciculate; leaf blade elliptic to oblong, papery, glabrous, margin entire or slightly undulate; secondary veins 7-9 on each side of midvein. Figs axillary on leafy branchlets, paired or solitary; involucral bracts broadly ovate, basally leathery; male, gall, and female flowers within same fig; male flowers few, gall flowers many, ovary with a thick long stalk; female flowers sessile or pedicellate.

Flower and Fruit: Fl. Apr. -Jun.

Distribution: Japan, Southeast Asia. East China to South China. Main island and Daleishan Island, usually wild or cultivated.

Habitat: Plains or villages mainly near seacoast.

A150. 桑科
Moraceae

桑
Morus alba
桑属 *Morus*

形态特征：落叶乔木或灌木。树皮厚，灰色，具不规则浅纵裂。叶互生，卵形或广卵形，边缘锯齿或分裂，脉腋有簇毛；叶柄具柔毛；托叶披针形，早落。花单性，雌雄异株，腋生；雄花总花梗长，雌花无花梗。聚花果卵状椭圆形，成熟时暗紫色或绿白色。

花果期：花期 4 ~ 5 月，果期 5 ~ 6 月。

分布：原产我国中部和北部，现全世界广泛栽培。我国各地广泛种植。南麂各岛常见，常栽培。

生境：村旁、田间、滩地或山坡上。

Description: Shrubs or trees, deciduous. Bark thick, gray, shallowly furrowed. Leaves alternate, leaf blade ovate to broadly ovate, margin serrate or irregularly lobed, abaxially sparsely pubescent along midvein; petiole pubescent; stipules lanceolate, caducous. Flowers unisexual, dioecism, axillary; male flower with long peduncle, female flower sessile. Syncarp ovoid-ellipsoid, blackish purple, or greenish white when mature.

Flower and Fruit: Fl. Apr. -May, fr. May-Jun.

Distribution: Originally endemic to central and northern China, widely cultivated throughout the world. Cultivated throughout China. Throughout main island of Nanji, usually cultivated.

Habitat: Near village, field, shoaly land or hillside.

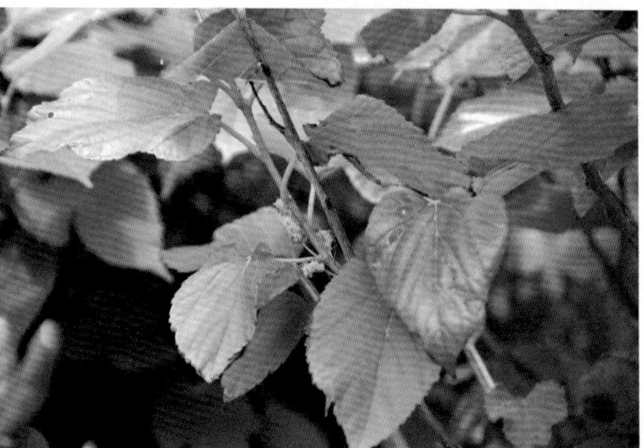

A151. 荨麻科
Urticaceae

野线麻（大叶苎麻）
***Boehmeria japonica*（*Boehmeria longispica*）**
苎麻属 *Boehmeria*

形态特征：亚灌木或多年生草本，不分枝或者少分枝，茎上部通常有较密的开展或贴伏的糙毛。叶纸质，对生，近圆形、圆卵形或卵形，顶端骤尖，基部宽楔形或截形，边缘在基部之上有锯齿。穗状花序单生叶腋，雌雄异株，不分枝；雄花花被片 4，椭圆形；雌花花被倒卵状纺锤形，顶端有 2 小齿，上部密被糙毛。瘦果倒卵球形，光滑。

花果期：花期 6 ～ 8 月，果期 9 ～ 11 月。

分布：日本。我国南北多地有分布。南麂主岛（百亩山、大山、门屿尾），偶见。

生境：林缘、灌丛，山沟小溪。

Description: Subshrubs or herbs perennial, simple or few branched, upper stems and branchlets densely appressed or patent strigose. Leaves papery, opposite; leaf blade suborbicular, orbicular-ovate, or ovate, apex cuspidat, base broadly cuneate, or truncate, margin dentate from base. Glomerules on axillary unbranched; Male flowers 4-merous, perianth lobes elliptic; female flowers perianth rhomboid-ellipsoidal, apex, 2-toothed, upper part densely hispid. Achenes subovoid, smooth.

Flower and Fruit: Fl. Jun. -Aug. , fr. Sep. -Nov.

Distribution: Japan. Most areas of south and north China. Main island (Baimushan, Dashan, Menyuwei), occasionally.

Habitat: Forest margins, thickets, along streams in hills and mountains.

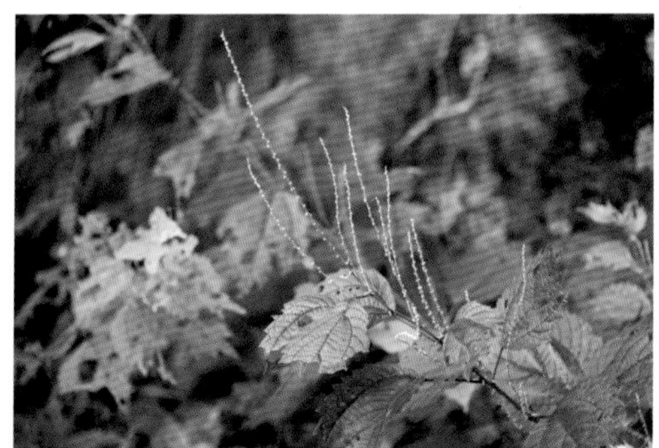

A151. 荨麻科
Urticaceae

苎麻
Boehmeria nivea
苎麻属 *Boehmeria*

形态特征：亚灌木或灌木，高 0.5 ～ 1.5 m；茎上部与叶柄均密被开展的长硬毛和贴伏的短糙毛。叶草质，互生，通常圆卵形或宽卵形，顶端骤尖，基部近截形或宽楔形，边缘具粗锯齿。圆锥花序腋生，上雌下雄，或全雌。瘦果近球形，光滑，基部突缩成细柄。

花果期：花期 5 ～ 8 月，果期 9 ～ 11 月。

分布：亚洲东部。我国秦淮以南各地。南麂各岛屿常见。

生境：林缘、灌丛，沿小溪、路旁的潮湿处。

Description: Subshrubs or shrubs, 0.5-1.5 m tall; upper stems, branchlets, and petioles densely patent hirsute, appressed strigose or only strigose. Leaves alternate, herbaceous, leaf blade often orbicular or broadly ovate, apex cuspidate, base subtruncate, or cuneate; margin dentate from base. Panicles axillary, with female flowers at upper part, male flowers at lower part, or only female flowers. Achene subglobose, smooth, base constricted and stipitate.

Flower and Fruit: Fl. May-Aug. , fr. Sep. -Nov.

Distribution: Eastern Asia. To the south of Qinling and Huaihe. Throughout Nanji Islands, commonly.

Habitat: Forest margins, thickets, moist places along streams, roadsides.

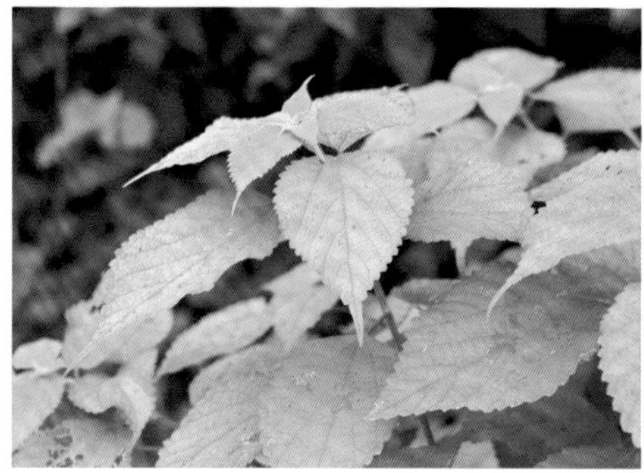

A151. 荨麻科
Urticaceae

波缘冷水花
Pilea cavaleriei
冷水花属 *Pilea*

形态特征：多年生草本，具匍匐茎，无毛。茎直立，多分枝，密被钟乳体。叶对生，集生于枝顶，宽卵形或菱状卵形，在近叶柄处常有不对称的小耳突，边缘全缘或稀波状，钟乳体密生于叶上面，基出脉不明显；托叶小，三角形，宿存。雌雄同株，聚伞花序常密集成近头状，有时具少数分枝；苞片三角状卵形。瘦果卵形，稍扁，光滑。

花果期：花期 5 ~ 8 月，果期 8 ~ 10 月。

分布：不丹。我国长江流域以南地区。南麂主岛（关帝岙、三盘尾），常见。

生境：林下石上或阴湿处。

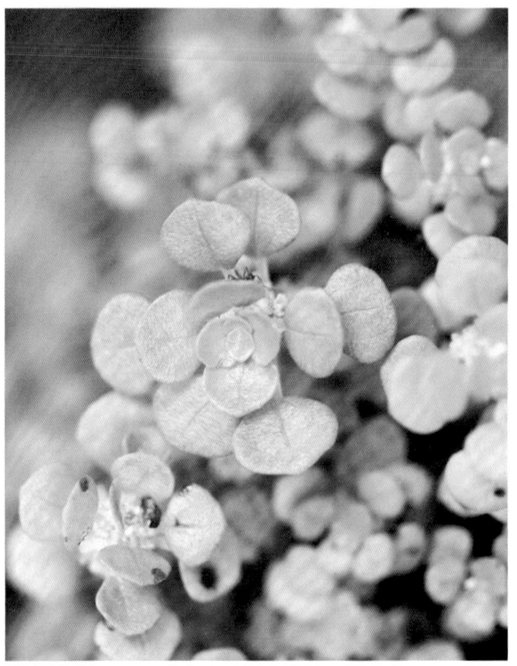

Description: Herbs perennial, stoloniferous, glabrous. Stems erect, much branched, fleshy, with dense cystoliths. Leaves opposite, crowded on apex of stem and branches; leaf blade broad-ovate, rhombic-ovate, or suborbicular, with asymmetric small auricular near the petiole, margin entire, rarely undulate or bluntly crenate, with dense cystoliths on the surface, basinerves inconspicuous; stipule small, triangular, persistent. Monoecious, cymes often dense into subcapitate, sometimes with few branches; bracts triangular ovate. Achene ovoid, compressed, smooth.

Flower and Fruit: Fl. May-Aug. , fr. Aug. -Oct.

Distribution: Bhutan. To the south of Yangtze River Basin. Main island (Guandiao, Sanpanwei), commonly.

Habitat: Shaded moist places, rocks in forests.

A151. 荨麻科
Urticaceae

小叶冷水花
Pilea microphylla
冷水花属 *Pilea*

形态特征：柔弱草本，无毛，雌雄同株。茎干燥时蓝绿色，肉质的，钟乳体紧密。托叶宿存，三角形，膜质；叶柄纤细，不等长；叶片倒卵形或者匙形，大小不等。花序通常雄雌同序，紧密聚伞状头状；团伞花序少数开花。瘦果卵圆形，压缩，光滑，藏于宿存花被内。

花果期：花期 6 ～ 8 月，果期 9 ～ 10 月。

分布：原产南美热带地区。我国南方地区广泛归化。南麂主岛（大山、国姓岙、美龄居），偶见。

生境：路边石缝和墙上阴湿处。

Description: Herbs weak, glabrous, monoecious. Stems blue-green when dry, succulent, cystoliths dense. Stipules persistent, triangular, membranous; petiole slender, unequal in length; leaf blade obovate or spatulate, unequal in size. Inflorescences often androgynous, compactly cymose-capitate; glomerules few flowered. Achene ovoid, compressed, smooth, enclosed by persistent perianth.

Flower and Fruit: Fl. Jun. -Aug. , fr. Sep. -Oct.

Distribution: Native to tropical South America. Commonly naturalized in Southern China . Main island (Dashan, Guoxingao, Meilingju), occasionally.

Habitat: Roadside crevices and damp walls.

A154. 杨梅科 Myricaceae

杨梅
Myrica rubra
香杨梅属 *Myrica*

形态特征：常绿乔木，雌雄异株，高 15 m，树皮灰色。小枝和芽无毛。叶革质，常密集于枝顶，楔状倒卵形或长椭圆状倒卵形，背面苍绿色，疏生金色腺体，全缘或中上部有锯齿。雄花序单生或数条丛生于叶腋，圆柱状；雌花序常单生于叶腋，多花，顶端 1（稀 2）雌花发育成果实。核果球状，外表面具乳头状凸起，成熟时深红色或紫红色。

花果期：花期 3 ~ 4 月，果期 5 ~ 7 月。

分布：日本、朝鲜。我国长江流域以南，常栽培。南麂主岛（后隆、三盘尾），常栽培。

生境：林中山坡、山谷，酸性土壤。

Description: Trees evergreen, dioecious, to 15 m tall, bark gray. Branchlets and buds glabrous. Leaf leathery, densely in branchlets apex, blade cuneate-obovate or narrowly elliptic-obovate, abaxially pale green and sparsely golden glandular, margin entire or serrate in apical 1/2. Male spikes solitary or sometimes few together in leaf axils, cylindrical; female spikes solitary in leaf axils, many flowered, upper 1 or 2 matured. Drupe globose, papilliferous, dark red or purple-red at maturity.

Flower and Fruit: Fl. Mar. -Apr. , fr. May-Jul.

Distribution: Japan, Korea. To the south of Yangtze River Basin, commonly cultivated, Main island (Houlong, Sanpanwei), usually cultivated.

Habitat: Forests in mountain slopes, valleys, acid soil.

A155. 胡桃科
Juglandaceae

化香树
Platycarya strobilacea
化香树属 *Platycarya*

形态特征：落叶乔木或灌木，高可达 15m。小叶 1 ~ 15，侧生小叶无柄，卵状披针形，背面基部和沿中脉簇生柔毛，基部偏斜至楔形；顶生小叶基部圆形或阔楔形。穗状花序雌雄同序，基部为雌花序，雄花序位于上部或有时缺失。穗状果序卵状椭圆形或椭圆状圆柱形至近球形，宿存苞片披针形；小坚果近圆形至倒卵形。

花果期：花期 5 ~ 7 月，果期 7 ~ 10 月。

分布：日本、朝鲜、越南。我国黄河流域以南地区。南麂主岛（百亩山、国姓岙、门屿尾），常见。

生境：山坡的混交林中，石灰岩。

Description: Deciduous trees or shrubs, to 15 m tall. Leaflet 1-15, lateral leaflets sessile, blade ovate-lanceolate, abaxially densely cluster hairs at base and along midvein abaxially, base oblique to cuneate; terminal leaflet base rounded or broadly cuneate. Spike androgynous, central spike female in basal, male in apical, or sometimes absent. Fruiting spike ovoid-ellipsoid or ellipsoid-cylindric to subglobose, bracts lanceolate; nutlets suborbicular to obovate.

Flower and Fruit: Fl. May-Jul. , fr. Jul. -Oct.

Distribution: Japan, Korea, Vietnam. South of the Yellow River Basin. Main island (Baimushan, Guoxingao, Menyuwei), commonly.

Habitat: Mixed forests on mountain slopes, limestone.

A156. 木麻黄科 Casuarinaceae

木麻黄

Casuarina equisetifolia

木麻黄属 *Casuarina*

形态特征：常绿乔木，大树根部无萌蘖。树冠狭长圆锥形；树皮有鳞，在幼树上较薄，皮孔密集排列为条状或块状，老树的树皮具不规则纵裂；枝有密集的节。花雌雄同株或异株；雄花序棒状圆柱形；雌花序通常顶近枝顶的侧生短枝上。球果椭圆形，幼嫩时外被灰绿色或黄褐色茸毛，成长时毛常脱落。

花果期：花期 4 ～ 5 月，果期 7 ～ 10 月。

分布：原产东南亚至大洋洲岛屿。我国华东和华南地区普遍栽植。南麂各岛屿常见。

生境：海岸沙地。

Description: Evergreen trees, not suckering from roots. Trunk straight, crown conical; bark scaly, thin when young, with irregular longitudinal fissures for old trees; lenticels densely arranged in strips or clumps; branches with dense nodes. Flowers monoecious or heteroecious; male inflorescences ellipsoid; female inflorescences usually apical on short lateral branches. Cones ellipsoid, covered with gray-green or yellow-brown fuzz when young, deciduous, shed when growing.

Flower and Fruit: Fl. Apr. -May, fr. Jul. -Oct.

Distribution: Native to Southeast Asia and Oceania islands. Widely cultivated in East and South China. Throughout Nanji Islands, commonly.

Habitat: Coastal sands.

A156. 木麻黄科
Casuarinaceae

粗枝木麻黄
Casuarina glauca
木麻黄属 *Casuarina*

形态特征：常绿乔木，树皮块状剥裂及浅纵裂。侧枝多，近直立而疏散。嫩梢具环裂反卷的鳞片状叶。小枝颇长，上举，末端弯垂，圆柱形，具浅沟槽，两端近节处略肿胀。花雌雄同株；雄花序密生于小枝顶；雌花序侧生，球形或椭圆形。球果状果序广椭圆形至近球形；小坚果淡灰褐色，有光泽。

花果期：花期 3 ～ 4 月，果期 6 ～ 9 月。

分布：原产澳大利亚。我国华东、华南地区有栽培。南麂各岛屿常见。

生境：海岸沼泽地至内陆地区。

Description: Evergreen trees, bark finely fissured and scaly. Many branchlets, upright and scattered. Leaves scaly, cyclorrhaphous, strongly recurved at new shoots. Ultimate branchlets ascending to pendulous, rather long, cylindrical, with shallow grooves, slightly swollen at apex. Flowers monoecious;male inflorescences densely borne on top of branchlets;female flowers lateral, globose or elliptic. Cones broadly ellipsoid to subglobose;samaras ficelle, lustrous.

Flower and Fruit: Fl. Mar.-Apr. , fr. Jun. -Sep.

Distribution: Native to Australia. Cultivated in East and South China. Throughout Nanji Islands, commonly.

Habitat: Coastal marshes to inland areas.

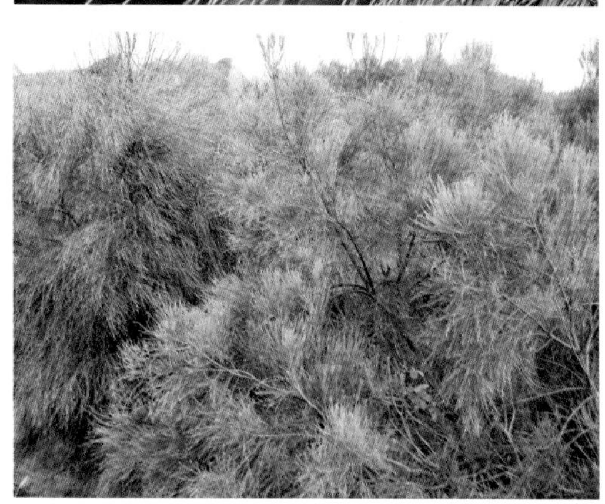

A163. 葫芦科
Cucurbitaceae

栝楼
Trichosanthes kirilowii
栝楼属 *Trichosanthes*

形态特征：攀援草本，长达 10 m。块根圆柱状，粗大肥厚。茎较粗，多分枝，具纵棱及槽。叶片纸质，近圆形，3 ~ 5 浅裂。花雌雄异株，雄花总状花序，雌花单生，花冠白色。果实椭圆形或圆形，成熟时黄褐色或橙黄色。种子狭卵形，淡黄褐色，近缘处具棱。

花果期：花期 5 ~ 8 月，果期 8 ~ 10 月。

分布：日本、朝鲜。华东至西北地区。南麂主岛各处常见。

生境：开阔的山坡林下、灌丛中、草地和村旁田边。

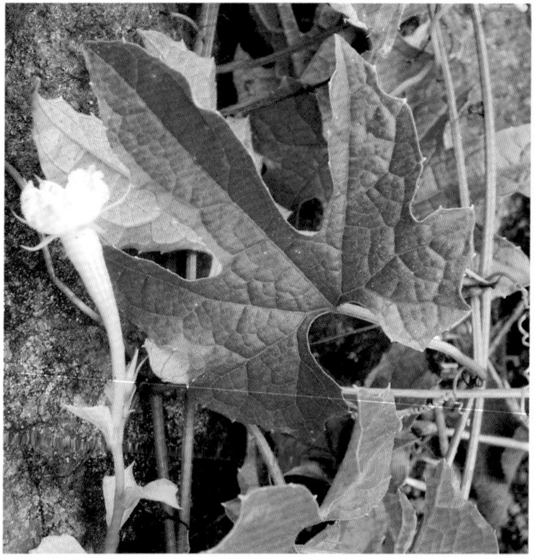

Description: Herbs climbing , to 10 m long. Tuber terete, thick and hypertrophic. Stem robust, much branched, with longitudinal ribs and grooved. Leaf blade papery, suborbicular, 3-5 lobed. Flowers dioecious, male raceme, female flowers solitary, corolla white. Fruit oval or round, brown-yellow or orange-yellow when ripe. Seeds oblong-ovate, light yellow brown, marginate.

Flower and Fruit: Fl. May-Aug. , fr. Aug. -Oct.

Distribution: Japan, Korea. East China to Northwest China. Main island of Nanji, commonly.

Habitat: Open forests, shrublands, grasslands and fields beside villages.

A168. 卫矛科
Celastraceae

扶芳藤
Euonymus fortunei
卫矛属 *Euonymus*

形态特征：常绿亚灌木，上升或平卧；分枝和小枝圆形，有时具条纹，常褐色或绿褐色。叶密生枝上，叶形多变，卵形或卵状椭圆形，无毛，基部近截形，边缘具细圆齿或锯齿，先端钝至急尖。花序梗通常少花；花4基数，淡绿色或带白色。蒴果褐色至红褐色，近球形，假种皮红色。

花果期：花期4～7月；果期9～12月。

分布：东亚至东南亚，各地广泛栽培。我国南北各地广布。南麂各岛屿常见。

生境：林地、灌丛，常栽培。

Description: Evergreen subshrubs, ascending or procumbent; branches and twigs rounded, sometimes striate, usually brown or green-brown. Leaves densely arranged on branches; leaf blade variously ovate or ovate-elliptic, glabrous, base nearly truncate, margin crenulate to serrate, apex obtuse to acute. Peduncle usually with few flowers; flowers 4-merous, greenish or whitish. Capsule brown to red-brown, subglobose, aril red.

Flower and Fruit: Fl. Apr. -Jul. , fr. Sep.-Dec.

Distribution: East to Southeast Asia, widely cultivated. Throughout the north and south areas of China. Throughout Nanji Islands, commonly.

Habitat: Woodlands, scrub, and forests, often cultivated in gardens.

A168. 卫矛科
Celastraceae

冬青卫矛（大叶黄杨）
Euonymus japonicus
卫矛属 *Euonymus*

形态特征：常绿灌木或小乔木。小枝绿色或淡绿，无毛。叶片革质或厚革质，卵形，倒卵形，边缘上部具细圆齿，下部近全缘，侧脉 6 ~ 8 对。聚伞花序腋生，多分枝；花 4 基数，萼片近圆形，花瓣绿色或黄绿色，近圆形。蒴果球形或近球形，黄棕色，4 裂。

花果期：花期 4 ~ 8 月，果期 8 月至翌年 1 月。

分布：原产日本，各地广泛栽植。我国大部分地区有分布。南麂各岛屿常见。

生境：花园和植物园里作栽培。

Description: Evergreen shrubs or small trees. Twigs green to light green, glabrous. Leaf blade leathery or thickly leathery, ovate, obovate, margin crenulate distally, nearly entire proximally; lateral veins 6-8 pairs. Cymes usually axillary, many branched with many flowers; flowers 4-merous, sepals nearly orbicular, petals green or yellowish green, nearly orbicular. Capsule globose or subglobose, yellow-brown, 4-lobed.

Flower and Fruit: Fl. Apr. -Aug. , fr. Aug. -Jan.

Distribution: Native to Japan, widely cultivated. Most areas of China. Throughout Nanji Islands, commonly.

Habitat: Cultivated, especially in gardens and arboreta.

A168. 卫矛科
Celastraceae

变叶裸实（变叶美登木）
Gymnosporia diversifolia（*Maytenus diversifolius*）
裸实属 *Gymnosporia*

形态特征：灌木或小乔木。一二年生小枝刺状，灰棕色，常被密点状锈褐色短刚毛。叶纸质，果时近革质，倒卵形、近阔卵圆形或倒披针形，先端圆或钝，基部楔形或渐窄下延成窄长楔形。圆锥聚伞花序，花序梗二歧分枝；花白色或淡黄色，萼片三角卵形，花盘扁圆，雄蕊着生花盘之外。蒴果 2 裂，扁倒心形，红色或紫色。种子椭圆状，黑褐色，基部有白色假种皮。

花果期：花期 5 ～ 6 月；果期 10 ～ 11 月。

分布：日本、东南亚。我国华东、华南地区。南麂各岛屿常见。

生境：疏林、山坡、海岸、路旁。

Description: Shrubs or small trees. Twigs spiny, pallid brown, densely covered with rust-colored punctiform short bristles, glabrescent with age. Leaf blade obovate, broadly obovate, or obovate-lanceolate, papery, or leathery with age, base cuneate. Cymes axillary, dichotomous; flowers white or light yellow, sepals triangular-ovate; disk oblate, stamens outside the disk. Capsule subobovoid, red or purple, 2-valved. Seeds ellipsoid, black-brown, basally covered by white aril.

Flower and Fruit: Fl. May-Jun. , fr. Oct. -Nov.

Distribution: Japan, Southeast Asia. East China, South China. Throughout Nanji Islands, commonly.

Habitat: Sparse forests, mountain slopes, seashores, roadsides.

A171. 酢浆草科
Oxalidaceae

酢浆草
Oxalis corniculata
酢浆草属 *Oxalis*

形态特征：一年生或短暂多年生植物。茎匍匐，上升到半直立。叶互生或假轮生；叶片倒心形，绿色或枣红色，正面和背面短柔毛，先端深凹。伞形花序 1 ~ 5 花，花序梗稍长于叶柄；苞片线状披针形，花梗密被糙伏毛，萼片长圆状披针形，具缘毛，花瓣明黄色，长圆倒卵形。蒴果长圆筒状，5 棱，被毛。

花果期：花果期 2 ~ 10 月。

分布：几乎世界广布。我国大部分区域均有分布。南麂主岛各处常见。

生境：山坡，森林，牧场，河边，路旁，田野，荒地。

Description: Herbs annual or short-lived perennial. Stems creeping, ascending to semierect. Leaves alternate or pseudoverticillate; leaflet blades obcordate, green or suffused purplish red, variably adaxially and abaxially pubescent, apex deeply emarginate. Inflorescences umbellate, 1-5-flowered, peduncle usually slightly longer than petioles; bracts linear-lanceolate; pedicel densely strigose, sepals oblong-lanceolate, margin ciliate, petals bright yellow, oblong-obovate. Capsule long cylindric, 5-sided, strigose.

Flower and Fruit: Fl. and fr. Feb. -Oct.

Distribution: Almost cosmopolitan. Most areas of China. Main island of Nanji, commonly.

Habitat: Mountain slopes, forests, grasslands, riversides, roadsides, fields, wastelands.

A171. 酢浆草科
Oxalidaceae

红花酢浆草
Oxalis corymbosa
酢浆草属 *Oxalis*

形态特征：多年生草本，高 10 ～ 25 cm，无茎，被短柔毛。叶基生；叶柄有平展白色毛；小叶片倒心形，两面被毛，正面边缘具暗色小腺点，先端深凹。伞房状聚伞花序，不规则分枝，花 8 ～ 15 朵；苞片披针形，膜质；萼片披针形，先端具 2 个红棕色小腺体；花瓣粉红，略带紫，具黑色脉，倒心形。蒴果少见。

花果期：花期 3 ～ 12 月。

分布：原产热带南美洲。我国大部分有栽培归化。南麂主岛（关帝岙、后隆、三盘尾），常见。

生境：耕地，开阔的生境。

Description: Herbs perennial, 10-25 cm tall, stemless, pubescent. Leaves basal; petiole with spreading white trichomes; leaflet blades obcordate, both surfaces covered with trichomes, adaxial surface punctate with dark calli especially near margin, apex deeply emarginate. Inflorescences corymbose cymes, irregularly branched, 8-15-flowered; bracts lanceolate, membranous, sepals lanceolate; apex with 2 reddish brown calli; petals purplish pink with darker veins, obcordate. Capsule rarely formed.

Flower and Fruit: Fl. Mar. -Dec.

Distribution: Native to tropical South America. Cultivated and naturalized in most areas of China. Main island (Guandiao, Houlong, Sanpanwei), commonly.

Habitat: Cultivated grounds and open habitats.

A186. 金丝桃科 Hypericaceae

地耳草（田基黄）
Hypericum japonicum Thunberg
金丝桃属 *Hypericum*

形态特征：一年生草本，高 2 ～ 45 cm。茎具 4 纵棱。叶对生，无柄，坚纸质，卵形或椭圆形，全缘，散布透明腺点。花序顶生，花 1 ～ 30，花瓣白色、淡黄至橙黄色。蒴果短圆柱形至圆球形。种子淡黄色，表面有细蜂窝纹。

花果期：花期 3 ～ 10 月，果期 4 ～ 11 月。

分布：亚洲，大洋洲至夏威夷群岛。我国南北各地广布。南麂主岛各处常见。

生境：田边、沟边、草地及荒地。

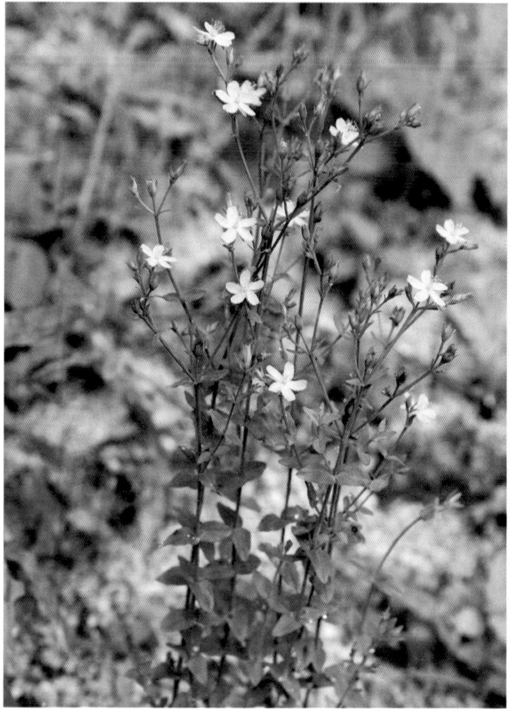

Description:Herbs annual, 2-45 cm. Stem 4-lined. Leaves opposite, sessile, thickly papery, margin entire, with transparent gland. Inflorescence 1-to 30-flowered, terminal, petals pale to bright yellow or orange. Capsule cylindric to globose. Seeds straw-yellow, testa finely linear-scalariform.

Flower and Fruit: Fl. Mar. -Oct, fr. Apr. -Nov.

Distribution: Asia, Oceania to Hawaiian Islands. Throughout the north and south areas of China. Main island of Nanji, commonly.

Habitat: Fields, ditches, marshes, grasslands or waste places.

A186. 金丝桃科
Hypericaceae

金丝桃
Hypericum monogynum
金丝桃属 *Hypericum*

形态特征：灌木，高 0.5 ～ 1.3 m，丛生或通常具疏松的枝，平展。叶长圆形至椭圆形，厚纸质，背面苍白；侧脉 2 ～ 3 对；基部楔形，先端锐尖到圆形。花序 1 ～ 15 花；苞片早落，线状披针形；花星状，萼片开展，椭圆形或倒披针形，花瓣金黄至柠檬黄，三角状倒卵形。蒴果宽卵形。

花果期：花期 5 ～ 8 月，果期 8 ～ 9 月。

分布：中国大部分区域有分布，世界广泛栽培。南麂主岛（大沙岙、美龄居、镇政府），栽培。

生境：山坡，路旁，干燥生境的灌丛。

Description: Shrubs, 0.5-1.3 m tall, bushy or usually with branches lax, spreading. Leaves blade oblong to elliptic, thickly papery, abaxially paler; main lateral veins 2-or 3-paired; base cuneate to subangustate, apex acute to rounded. Inflorescence 1-15-flowered; bracts caducous, linear-lanceolate; flowers stellate, sepals spreading, elliptic or oblanceolate, petals golden yellow to lemon yellow, triangular-obovate. Capsule broadly ovoid.

Flower and Fruit: Fl. May-Aug. , fr. Aug. -Sep.

Distribution: Most area of China, widely cultivated all over the world. Main island (Dashaao, Meilingju, Zhenzhengfu), cultivated.

Habitat: Mountain slopes, roadsides, thickets in dry habitats.

A200. 堇菜科
Violaceae

七星莲
Viola diffusa
堇菜属 *Viola*

形态特征：一年生草本，全株有毛。匍匐茎先端具莲座状叶丛，有不定根。基生叶多数，丛生和莲座状，或在匍匐茎上互生；叶柄显著具翅，常被柔毛；叶片卵形，两面密被白色柔毛，后逐渐稀少，基部宽楔形，明显下延，边缘钝齿和缘毛。花略带紫色或淡黄，小；花梗细长，无毛或疏生微柔毛，侧花瓣倒卵形。蒴果长圆形，无毛，花柱宿存。

花果期：花期 3 ~ 5 月，果期 5 ~ 10 月。

分布：东亚至东南亚。我国南北各地广布。南麂主岛（百亩山、国姓岙、后隆、门屿尾），常见。

生境：山地森林，林缘，草坡，河谷，岩石裂缝。

Description: Herbs annual, stiffly hairy throughout. Stolon with rosulate leaves at top, usually producing adventitious roots. Basal leaves numerous, fasciculate and rosulate, or alternate on stolon; petiole conspicuously winged, usually puberulous; leaf blade ovate, both surfaces densely white puberulous on young leaves, later gradually sparsely so, base broadly cuneate, conspicuously decurrent to petiole, margin obtusely dentate and ciliate. Flowers purplish or yellowish, small, long pedicellate; pedicels slender, glabrous or sparsely puberulous, lateral petals obovate, glabrous or shortly bearded. Capsule oblong, glabrous, often with persistent styles.

Flower and Fruit: Fl. Mar. -May , fr. May-Oct.

Distribution: East Asia to Southeast Asia. Throughout the north and south areas of China. Main island (Daimushan, Guoxingao, Houlong, Menyuwei), commonly.

Habitat: Mountain forests, forest margins, grassy slopes, stream valleys, rock crevices.

A200. 堇菜科
Violaceae

紫花地丁
Viola philippica (Viola yedoensis)
堇菜属 *Viola*

形态特征：多年生草本，无地上茎，高 4 ~ 14 cm。叶纸质，基生，莲座状，下部叶片小，三角状卵形，上部叶长圆形或长圆状卵形；托叶 2/3 ~ 4/5 与叶柄合生，离生部分线状披针形，边缘疏生具腺体的流苏状细齿或近全缘。花紫堇色，稀白色，喉部有紫色条纹。蒴果长圆形。种子卵球形，淡黄色。

花果期：花期 4 ~ 5 月，果期 5 ~ 9 月。

分布：东亚至东南亚。分布几遍我国。南麂主岛各处常见。

生境：田间、荒地，山坡、林缘、灌丛或路旁草丛。

Description: Herbs perennial, acaulescent, 4-14 cm tall. Rhizome short, erect. Leaves papery, basal, rosulate, lower blades small, triangular-ovate, upper blades oblong or oblong-ovate; stipules 2/3-4/5 adnate to petioles, free part linear-lanceolate, margin remotely glandular fimbriate-denticulate or subentire. Flowers purple-violet, rarely white, purple-striate at throat. Capsule ellipsoid. Seeds ovoid-globose, yellowish.

Flower and Fruit: Fl. Apr. -May, fr. May-Sep.

Distribution: East Asia to Southeast Asia. Almost throughout China. Main island of Nanji, commonly.

Habitat: Fields, waste places, grassy places on mountain slopes, forest margins, thickets, roadsides.

A204. 杨柳科
Salicaceae

柞木
Xylosma congesta
柞木属 *Xylosma*

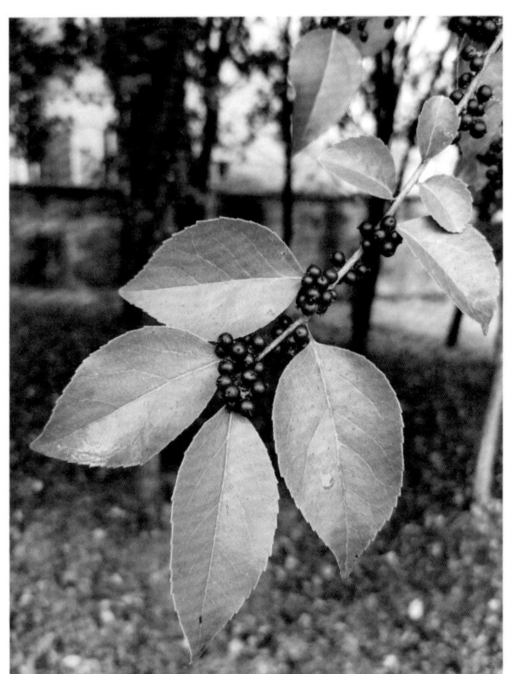

形态特征：常绿灌木或小乔木，高 4 ～ 15 m；树皮棕灰色；幼时具枝刺，老时无刺。叶柄短，叶片宽卵形，革质，边缘锯齿，先端锐尖。总状花序腋生，短；花淡黄；花梗极短，具短柔毛；苞片卵形到狭披针形，背面被短柔毛；萼片宽卵形具圆形先端，具短柔毛，具缘毛。浆果深红色至黑色，球形。种子干燥时红棕色，卵形。

花果期：花期 7 ～ 11 月，果期 8 ～ 12 月。

分布：日本、朝鲜。我国长江流域以南广泛分布。南麂主岛（国姓岙、后隆、门屿尾），常见。

生境：林缘，丘陵，平原或村庄周围的灌丛。

Description: Shrubs or small trees, evergreen, 4-15 m tall; bark brown-gray; branches spiny when young, unarmed when old. Petiole short, leaf blade broadly ovate, leathery, margin serrate, apex acute. Inflorescence axillary, racemose, short; flowers yellowish; pedicels very short, pubescent; bracts ovate to narrowly lanceolate, abaxially pubescent; sepals broadly ovate with rounded apex, outside pubescent, ciliate. Berry dark red to black, globose. Seeds reddish brown when dry, ovoid.

Flower and Fruit: Fl. Jul. -Nov. , fr. Aug. -Dec.

Distribution: Japan, Korea, widely distributed in the south of Yangtze River Basin. Main island (Guoxingao, Houlong, Menyuwei), commonly.

Habitat: Forest margins, thickets on hills, plains, surrounding villages.

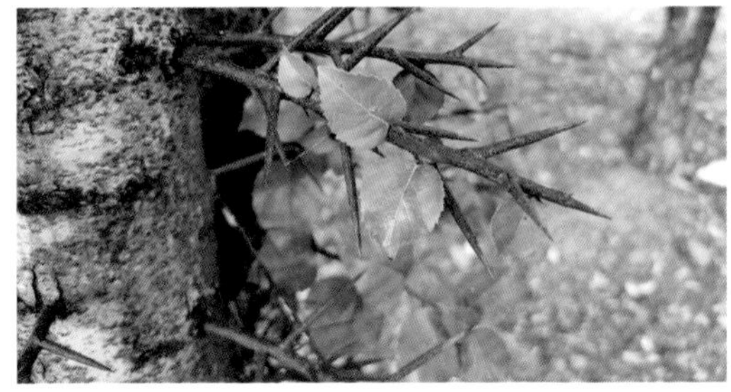

A207. 大戟科
Euphorbiaceae

铁苋菜
Acalypha australis
铁苋菜属 *Acalypha*

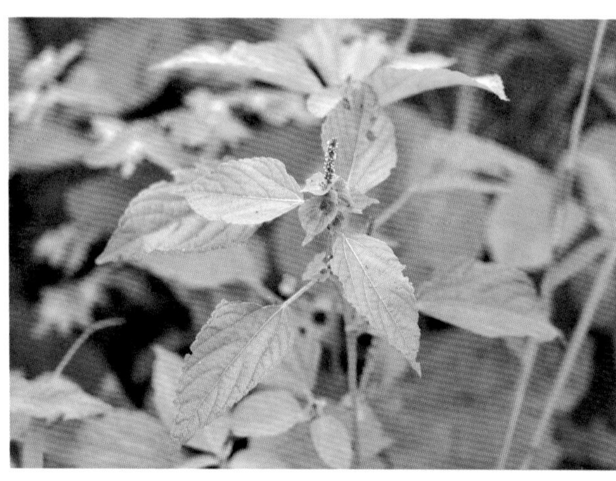

形态特征：一年生草本，高 0.2 ～ 0.5 m，雌雄同株。小枝被长柔毛。托叶披针形；叶片长圆状卵形，菱状卵形或宽披针形，膜质，背面沿脉具柔毛，正面无毛，基部楔形，具圆齿，先端短渐尖。花序腋生，少顶生，不分枝，具柔毛，两性；萼片 3，狭卵形，具柔毛。蒴果 3 室，具柔毛和小瘤。种子近卵形，光滑。

花果期：花果期 4 ～ 12 月。

分布：亚洲东部。除内蒙古、新疆外广布我国。南麂主岛各处常见。

生境：草地，斜坡，耕地。

Description: Herbs annual, 0.2-0.5 m tall, monoecious. Branchlets pilose. Stipules lanceolate; leaf blade oblong-ovate, rhombic-ovate, or broadly lanceolate, membranous, abaxially pilosulose along veins, adaxially glabrous, base cuneate, crenate, apex shortly acuminate. Inflorescences axillary, rarely terminal, unbranched, pilosulose, bisexual; sepals 3, narrowly ovate, pilose. Capsule 3-locular, pilose and tuberculate. Seeds subovoid, smooth.

Flower and Fruit: Fl. and fr. Apr. -Dec.

Distribution: Eastern Asia. Throughout China except Nei Mongol and Xinjiang. Main island of Nanji, commonly.

Habitat: Grasslands, slopes, cultivated areas.

A207. 大戟科
Euphorbiaceae

喙果黑面神
Breynia rostrata
黑面神属 *Breynia*

形态特征：常绿灌木或乔木，全株无毛。小枝和叶片干后呈黑色。托叶三角状披针形；叶片卵状或长圆状披针形，纸质或薄革质，背面灰绿色，正面绿色，侧脉3～5对。雌花与雄花簇生叶腋；雄花花萼漏斗状，先端具6齿；雌花花萼6裂，花瓣不等长，二列，柱头3，先端2裂。蒴果圆球状，具宿存喙状花柱。

花果期：花期5～12月；果期8～12月。

分布：越南。我国浙江、华南和云南地区。南麂各岛屿常见。

生境：山地林中或灌丛山坡。

Description: Evergreen shrubs or trees, glabrous throughout. Branches and leaf blade black when dry. Stipules triangular-lanceolate; leaf blade ovate- or oblong-lanceolate, papery or thinly leathery, abaxially gray-green, adaxially green, lateral veins 3-5 pairs. Flowers male and female mixed in axillary clusters; male flowers calyx funnelform, 6-dentate at apex; female flowers calyx 6-lobed, sepals unequal, biseriate, stigmas 3, bifid at apex. Capsules globose, apex beaked, with persistent stigmas.

Flower and Fruit: Fl. May-Dec. , fr. Aug. -Dec.

Distribution: Vietnam. Zhejiang, South China and Yunnan. Throughout Nanji Islands, commonly.

Habitat: Montane forests or scrub-covered slopes.

A207. 大戟科
Euphorbiaceae

泽漆
Euphorbia helioscopia
大戟属 *Euphorbia*

形态特征：草本，常一年生，直立或上升，高 10 ～ 30 cm。茎单生或近基部分枝上升，有时中空，光滑无毛。叶互生；无托叶、叶柄；叶片匙形或倒卵形，基部楔形，边缘具齿，先端圆形。复合假伞形花序，杯状聚伞花序近无柄；总苞钟状，光滑无毛，裂片圆形，边缘和里面具柔毛。蒴果三角圆柱状，具 3 垂直沟，光滑，无毛。种子深棕色。

花果期：花果期 4 ～ 10 月。

分布：欧亚大陆及北非有分布。分布我国各地。南麂主岛各处常见。

生境：田野，路旁，混交林的灌丛，林缘。

Description: Herbs, usually annual, erect or ascending, 10-30 cm tall. Stems single or with ascending branches from near base, sometimes slightly fistulose, smooth and glabrous. Leaves alternate; stipules and petiole absent; leaf blade obovate to spoon-shaped, base cuneate, margin dentate, apex rounded. Inflorescence a compound pseudumbel, cyathium subsessile; involucre campanulate, smooth and glabrous, lobes rounded, pilose at margin and inside. Capsule trigonous-terete, with 3 vertical furrows, smooth, glabrous. Seeds dark brown.

Flower and Fruit: Fl. and fr. Apr. -Oct.

Distribution: North Africa, Asia and Europe. Throughout China. Main island of Nanji, commonly.

Habitat: Fields, roadsides, scrub, margins of mixed forests.

A207. 大戟科
Euphorbiaceae

飞扬草
Euphorbia hirta
大戟属 *Euphorbia*

形态特征：一年生草本，高 30 ~ 60 cm，少分枝。茎中部以上分枝，上升到直立，具毛。叶对生，托叶膜质，三角形，早落；叶片披针形，正面绿色到红色，背面灰绿色，两面具柔毛，叶全缘或有细锯齿。聚伞花序密集，头状；总苞钟状，具柔毛，三角状卵形。蒴果三棱状，光滑，被短柔毛。种子带红色。

花果期：花果期 6 ~ 12 月。

分布：世界热带亚热带的地区。我国长江以南各地。南麂主岛（大沙岙、关帝岙、三盘尾），偶见。

生境：路旁，田野，灌丛，疏林。

Description: Herbs annual, 30-60 cm tall, usually few branched. Stem branched from middle or above, ascending to erect, with hairs. Leaves opposite, stipules membranous, triangular, caducous; leaf blade lanceolate-oblong, adaxially green to red, abaxially gray-green, both surfaces pilose, margin entire or serrulate. Cyathia in dense, often headlike; involucre campanulate, pilose, triangular-ovate. Capsule 3-angular, smooth, shortly pilose. Seeds reddish.

Flower and Fruit: Fl. and fr. Jun. -Dec.

Distribution: Tropical and subtropical regions of the world. To the south of Yangtze River. Main island (Dashaao, Guandiao, Sanpanwei), occasionally.

Habitat: Roadsides, fields, scrub, open forests.

A207. 大戟科
Euphorbiaceae

地锦草（地锦）
Euphorbia humifusa
大戟属 ***Euphorbia***

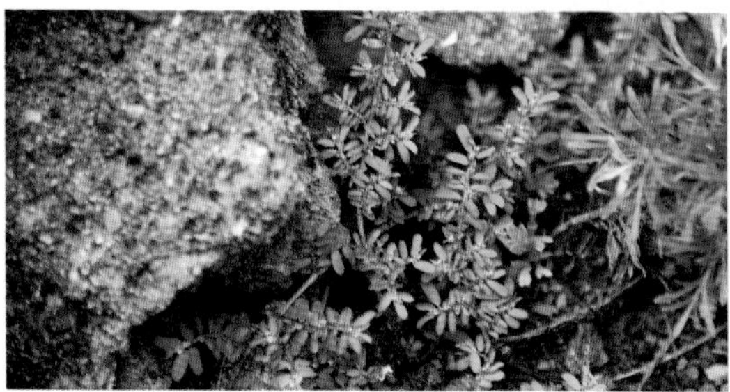

形态特征：一年生草本，高 20 cm，偶见 30 cm。茎匍匐，自基部多分枝，常红或淡红色，被柔毛或疏柔毛。叶对生；托叶线性深裂，早落；叶片长圆形或椭圆形，先端钝圆，基部常偏斜，边缘中部以上具细锯齿，先端钝。杯状聚伞花序单生叶腋，总苞陀螺状，边缘裂片 4，三角形；腺体 4，附属物白色。蒴果三棱状卵球形，光滑无毛。

花果期：花果期 5 ~ 10 月。

分布：广布欧亚大陆和非洲温带地区。除海南外广布我国。南麂主岛各处常见

生境：田野、路旁、沙丘，海岸，斜坡，常见于开阔地。

Description: Herbs annual, 20(-30) cm tall. Stems many from base, prostrate or ascending, often red or pinkish red, glabrous or pilose. Leaves opposite; stipules deeply divided into often threadlike lobes, caduceus; leaf blade oblong or elliptic, base obliquely truncate, margin finely serrulate above middle, apex obtuse. Cyathia single, axillary, involucre turbinate, marginal lobes 4, triangular; glands 4, appendages white. Capsule 3-angular-ovoid-globose, smooth.

Flower and Fruit: Fl. and fr. May -Oct.

Distribution: Widely distributed in temperate regions of Africa, Asia, and Europe. Throughout China except Hainan.Main island of Nanji, commonly.

Habitat: Fields, roadsides, sandy hills, seashores, slopes, usually in open situations.

A207. 大戟科
Euphorbiaceae

斑地锦
Euphorbia maculata
大戟属 *Euphorbia*

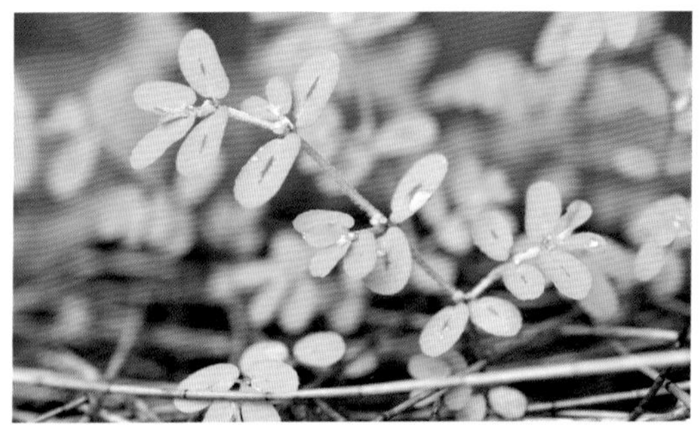

形态特征：一年生草本，高 10 ～ 17 cm。茎多基生，平卧，疏生白色柔毛。托叶成刺，具缘毛；叶对生；叶片长椭圆形，正面绿色，通常中部具一椭圆形紫斑，背面淡绿色，两面无毛，基部偏斜，略狭圆形先端钝。杯状聚伞花序生于节上，总苞狭杯状，外面具白色短柔毛，三角状圆形。蒴果三棱球形，光滑，疏生柔毛。

花果期：花果期 4 ～ 9 月。

分布：亚洲、欧洲、北美洲。我国华北至华东地区。南麂主岛各处常见。

生境：草地、路旁。

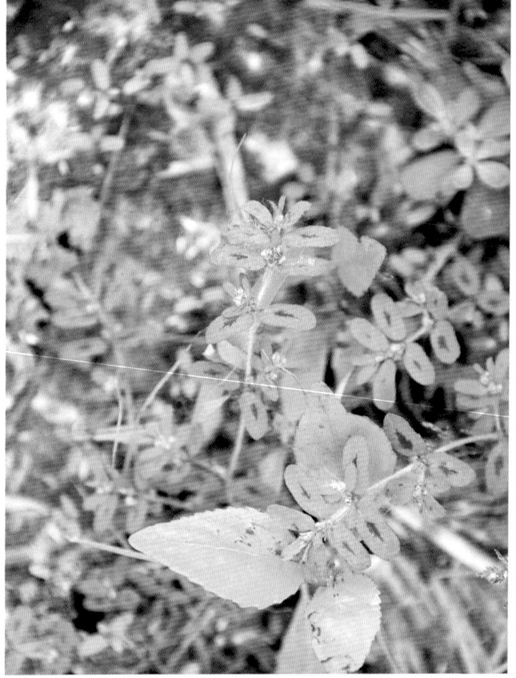

Description: Herbs annual, 10-17 cm tall. Stems many from base, prostrate, sparsely white pilose. Stipules forming prickles, ciliate; leaves opposite; leaf blade long elliptic, adaxially green, often with an oblong purple spot in middle, abaxially light green, both surfaces glabrous, base obliquely slightly attenuate-rounded, apex obtuse. Cyathia from nodes; involucre narrowly cuplike, white pubescent outside, triangular-rounded. Capsule 3-angular-ovoid, smooth, sparsely pilose.

Flower and Fruit: Fl. and fr. Apr. -Sep.

Distribution: Asia, Europe, North America. North China to East China. Main island of Nanji, commonly.

Habitat: Grasslands, roadsides.

A207. 大戟科
Euphorbiaceae

千根草
Euphorbia thymifolia
大戟属 *Euphorbia*

形态特征：一年生草本，高 10 ~ 20 cm。茎纤细，常呈匍匐状，自基部极多分枝，被稀疏柔毛。叶对生；托叶披针形或线形，易脱落；叶片圆形或心形，边缘常有锯齿，两面被毛。聚伞花序单生或数个腋生，总苞钟状至陀螺状，外被短柔毛，边缘 5 裂，裂片卵形；腺体 4，被白色附属物。蒴果三角状卵形，平滑，被短柔毛。

花果期：花果期 6 ~ 11 月。

分布：广布南北半球温暖地区。我国华东至华南地区。南麂主岛各处常见。

生境：常见于路旁、草地、灌丛、田野。

Description: Herbs annual, 10-20 cm tall. Stems slender and thin, many from base, usually prostrate, sparsely pilose. Leaves opposite; stipules lanceolate or linear, easily fallen; leaf blade rounded or cordate, margin usually finely serrulate, both surfaces pubescent. Cyathia single or numerous clustered and axillary, involucre campanulate to turbinate, outside shortly pilose, marginal lobes 5, ovate; glands 4, appendage white. Capsule 3-angular-ovoid, smooth, shortly pubescent.

Flower and Fruit: Fl. and fr. Jun. -Nov.

Distribution: widely spread in warm countries of both hemispheres. East China to South China. Main island of Nanji, commonly.

Habitat: Roadsides, grasslands, scrub, fields, very common.

A207. 大戟科
Euphorbiaceae

倒卵叶算盘子
Glochidion obovatum
算盘子属 *Glochidion*

形态特征：灌木或小灌木，高 0.5 ～ 1 m，小枝被短柔毛。托叶卵状三角形；叶片倒卵形或长圆状倒卵形，薄革质，顶端钝或短渐尖，基部楔形，干后棕色，无毛。聚伞花序腋生；雄花萼片 6，倒卵形，雄蕊 3，合生；雌花萼片 6，子房卵形，4 ～ 6 室，无毛，花柱合生呈圆柱状，顶端 6 裂。蒴果扁球状，具 8 ～ 12 条纵沟。

花果期：花期 5 ～ 6 月；果期 10 ～ 11 月。

分布：日本。我国华东地区。南麂各岛屿常见。

生境：山坡灌丛。

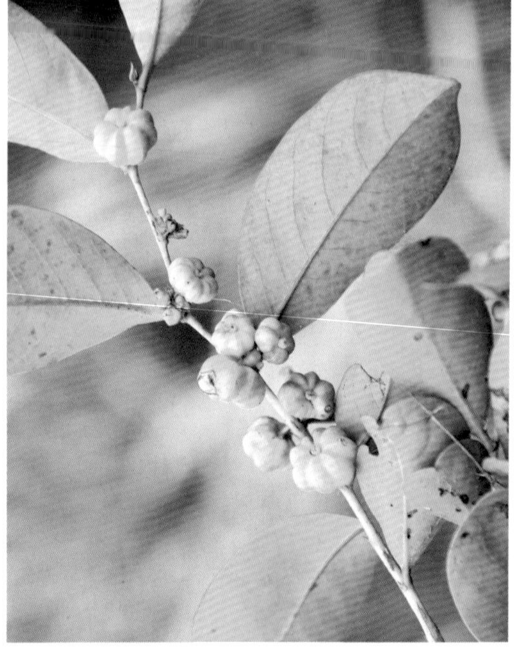

Description: Shrubs or shrublets, 0.5-1 m tall, branches pubescent. Stipules ovate-triangular; leaf blade obovate or oblong-obovate, thinly leathery, base cuneate, apex obtuse or shortly acuminate, brown when dry, glabrous. Flowers in cymes; male flowers sepals 6, obovate, stamens 3, connate; female flowers sepals 6, ovary ovoid, 4-6-locular, glabrous, style column cylindric, 6-lobed at apex. Capsules depressed globose, 8-12-grooved.

Flower and Fruit: Fl. May-Jun. , fr. Oct. -Nov.

Distribution: Japan. East China. Throughout Nanji Islands, commonly.

Habitat: Montane scrub on slopes.

A207. 大戟科
Euphorbiaceae

算盘子
Glochidion puberum
算盘子属 *Glochidion*

形态特征：直立灌木，高 1 ~ 5 米，小枝灰褐色，密被短柔毛。托叶三角形；叶片长圆形或倒卵状长圆形，纸质或近革质，基部宽楔形，先端短尖，网脉明显。花 2 ~ 5 朵簇生叶腋内，雄花束生小枝下部，雌花束在上部；萼片 6，绿色，雄蕊 3 枚合生呈圆柱状，雌花花柱合生呈环状，顶端短裂。蒴果扁球形，有 8 ~ 10 纵沟，密被毛，熟时带红色。

花果期：花期 4 ~ 8 月，果期 7 ~ 11 月。

分布：日本。我国华北以南地区。南麂主岛各处常见。

生境：溪岸旁的斜坡和灌丛，林缘。

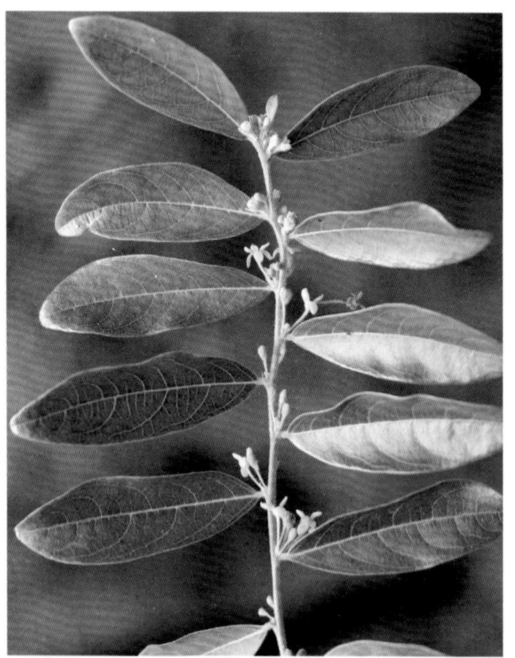

Description: Shrubs erct, 1-5 m tall, branchlets gray-brown, densely pubescent. Stipules triangular; leaf blade oblong, or obovate-oblong, papery or subleathery, base cuneate to obtuse, apex acute or rounded, reticulate nerves prominent. Flowers in axillary clusters, 2-5-flowered, proximal axils mostly to all male flowers, distal axils mostly to all female flowers; sepals 6, green, stamens 3, connate into a cylindric column, style column annular, shortly lobed in summit. Capsules depressed-globose, 8-10-grooved, densely pubescent, reddish when mature.

Flower and Fruit: Fl. Apr. -Aug. , fr. Jul. -Nov.

Distribution: Japan. To the south of North China. Main island of Nanji, commonly.

Habitat: Slopes, scrub on stream banks, forest margins.

A207. 大戟科
Euphorbiaceae

野梧桐
Mallotus japonicus
野桐属 *Mallotus*

形态特征：落叶灌木或小乔木。嫩枝、叶柄和花序轴均密被褐色星状毛。叶互生，宽卵形，纸质，基部楔形，或稍心形，具 2 腺体，边全缘或微裂，上面无毛，下面疏散橙红色腺点；基脉 3 ～ 5。顶生总状花序常成圆锥状；花雌雄异株，雄花花萼 3 裂，雄蕊多数；雌花密生，花萼 5 裂，花柱 3。蒴果球形，密生软刺及紫红色腺点，疏被星状毛。

花果期：花期 6 ～ 7 月；果期 8 ～ 10 月。

分布：日本、朝鲜。我国华东地区。南麂各岛屿常见。

生境：山谷、森林或林缘。

Description: Deciduous shrubs or small trees. Young branchlets, petiole and inflorescence axis dull brownish stellate-tomentulose. Leaf blade suborbicular-ovate, alternate, papery, base cuneate, or slightly cordate, with 2 glands, margin entire or lobed, adaxially glabrous, below evacuate orange-red gland spot; basal veins 3-5. Racemes terminal, usually branched into paniculate; flowers dioecious, male flowers calyx lobes 3, stamen majority; female flowers densely, calyx 5-lobed, styles 3. Capsule spherical, dense soft spines and purplish red glandular spot, sparse by stellate hair.

Flower and Fruit: Fl. Jun. -Jul. , fr. Aut. -Oct.

Distribution: Japan, Korea. East China. Throughout Nanji Islands, commonly.

Habitat: Valleys, forests, forest margins.

A207. 大戟科
Euphorbiaceae

蜜甘草（蜜柑草）
Phyllanthus ussuriensis
叶下珠属 ***Phyllanthus***

形态特征：一年生草本，高达 60 cm，雌雄同株，全株无毛。茎直立，通常在基部分枝；枝柔弱；小枝具角。托叶卵状披针形；叶片椭圆形至长圆形，纸质，背面白绿色，基部圆形，先端锐尖到钝。腋生束状花序，花 1 到多朵；花梗丝状，具几个小苞片；雄花萼片 4，雌花萼片 6，果期反折；花盘腺体 6，花柱 3，先端 2 裂。蒴果扁圆形，光滑。

花果期：花期 4 ~ 7 月，果期 7 ~ 10 月。

分布：东亚。我国东部湿润半湿润地区。南麂主岛（关帝岙、三盘尾），偶见。

生境：山地的斜坡，草地，荒地，路边。

Description: Herbs annual, up to 60 cm tall, monoecious, glabrous throughout. Stem erect, usually branched at base; branches delicate; branchlets angular. Stipules ovate-lanceolate; leaf blade elliptic to oblong, papery, abaxially white-green, base rounded, apex acute to obtuse. Inflorescence an axillary fascicle, 1-to several flowered; pedicels filamentous, with several bracteoles; male flowers sepals 4, female flowers sepals 6, reflexed in fruit; disk glands 6, styles 3, bifid at apex. Capsules oblate, smooth.

Flower and Fruit: Fl. Apr. -Jul. , fr. Jul. -Oct.

Distribution: East Asia. most areas of China. Humid and subhumid areas of eastern China. Main island (Guandiao, Sanpanwei), occasionally.

Habitat: Montane slopes, grasslands, wastelands, pathsides.

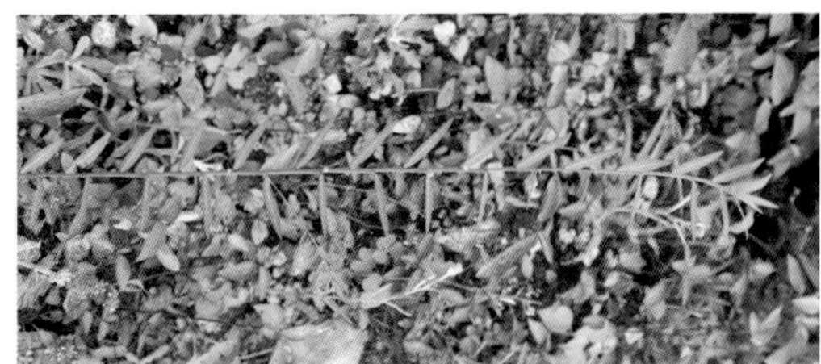

A207. 大戟科
Euphorbiaceae

蓖麻
Ricinus communis
蓖麻属 *Ricinus*

形态特征：一年生草本或草质灌木，高 2 ～ 5 m，茎直立，中空，幼嫩部分被白霜，全株常带红色或略带紫色。托叶合生，叶柄长，顶端具 2 枚盾状腺体；叶掌状 7 ～ 11 裂，边缘有锯齿。花序顶生，两性；苞片阔三角形，早落，雄花在下，雌花在上；花萼镊合状，雄蕊极多，子房 3 室，具软刺，柱头 3 枚，2 裂。蒴果椭圆形或卵球形，具刺。种子浅灰色，种阜扁平圆锥形。

花果期：花期 6 ～ 9 月或全年。

分布：全世界栽培。我国各地有分布。南麂各岛常见，常栽培。

生境：栽培，荒地杂草，河床上偶见归化。

Description: Annual herbs or herbaceous shrubs, 2-5m tall; stem erect, hollow, often single-stemmed but sometimes bushlike or treelike, younger parts glaucous, whole plant often reddish or purplish. Stipules connate, petiole long, apex with 2 glands, insertion peltate; leaf blade palmately 7-11-lobed, margin serrate. Inflorescences terminal, bisexual; bracts broadly triangular, deciduous; male flowers proximal, female flowers distal; calyx , valvate, stamens very many; ovary 3-locular, softly spiny, styles 3, 2-lobed. Capsule ellipsoid or ovoid, echinate. Seed grayish, caruncle depressed-conical.

Flower and Fruit: Fl. Jun. -Sep. , or year round.

Distribution: Cultivated worldwide. Throughout China. Throughout Nanji Islands, usually cultivated.

Habitat: Cultivated, ruderal weed, sometimes naturalized in riverbeds.

A207. 大戟科
Euphorbiaceae

乌桕
Triadica sebifera（*Sapium sebiferum*）
乌桕属 *Triadica*

形态特征：落叶乔木。树皮暗灰色，有深纵裂纹。叶互生，纸质，叶片菱形或菱状卵形，先端突尖，基部楔形或钝，全缘。总状花序顶生；雌雄同株；雄花多生花序轴上部，花萼杯状，3浅裂，花丝极短；雌花多生苞片内，花萼3深裂，裂片披针形；子房3室，基部合生。蒴果梨状球形，成熟时黑色。种子扁球形，黑色，外被白色蜡质的假种皮。

花果期：花期3～6月；果期8～10月。

分布：日本、越南。我国黄河以南各省区，北达陕西、甘肃，广泛栽培。南麂主岛各处常见，常栽培。

生境：旷野、塘边、疏林。

Description: Deciduous trees. Bark dark gray, with longitudinal stripes. Leaves alternate, papery, leaf blade rhomboid, apex acutely acuminate, base broadly rounded, truncate, margin entire; Flowers is terminal racemes; monoecious; male flowers on the upper part of inflorescence axis, calyx cup-shaped, shallowly 3-lobed, filaments very short; female flowers in bracts, calyx 3 deep partite, lobes lanceolate; ovary 3, connate at base. Capsules pyriform, black when mature. Seeds oblate, black, covered with white, waxy aril.

Flower and Fruit: Fl. Mar. -Jun. , fr. Aut. -Oct.

Distribution: Japan, Vietnam. To the south of Yellow River, north to Shanxi, Gansu, widely cultivated. Throughout main island of Nanji, usually cultivated.

Habitat: Open fields, around ponds, open forests.

A207. 大戟科
Euphorbiaceae

油桐
Vernicia montana
油桐属 *Vernicia*

形态特征：落叶乔木。枝无毛，散生突起皮孔。叶阔卵形，顶端短尖至渐尖，基部心形至截平，全缘或 2 ~ 5 裂；叶背面沿叶脉被短柔毛；叶柄顶端有 2 枚具柄的杯状腺体。花单性异株或同株，花萼无毛；花瓣白色或基部紫红色，倒卵形，基部爪状，被褐色短柔毛。核果卵球状，具 3 条纵棱，棱间有粗疏网状皱纹。种子扁球状，种皮厚，有疣突。

花果期：花期 3 ~ 4 月；果期 8 ~ 10 月

分布：越南，新旧世界广泛栽培。我国秦岭以南地区。南麂主岛偶见栽培。

生境：疏林、山谷等。

Description: Deciduous trees. Branches glabrous, with sparsely elevated lenticels. Leaf blade broadly ovate, apex acute to acuminate, base cordate to truncate, margin entire or 2-5-fid; abaxially pubescent along base of nerves; petiole apex with 2 stalked cupular glands. Inflorescences unisexual, dioecious or monoecious, calyx glabrous; petals white or purple-red at base, obovate, base clawed, covered with brown pubescence. Drupes ovoid, longitudinally 3-angular, between angles with sparsely reticulate wrinkles. Seeds compressed globose, coat thicker, verrucose.

Flower and Fruit: Fl. Mar. -Apr. , fr. Aut. -Oct.

Distribution: Vietnam, cultivated in the Old and New Worlds. To the south of Qinling. Cultivated on main island of Nanji, occasionally.

Habitat: Open forests, valleys.

A215. 千屈菜科
Lythraceae

紫薇
Lagerstroemia indica
紫薇属 *Lagerstroemia*

形态特征： 灌木或小乔木，高 7 m。小枝纤细，具四棱，略成翅状。叶无柄或具短柄；叶片椭圆形，长圆形，倒卵形或近圆形，纸质到稍革质，无毛，侧脉 3 ～ 7 对。圆锥花序近锥形，被微柔毛，花密集；花冠管 6 瓣，无毛，萼片正面无毛，花瓣紫色、粉色或白色，圆形，具爪。蒴果椭圆形，种子具翅。

花果期： 花期 6 ～ 9 月，果期 9 ～ 11 月。

分布： 原产亚洲，世界热带和温暖地区广泛栽培。我国南北各地有野生或栽培。南麂主岛（大沙岙、美龄居、镇政府），栽培。

生境： 半阴处，肥沃的田野。

Description: Shrubs or small trees, to 7 m tall. Branchlets slender, 4-angled or subalate. Leaves sessile or with petiole; leaf blade elliptic, oblong, obovate, or suborbicular, papery to slightly leathery, glabrous, lateral veins 3-7 pairs. Panicles subpyramidal, puberulous, densely flowered; floral tube 6-merous, glabrous, sepals adaxially glabrous, petals purple, pink or white, orbicular, clawed. Capsules ellipsoidal, seeds winged.

Flower and Fruit: Fl. Jun. -Sep. , fr. Sep. -Nov.

Distribution: Native to Asia, widely cultivated throughout the tropical and warm areas of the world. Wild or cultivated in the north and south areas of China. Main island (Dashaao, Meilingju, Zhenzhengfu), cultivated.

Habitat: Semishaded places, rich fields.

A216. 柳叶菜科
Onagraceae

假柳叶菜
Ludwigia epilobioides
丁香蓼属 *Ludwigia*

形态特征：一年生草本，高 10 ~ 60 cm。茎常淡红色，多分枝，近无毛。叶椭圆形至狭椭圆形，基部狭楔形，先端锐尖，侧脉每边 8 ~ 12 条。花腋生，萼片 4，三角状卵形，花瓣黄色，狭匙形。蒴果淡褐色，近圆柱状，稍 4 棱，无毛，果皮薄，熟时不规则开裂。

花果期：花期 8 ~ 10 月；果期 9 ~ 11 月。

分布：东亚。分布于我国各地。南麂主岛（关帝岙、门屿尾、三盘尾），偶见。

生境：常见于低地潮湿处，如稻田，沟渠，溪岸。

Description: Herbs annual, 10-60 cm tall. Stems often red tinged, often branched, subglabrous. Leaf blade elliptic to narrowly elliptic, base narrowly cuneate, apex acute, lateral veins 8-12 per side. Flowers axillary, sepals 4, deltate, puberulous, petals yellow, narrowly spatulate. Capsule pale brown, subcylindric, slightly 4-angled, glabrous, thinly walled, readily and irregularly dehiscent.

Flower and Fruit: Fl. Aug. -Oct. , fr. Sep. -Nov.

Distribution: East Asia. Throughout China. Main island (Guandiao, Menyuwei, Sanpanwei), occasionally.

Habitat: Often common in low moist places such as paddy fields, ditches, steam banks.

A218. 桃金娘科
Myrtaceae

桉（大叶桉）
Eucalyptus robusta
桉属 *Eucalyptus*

形态特征：乔木。树皮宿存，深褐色，有不规则斜裂沟。嫩枝有棱。幼态叶对生，有柄；成熟叶互生，卵状披针形，厚革质，两面均有腺点。伞形花序腋生，简单，4～8花，总梗扁平，萼管半球形或倒圆锥形，帽状体先端收缩成喙。蒴果卵状壶形，上半部略收缩，果瓣3～4，深藏于萼管内。

花果期：花期几乎全年。

分布：原产澳大利亚。我国长江以南各地栽培。南麂主岛（大沙岙、后隆、美龄居、镇政府）、大檑山屿，栽培。

生境：丘陵、路边、林缘。

Description: Trees. Bark persistent, dark gray, rough and subfibrous. Branchlets ridged. Young leaves opposite, petiolate; mature leaves twisted, leaf blade ovate-lanceolate, oblique, thickly leathery, both surfaces glandular. Inflorescences axillary, simple, umbels 4-8-flowered, peduncle compressed, hypanthium semiglobose to obconic, calyptra apex constricted into a beak. Capsule pot-shaped, constricted in middle, valves 3 or 4, included in hypanthium.

Flower and Fruit: Fl. almost year-round.

Distribution: Native to Australia. Cultivated to the south of Yangtze River. Main island (Dashaao, Houlong, Meilingju, Zhenzhengfu), Daleishan Island, cultivated.

Habitat: Hillsides, roadsides, forest margins.

A219. 野牡丹科 Melastomataceae

地菍（地稔）
Melastoma dodecandrum
野牡丹属 *Melastoma*

形态特征：小灌木，高 10 ～ 30cm。茎常匍匐；分枝多，幼时被糙伏毛，后无毛。叶片卵形至椭圆形，硬纸质，背面沿脉具稀疏糙伏毛，正面通常在边缘具糙伏毛，边缘密细锯齿或全缘，先端锐尖。聚伞花序顶生，（1）-3 花，基部具 2 叶状苞片；花梗具糙伏毛；萼裂片披针形，疏生糙伏毛；花瓣淡紫色至紫色，菱状倒卵形。果实坛状球形，肉质，具糙伏毛。

花果期：花期 5 ～ 7 月，果期 7 ～ 9 月。

分布：越南。我国华东至华南地区。南麂主岛（国姓岙、门屿尾、三盘尾），常见。

生境：空旷地带，灌丛，草地。

Description: Shrublets, 10-30 cm tall. Stems often repent; branchlets numerous, procumbent, strigose when young, later glabrous. Leaf blade ovate to elliptic, stiffly papery, abaxially very remotely strigose along veins, adaxially usually strigose at margin only, margin densely serrulate or entire, apex acute. Inflorescences terminal, cymose, (1-)3-flowered, with 2 leaflike bracts at base; pedicel strigose; calyx lobes lanceolate, sparsely strigose; petals lavender to purple, rhomboid-obovate. Fruit urceolate-globular, succulent, strigose.

Flower and Fruit: Fl. May-Jul. , fr. Jul. -Sep.

Distribution: Vietnam. East China to South China. Main island (Guoxingao, Menyuwei, Sanpanwei), commonly.

Habitat: Open fields, thickets, grasslands.

A239. 漆树科
Anacardiaceae

盐肤木
Pistacia chinensis
黄连木属 *Pistacia*

形态特征：灌木至乔木，高 2 ～ 10 m；小枝具皮孔，与叶轴、叶背、花序被铁锈色短柔毛。奇数羽状复叶，叶无柄；叶轴具宽翅至翅翼，叶片卵形至长圆形，正面深绿色，背面浅绿色，苍白。花序多分枝，花瓣倒卵形长圆形。核果球状，稍压扁，混合具柔毛和腺状短柔毛，成熟时红色。

花果期：花期 8 ～ 9 月，果期 10 月。

分布：东亚、东南亚。除青海、新疆外，分布几遍我国。南麂主岛各处常见。

生境：丘陵和山地森林，沿溪的森林、灌丛。

Description: Shrubs to trees, 2-10 meters tall; branchlets lenticellate, ferruginous pubescent with rachis, leaf abaxial and inflorescence. Leaf sessile, imparipinnately compound; rachis broadly winged to wingless; leaflets ovate to oblong, adaxially dark green, abaxially lighter green, glaucous. Inflorescences many branched, petals obovate and oblong. Drupe globose, slightly compressed, mixed with pilose and glandular pubescent, red at maturity.

Flower and Fruit: Fl. Aug. -Sep. , fr. Oct.

Distribution: East Asia, Southeast Asia. Throughout China except Qinghai and Xinjiang. Main island of Nanji, commonly.

Habitat: Hill, and mountain forests, forests along streams, thickets.

A239. 漆树科
Anacardiaceae

野漆
Toxicodendron succedaneum
漆树属 *Toxicodendron*

形态特征：乔木或灌木，高可达10m。小枝无毛或具柔毛。奇数羽状复叶互生，常集生枝顶，小叶4～7对，对生或近对生，叶片长圆状椭圆形至卵状披针形，坚纸质至薄革质，叶背常具白粉，基部偏斜，全缘；叶轴圆柱形或上部具狭翅。圆锥花序多分枝，无毛；花黄绿色，花萼裂片阔卵形，花瓣长圆形，开花时外卷。核果大，偏斜，压扁，外果皮薄，淡黄色。

花果期：花期5月，果期7～10月。

分布：印度、中南半岛、朝鲜和日本。我国华北至长江以南各地。南麂主岛各处常见。

生境：低地、丘陵森林和低地石灰岩灌丛。

Description: Trees or shrubs, up to 10 m tall. Branchlets glabrous to pubescent. Odd-pinnate alternate, usually terminal clustering; leaflets 4-7, opposite or subopposite, leaflet blade oblong-elliptic to ovate-lanceolate, papery or thinly leathery, glaucous abaxially, base oblique, margin entire; rachis terete or narrowly winged distally. Inflorescence paniculate, many branched, glabrous; flowers yellowish green, calyx lobes broadly ovate, petals oblong, revolute at anthesis. Drupe large, asymmetrical, compressed, epicarp thin, light yellow.

Flower and Fruit: Fl. May, fr. Jul. -Oct.

Distribution: India, Indochina Peninsula, Korea and Japan. North China to the south of Yangtze River. Main island of Nanji, commonly.

Habitat: Lowland and hill forests, lowland thickets on limestone.

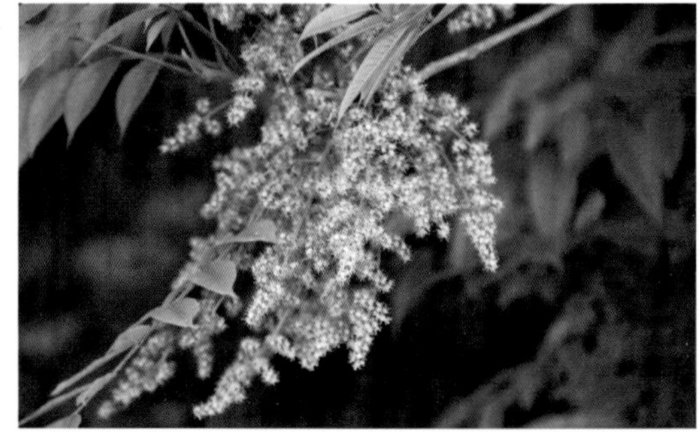

A241. 芸香科
Rutaceae

楝叶吴萸（臭辣吴萸、臭辣树）
***Tetradium glabrifolium*（*Evodia fargesii*）**
吴茱萸属 *Tetradium*

形态特征：落叶乔木或灌木。树皮平滑，嫩枝紫褐色。奇数羽状复叶互生，对生，小叶 5 ~ 19；阔卵形至披针形，背面常具白霜，无乳突，侧脉每边 8 ~ 18，网脉清晰密集。聚伞圆锥花序顶生，花单性，细小，5 基数；花瓣绿色、黄色或白色，干后发白或褐色，无毛或有毛。蓇葖果三棱形，成熟时紫红色或淡红色，微皱褶。种子黑色。

花果期：花期 6 ~ 8 月；果期 9 ~ 10 月。

分布：东南亚。我国长江流域以南各地。南麂主岛各处常见。

生境：森林、灌丛、开阔地。

Description: Deciduous trees or shrubs. Bark smooth, twigs purplish brown. Imparipinnate leaf alternate, 5-19-foliolate; leaflet blades opposite, broadly ovate to lanceolate, abaxially usually glaucous and not papillate, secondary veins 8-18 on each side of midvein, reticulate veinlets abaxially clearly defined and dense. Thyrse terminal, flowers unisexual, tiny, 5-merous; petals green, yellow, or white but drying whitish to brown, glabrous to villous. Follicles trigonous, purplish red or pale red when ripe, slightly creased. Seeds black.

Flower and Fruit: Fl. Jun. -Aug. , fr. Sep. -Oct.

Distribution: Southeast Asia. To the south of the Yangtze River Basin. Main island of Nanji, commonly.

Habitat: Forests, thickets, open places.

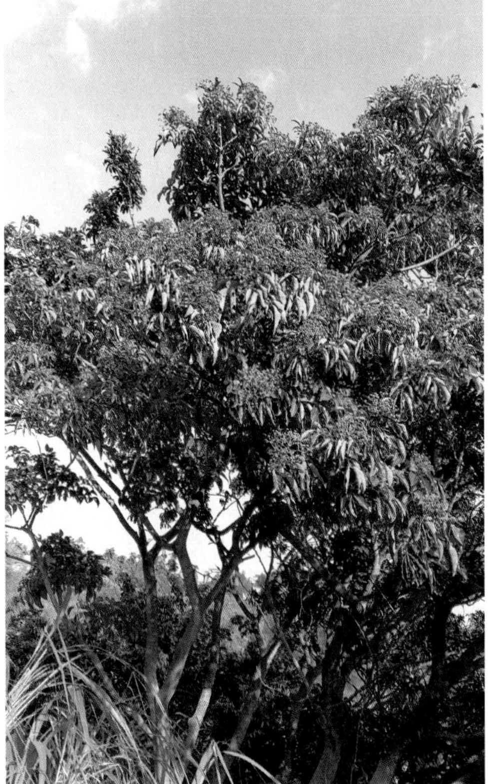

A241. 芸香科
Rutaceae

两面针
Zanthoxylum nitidum
花椒属 *Zanthoxylum*

形态特征：灌木，直立或攀缘，或有时为木质藤本，无毛。树干具翅。茎，小枝和叶轴通常具皮刺。奇数羽状复叶互生，小叶 5 ～ 11；对生，阔卵形、近心形或椭圆形，革质，全缘或具圆齿，先端渐尖。花序腋生，花 4 数，花被片 2 轮，萼片顶部紫绿色，花瓣淡黄绿色。蓇葖果红褐色，顶端具短喙。种子亮黑色。

花果期：花期 3 ～ 5 月，果期 9 ～ 11 月。

分布：东南亚至太平洋岛屿。我国华东至华南、西南地区。南麂各岛屿常见。

生境：山坡灌丛。

Description: Shrubs, erect or scrambling, or sometimes woody climbers, glabrous. Trunk winged. Stems, branchlets, and leaf rachises usually with prickles. Imparipinnate leaf alternate, 5-11-foliolate; leaflet blades opposite, broadly ovate, subcordate, or elliptic, leathery, margin crenate at least toward apex or entire, apex acuminate. Inflorescences axillary, flowers 4-merous, perianth in 2 series, sepals apically purplish green, petals pale yellowish green. follicles reddish brown, apex beaked. Seeds light black.

Flower and Fruit: Fl. Mar. -May, fr. Sep. -Nov.

Distribution: Southeast Asia to Pacific islands. East China to South and Southwest China. Throughout Nanji Islands, commonly.

Habitat: Slopes, thickets.

A243. 楝科
Meliaceae

楝（苦楝）
Melia azedarach
楝属 *Melia*

形态特征：落叶乔木。树皮灰褐色，纵裂。小枝有叶痕，2 ～ 3 回奇数羽状复叶，小叶对生，卵形、椭圆形至披针形，顶生一片通常略大，边缘有钝锯齿。圆锥花序，无毛或幼时被鳞片状短柔毛；花萼 5 深裂，裂片卵形或长圆状卵形，花瓣淡紫色，倒卵状匙形。核果球形至椭圆形，内果皮木质。种子椭圆形。

花果期：花期 3 ～ 5 月，果期 10 ～ 12 月。

分布：东亚和东南亚。我国华北地区以南各地。南麂各岛常见，常栽培。

生境：疏林，田缘，路旁。

Description: Deciduous trees. Bark brownish gray, longitudinally exfoliating. Branchlets with leaf scars. Leaves odd-pinnate, 2-or 3-pinnate, leaflets opposite; leaflet blades ovate, elliptic, or lanceolate, but terminal one usually slightly larger, margin crenate. Thyrses glabrous or covered with short lepidote pubescence; calyx 5-parted, sepals ovate to oblong-ovate, petals lilac-colored, obovate-spatulate. Drupe globose to ellipsoid, endocarp ligneous. Seed ellipsoid.

Flower and Fruit: Fl. Mar. -May, fr. Oct. -Dec.

Distribution: East and Southeast Asia. To the south of North China. Throughout main island of Nanji, usually cultivated.

Habitat: Sparse forests, field margins, roadsides.

A247. 锦葵科
Malvaceae

田麻
Corchoropsis crenata (Corchoropsis tomentosa)
田麻属 *Corchoropsis*

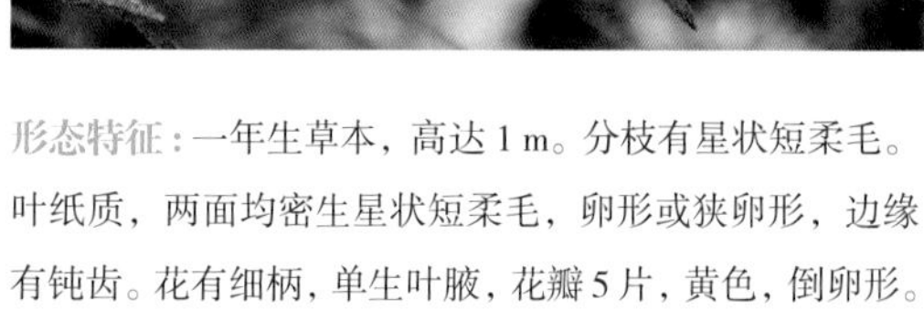

形态特征：一年生草本，高达 1 m。分枝有星状短柔毛。叶纸质，两面均密生星状短柔毛，卵形或狭卵形，边缘有钝齿。花有细柄，单生叶腋，花瓣 5 片，黄色，倒卵形。蒴果角状圆筒形，有星状柔毛。种子长卵形，略具横纹。

花果期：花期 4 ～ 6 月，果期秋季。

分布：日本、朝鲜。除西北地区外遍布我国。南麂主岛（关帝岙、门屿尾、三盘尾），偶见。

生境：丘陵或低山干山坡或多石处。

Description: Herbs annual, to 1 m tall. Branches stellate pilose. Leaf blade papery, both surfaces densely stellate pubescent, ovate or narrowly ovate, margin obtuse serrate. Pedicels slender, solitary in leaf axils, petals 5, yellow, obovate. Capsule narrowly cylindric, stellate pilose. Seeds ovoid, obscurely transversely ridged.

Flower and Fruit: Fl. Apr. -Jun. , fr. autumn.

Distribution: Japan, Korea. Throughout China except Northwest China. Main island (Guandiao, Menyuwei, Sanpanwei), occasionally.

Habitat: Hill, low mountain slopes, rocky lands.

A247. 锦葵科
Malvaceae

甜麻
Corchorus aestuans
黄麻属 *Corchorus*

形态特征：一年生草本，高约 1 m。茎红褐色；枝细长，披散。叶纸质，两面被长粗毛，互生，卵形或阔卵形，边缘有锯齿。花单生或成聚伞花序；萼片 5 片，外面紫红色，花瓣 5 片，黄色。蒴果长筒形，具 6 条纵棱，成熟时 3 ～ 5 瓣裂。种子多数。

花果期：花期 7 ～ 8 月，果期 9 ～ 10 月。

分布：热带亚洲、非洲和中美洲。我国长江以南各地。南麂主岛（关帝岙、门屿尾、三盘尾），偶见。

生境：荒地、旷野、村旁，为南方各地常见杂草。

Description: Herbs annual, to 1 m tall. Stem red-brown; branches slender, scattered. Leaf blade papery, pilose on both surfaces, alternate, ovate or broadly ovate, margin serrate. Flowers solitary or several together in cymes; sepals 5, purple-red abaxially, petals 5, yellow. Capsule cylindrical, 6 -ribbed, 3-5-valved at maturity. Seeds many.

Flower and Fruit: Fl. Jul. -Aug. , fr. Sep. -Oct.

Distribution: Tropical Asia, Africa, Central America. To the south of Yangtze River. Main island (Guandiao, Menyuwei, Sanpanwei), occasionally.

Habitat: Waste places, fields, village, a common weed throughout southern China.

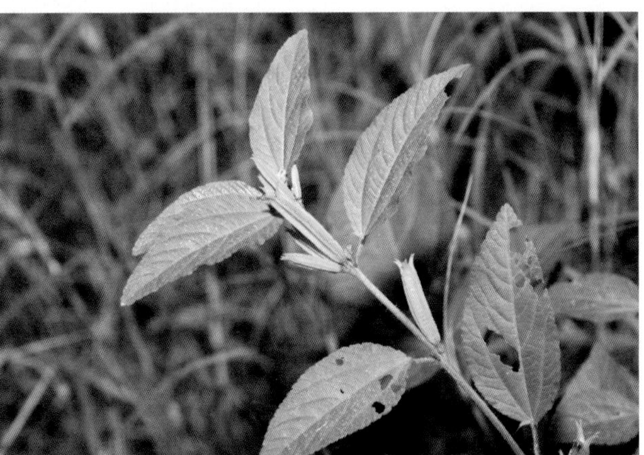

A247. 锦葵科
Malvaceae

小花扁担杆
Grewia biloba var. *parviflora*
扁担杆属 *Grewia*

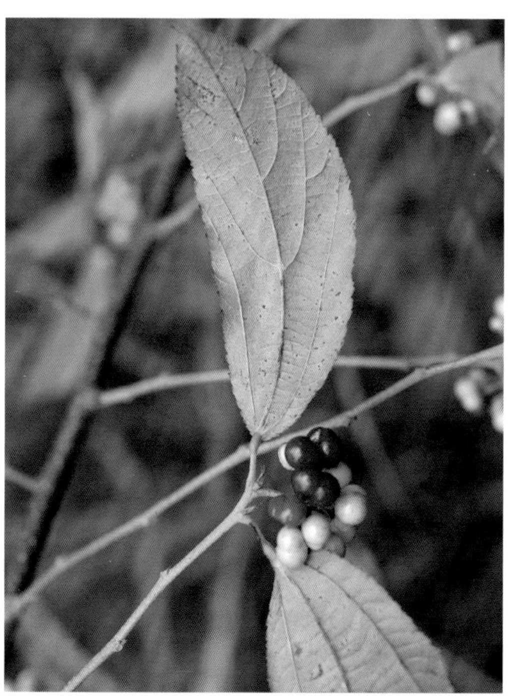

形态特征：灌木或小乔木，高 1 ~ 4 m，多分枝。小枝具短柔毛。叶椭圆形或倒卵状椭圆形，薄革质，叶下面密被星状茸毛，基出 3 脉。聚伞花序腋生，多花；花朵较短小，苞片钻形，萼片狭长圆形，花瓣白色，柱头盘状，浅裂。核果红色，有 2 ~ 4 分核。

花果期：花期 5 ~ 7 月，果期 8 ~ 10 月。

分布：我国大部分地区有分布。南麂主岛（百亩山、国姓岙、门屿尾、三盘尾），常见。

生境：山坡、沟边、灌丛及林下。

Description: Shrubs or small trees, 1-4 m tall, many branched. Branchlets pubescent. Leaf blade elliptic or obovate-elliptic, thinly leathery, abaxially densely softly stellate tomentose, basal veins 3. Cymes axillary, many flowered; flowers short, bracts subulate, sepals narrowly oblong, petals white, stigma disk, lobed. Drupe red, 2-4-lobed.

Flower and Fruit: Fl. May-Jul. , fr. Aug. -Oct.

Distribution: Most areas of China. Main island (Baimushan, Guoxingao, Menyuwei, Sanpanwei), commonly.

Habitat: Mountain slopes, ditches, thickets or forests.

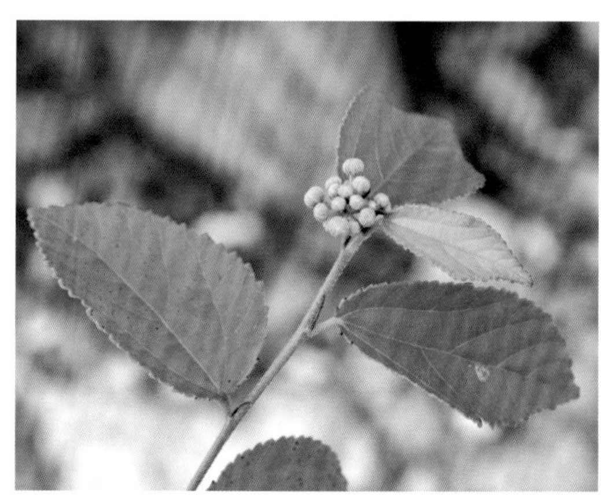

A247. 锦葵科 Malvaceae

海滨木槿
Hibiscus hamabo
木槿属 *Hibiscus*

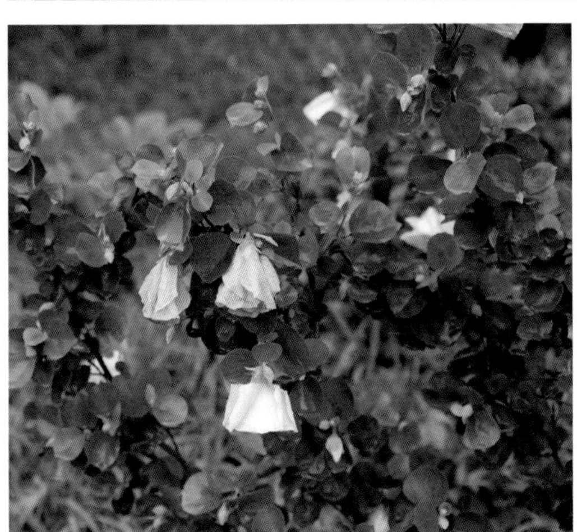

形态特征：落叶灌木或小乔木，高 1 ~ 5 m。嫩枝、叶背、花梗和花瓣被星状毛。托叶叶状，长圆状卵形，脱落；叶片圆形至阔倒卵形，不裂，基脉 5 ~ 7，基部心形，边缘具不规则细齿至近全缘。单花腋生，副萼杯状，裂片狭三角形，花萼钟状，深 5 裂，长于副萼，花冠黄色，后变橘红，内面基部具暗红色斑点。蒴果卵球形，密被褐色硬毛。种子肾形。

花果期：花期 7 ~ 10 月，果熟期 10 ~ 11 月。

分布：日本、朝鲜。浙江、福建沿海岛屿。南麂主岛（美龄居、门屿尾、三盘尾、镇政府），栽培。

生境：海滨沙地。

Description: Trees or shrubs, deciduous, 1-5 m tall. Young stems, leaf abaxially, pedicel and petal abaxially with stellate hairs. Stipules foliaceous, oblong-ovate, deciduous; leaf blade orbicular to broadly obovate, not lobed, basal veins 5-7, base cordate, margin irregularly crenulate to subentire. Flowers solitary, axillary, epicalyx cup-shaped, lobes narrowly triangular, calyx campanulate, deeply 5-lobed, longer than epicalyx, corolla yellow later turning orange-red, with dark red spots in center. Capsule ovoid, densely brownish hirsute. Seeds reniform.

Flower and Fruit: Fl. Jul. -Oct. , fr. Oct. -Nov.

Distribution: Japan, Korea. Costal islands of Zhejiang and Fujian. Main island (Meilingju, Menyuwei, Sanpanwei), cultivated.

Habitat: Coastal sands.

A247. 锦葵科
Malvaceae

木芙蓉
Hibiscus mutabilis
木槿属 *Hibiscus*

形态特征：落叶灌木或小乔木，直立，高 2 ～ 5 m。小枝，叶柄，花梗，副萼和花萼密被星状短柔毛。托叶披针形，通常早落；叶片宽卵形到圆形，被星状毛，裂片三角形，边缘有钝齿。花单生上部枝腋；花冠白色或带红色，花瓣近圆形，背面有毛，基部具髯毛。蒴果扁平球状，淡黄色具糙硬毛和绵状毛，分果爿 5。种子肾形，具长柔毛背面。

花果期：花期 8 ～ 10 月。

分布：原产我国东南部，世界各地广泛栽培，且偶有归化。南麂主岛（美龄居、门屿尾、三盘尾、镇政府），栽培。

生境：溪边的灌丛。

Description: Shrubs or small trees, erect, 2-5 m tall, deciduous. Branchlets, petioles, pedicel, epicalyx, and calyx densely stellate and woolly pubescent. Stipules lanceolate, usually caducous; leaf blade broadly ovate to round-ovate, papery, stellate minutely tomentose, lobes triangular, margin obtusely serrate. Flowers solitary, axillary on upper branches; corolla white or reddish, petals nearly orbicular, hairy abaxially, barbate at base. Capsule flattened globose, yellowish hispid and woolly; mericarps 5. Seeds reniform, villous abaxially.

Flower and Fruit: Fl. Aug. -Oct.

Distribution: Native to southeastern China, cultivated throughout the world and occasionally naturalized. Main island (Meilingju, Menyuwei, Sanpanwei), cultivated.

Habitat: Thickets along streams.

A247. 锦葵科
Malvaceae

木槿
Hibiscus syriacus
木槿属 *Hibiscus*

形态特征： 灌木落叶，直立，高 1.5 ~ 4 m。小枝被黄色星状微柔毛。托叶丝状钻形，具柔毛；叶片菱形至三角状卵形，3 裂或全缘，纸质，基部楔形，边缘不规则缺刻，先端钝到稍尖。花单生，花梗被星状毛，花萼钟状，裂片 5，花冠钟状，有时重瓣，花瓣倒卵形，背面具长柔毛。蒴果卵球形，密被黄色星状毛。种子肾形，毛背面黄白色。

花果期： 花期 7 ~ 10 月。

分布： 原产我国东南部，热带和温带地区广泛栽培。南麂主岛（美龄居、门屿尾、三盘尾、镇政府），栽培。

生境： 海边悬崖，山腰，溪边，路旁。

Description: Shrubs deciduous, erect, 1.5-4 m tall. Branchlets yellow stellate puberulent. Stipules filiform-subulate, pilose; leaf blade rhomboid to triangular-ovate, variously 3-lobed or entire, papery, base cuneate, margin irregularly incised, apex obtuse to subacute. Flowers solitary, pedicel stellate puberulent, calyx campanulate, lobes 5, corolla petals obovate, pilose abaxially. Capsule ovoid-globose, densely yellow stellate puberulent. Seeds reniform, with yellow-white hairs abaxially.

Flower and Fruit: Fl. Jul. -Oct.

Distribution: Native to southeastern China, cultivated Main island (Meilingju, Menyuwei, Sanpanwei), cultivated.

Habitat: Sea cliffs, hillsides, along streams, roadsides.

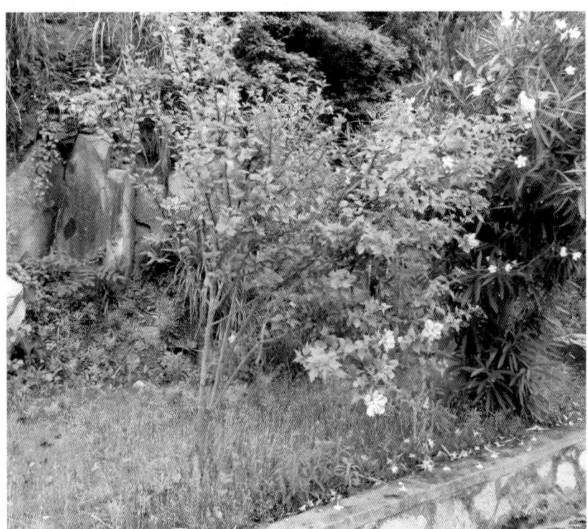

A247. 锦葵科
Malvaceae

马松子
Melochia corchorifolia
马松子属 *Melochia*

形态特征：草本或亚灌木，高不足 1 m，直立或外倾。枝黄棕色，疏生星状微柔毛。托叶线性；叶片卵形，很少不明显 3 浅裂，薄纸状，背面疏生星状被微柔毛，基部圆形，边缘具齿，先端锐尖或钝。花序呈密集的聚伞花序或团伞花序；花萼钟状，5 浅裂，背面具长柔毛和具糙硬毛，裂片三角形。蒴果球形，5 棱，具长柔毛。种子棕黑色，卵形，近三角形。

花果期：花期夏秋季。

分布：泛热带地区。我国长江以南各地。南麂主岛各处常见。

生境：田野或低山丘陵。

Description: Herbs or subshrubs, less than 1 m tall, erect or decumbent. Branches yellow-brown, sparsely stellate puberulent. Stipules linear; leaf blade ovate, rarely obscurely 3-lobed, thinly papery, abaxially sparsely stellate puberulent, base rounded, margin dentate, apex acute or obtuse. Inflorescence a dense cyme or glomerule; calyx campanulate, 5-lobed, abaxially villous and hispid, lobes triangular. Capsule globose, 5-angular, villous. Seeds brown-black, ovoid, slightly triangular.

Flower and Fruit: Fl. summer-autumn.

Distribution: Pantropical regions. To the south of Yangtze River. Main island of Nanji, commonly.

Habitat: Fields or low hills.

A247. 锦葵科
Malvaceae

小叶黄花稔
Sida alnifolia var. *microphylla*
黄花稔属 *Sida*

形态特征：亚灌木或灌木，匍匐或斜升。小枝细瘦，被星状柔毛。叶小，叶片长圆形至卵圆形，边缘具齿，上面被星状柔毛，下面密被星状长柔毛。花单生于叶腋；花瓣黄色，倒卵形，花丝筒被长硬毛。分果近球形；分果爿6～8，顶端2芒，被长柔毛。

花果期：花期7～12月。

分布：印度。我国浙江、福建、广东、广西和云南等省区。南麂主岛各处常见。

生境：山麓沟边灌丛或路边草丛。

Description: Subshrub or shrubs, creeping or ascending. Branchlets thin, stellate pilose. Leaf blade ovate to oblong, margin dentate, abaxially stellate velutinous, adaxially stellate pilose. Flowers solitary, axillary; corolla yellow, petals obovate, filament tube hirsute. Schizocarp subglobose; mericarps 6-8, apex 2-awned, with long hairy.

Flower and Fruit: Fl. Jul. -Dec.

Distribution: India. Zhejiang, Fujian, Guangdong, Guangxi, and Yunnan. Main island of Nanji, commonly.

Habitat: Foothill gully scrub or roadside grass.

A247. 锦葵科
Malvaceae

地桃花
Urena lobata
梵天花属 Urena

形态特征：直立亚灌木状草本，高达 1 m，小枝被星状绒毛。托叶丝状，早落；下部叶近圆形，基部圆形或近心形，边缘具锯齿，中部叶卵形，上部叶长圆形至披针形，叶上面被柔毛，下面被灰白色星状绒毛。花单生或稍丛生，腋生，花梗被绵毛；花冠淡红色，花瓣 5，倒卵形，外面被星状柔毛。果扁球形，分果爿被星状短柔毛和锚状刺。

花果期：花期 7 ~ 10 月。

分布：中南半岛、印度和日本。我国长江以南各省区。南麂主岛各处常见。

生境：草地，灌丛，路旁。

Description: Subshrublike herbs, erect, to 1 m tall, branchlets stellate tomentose. Stipules filiform, caducous; leaf blades on proximal part of stem nearly orbicular, base rounded or nearly cordate, margin serrate, blades on middle part of stem ovate; those on distal part of stem oblong to lanceolate, abaxially gray stellate puberulent, adaxially puberulent. Flowers solitary or slightly aggregated, axillary, pedicel woolly; corolla reddish, petals 5, obovate, abaxially stellate puberulent. Fruit flattened globose. mericarps stellate puberulent and spiny with hooked spines.

Flower and Fruit: Fl. Jul. -Oct.

Distribution: Indo-China Peninsula, India and Japan. To the south of Yangtze River. Main island of Nanji, commonly.

Habitat: Grasslands, scrub, roadsides.

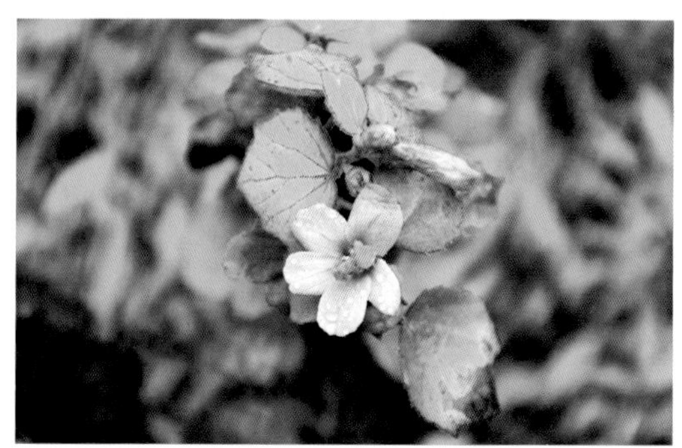

A249. 瑞香科 Thymelaeaceae

芫花
Daphne genkwa
瑞香属 *Daphne*

形态特征：落叶灌木，高 0.3 ～ 1 m。多分枝，幼枝纤细，黄绿色，密被淡黄色丝状毛，老枝褐色或带紫红色，无毛。叶对生，稀互生，纸质，卵形、卵状披针形或椭圆形。花 3 ～ 7 朵簇生叶腋，淡紫红或紫色，先叶开花。果肉质，白色，椭圆形，包于宿存花萼下部。

花果期：花期 3 ～ 5 月，果期 6 ～ 7 月。

分布：朝鲜。长江流域及华北地区。南麂各岛屿常见。

生境：林下，山地灌丛。

Description: Shrubs deciduous, 0.3-1 m tall. Branches many, young branches slender, yellow green, densely yellowish sericeous. Old branches brown or purplish red, glabrous. Leaves opposite, rarely alternate, papery, ovate, ovate-lanceolate, or elliptic. Flowers 3-7, produced before leaves, clusters axillary, lilac or purple. Drupe fleshy, white, ellipsoid, enclosed in persistent calyx.

Flower and Fruit: Fl. Mar. -May, fr. Jun. -Jul.

Distribution: Korea. The Yangtze River Basin and North China. Throughout Nanji Islands, commonly.

Habitat: Forests, shrubby slopes.

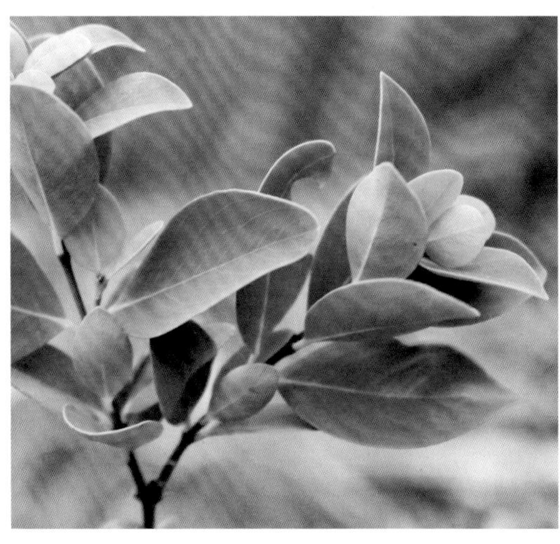

A249. 瑞香科
Thymelaeaceae

了哥王（南岭荛花）
Wikstroemia indica
荛花属 ***Wikstroemia***

形态特征：灌木，高达 2 m。枝红褐色，无毛。叶对生，纸质或近革质，倒卵形、长圆形或披针形。顶生短总状花序，花数朵组成顶生头状总状花序；花冠黄绿色。果椭圆形，无毛，成熟时暗紫黑或鲜红色。

花果期：花期 6 ~ 7 月，果期 9 ~ 10 月。

分布：印度、越南、菲律宾。我国长江以南各地。南麂各岛屿常见。

生境：开旷林下或石山上。

Description: Shrubs, to 2 m tall. Branches reddish brown, glabrous. Leaves oppssite, papery to thinly leathery, obovate, elliptic-oblong, or lanceolate. Inflorescences terminal, capitate, several flowered; Corolla yellow green. Drupe red to dark purple, ellipsoid, glabrous.

Flower and Fruit: Fl. Jun. -Jul, fr. Sep. -Oct.

Distribution: India, Vietnam, Philippines. To the south of Yangtze River. Throughout Nanji Islands, commonly.

Habitat: Forests, rocky shrubby slopes.

A270. 十字花科
Brassicaceae

碎米荠
Cardamine hirsuta
碎米荠属 *Cardamine*

形态特征：一年生草本，高 15 ～ 35 cm。基生叶具叶柄，小叶 2 ～ 5 对，顶生小叶肾形，边缘有 3 ～ 5 圆齿，侧生小叶长卵形至线形，全缘。总状花序顶生，花冠白色。长角果线形，稍扁，无毛。种子椭圆形，顶端有的具明显的翅。

花果期：花期 2 ～ 5 月，果期 4 ～ 7 月。

分布：全球温带地区。分布几遍我国。南麂主岛（关帝岙、门屿尾、三盘尾），常见。

生境：山坡、路旁、荒地及耕地的草丛中。

Description: Herbs annual, 15-35 cm tall. Basal leaves petiolate, lobules 2 to 5 pairs, terminal lobules renal, margin 3-5 crenate, lateral lobules long ovate to linear, entire. Racemes on top of branches, corolla white. Siliques linear, slightly flattened, glabrous. Seeds elliptic, apical some with distinct wings.

Flower and Fruit: Fl. Feb. -May, fr. Apr. -Jul.

Distribution: Temperate regions. Almost throughout China. Main island (Guandiao, Menyuwei, Sanpanwei), commonly.

Habitat: Mountain slopes, roadsides, wastelands, grassy areas of farmland.

A270. 十字花科
Brassicaceae

臭独行菜（臭荠）
Lepidium didymium（*Coronopus didymus*）
独行菜属 *Lepidium*

形态特征：一年或二年生匍匐草本，高 5 ~ 30 cm，全株有臭味。主茎短且不明显。叶一回或二回羽状全裂，裂片 3 ~ 5 对，线形或窄长圆形，全缘，两面无毛。花极小，萼片具白色膜质边缘，花瓣白色，长圆形。短角果肾形，2 裂，果瓣半球形。种子肾形，红棕色。

花果期：花期 3 月，果期 4 ~ 5 月。

分布：亚洲、欧洲、北美洲。我国长江流域以南地区。南麂各岛屿常见。

生境：路旁或荒地。

Description: Creeping Herbs annual or biennial, 5-30cm tall, all smelly. Main stem short and inconspicuous. Leaves once or twice pinnate, lobes 3-5 pairs, linear or narrowly oblong, entire, glabrous on both surfaces. Flowers extremely small, sepals with white membranous margin, petals white, oblong. Silicles reniform, 2-lobed, fruit petals hemispherical. Seeds reniform, reddish brown.

Flower and Fruit: Fl. Mar. , fr. Apr. -May.

Distribution: Asia, Europe, North America. To the south of Yangtze River Basin. Throughout Nanji Islands, commonly.

Habitat: Roadsides, wastelands.

A270. 十字花科
Brassicaceae

北美独行菜
Lepidium virginicum
独行菜属 *Lepidium*

形态特征：一年生或二年生草本，高达 20 ～ 50 cm。上部分枝。基生叶倒披针形，边缘羽状分裂或大头羽裂，茎生叶有短柄，倒披针形或线形。总状花序顶生；萼片椭圆形，花瓣白色，倒卵形。短角果近圆形，顶端微缺，有窄翅。种子卵形，红棕色，边缘有窄翅。

花果期：花期 4 ～ 6 月，果期 5 ～ 9 月。

分布：原产北美，世界广布。我国华北及长江流域以南地区。南麂主岛各处常见。

生境：田野，路旁，荒地，杂草丛中。

Description: Herbs annual or biennial, 20-50 cm tall. Stems branched above. Basal leaves oblanceolate, margin pinnatifid or lyrate, cauline leaves shortly petiolate, leaf blade oblanceolate or linear. Racemes terminal; sepals elliptic, petals white, obovate. Silicles suborbicular, apex emarginate apically, narrowly winged. Seeds ovate, reddish brown, margin with narrow wings.

Flower and Fruit: Fl. Apr. -Jun, fr. May-Sep.

Distribution: Native to North America; introduced elsewhere. North China and to the south of Yangtze River Basin. Main island of Nanji, commonly.

Habitat: Fields, roadsides, waste places, grassy areas.

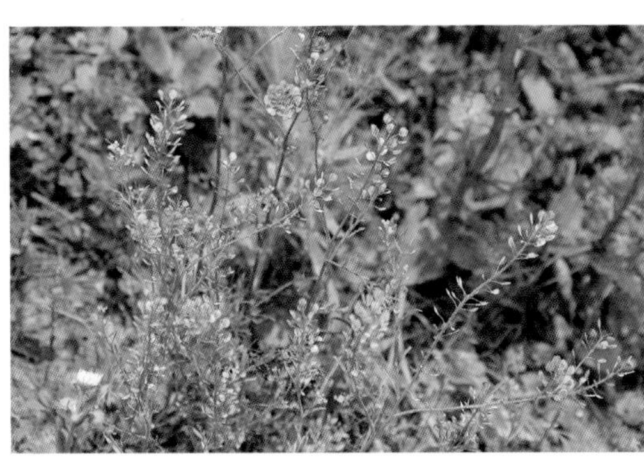

A270. 十字花科
Brassicaceae

蔊菜
Rorippa indica
蔊菜属 *Rorippa*

形态特征：一年生或二年生直立草本，高 20 ～ 40 cm。叶互生，基生叶及茎下部叶具长柄，叶形多变，常大头羽状；上部叶片宽披针形或匙形，边缘具疏齿，具短柄或基部耳状抱茎。总状花序顶生或侧生，花冠黄色。长角果线状圆柱形，短而粗。种子多数，细小。

花果期：花期 4 ～ 6 月，果期 6 ～ 8 月。

分布：亚洲东部和南部。我国东南沿海和华中、西北地区。南麂主岛（百亩山、国姓岙、后隆、门屿尾、三盘尾），常见。

生境：路旁、田边、园圃、河边。

Description: Herbs annual or biennial, erect, 20-40 cm tall. Leaves alternate, basal leaves and lower stem leaves with a long stalk, leaf shape, often lyrate-pinnatipartite; uppermost leaves broadly lanceolate or spatulate, margin sparsely serrate, cauline with short stipe or base auriculate. Racemes terminal or lateral, corolla yellow. Siliques linear cylindrical, short and thick. Seeds many, small.

Flower and Fruit: Fl. Apr. -Jun. , fr. Jun. -Aug.

Distribution: Eastern and southern Asia. Southeast coastal, Central China and Northwest China. Main island (Baimushan, Guoxingao, Houlong, Menyuwei, Sanpanwei), commonly.

Habitat: Roadsides, field margins, gardens, river banks.

A282. 白花丹科
Plumbaginaceae

补血草
Limonium sinense
补血草属 *Limonium*

形态特征：多年生草本，高 15 ~ 60cm。叶基生，花期宿存，叶柄宽，叶片倒卵状长圆形，长圆状披针形至披针形，基部渐狭，先端通常钝或急尖。花序伞房状或圆锥状，3 ~ 5 枚生于同一莲座叶，上升或直立，花序轴 4 棱或具 4 沟槽；穗状花序具 2 ~ 6 个小穗，穗轴 2 棱，每小穗具花 2 或 3 朵，苞片卵形，萼片漏斗状，萼檐白色，花冠黄色。

花果期：花期 4 ~ 12 月。

分布：日本、越南。我国滨海各地。南麂主岛（关帝岙、三盘尾），偶见。

生境：沿海潮湿盐土或砂土。

Description: Herbs perennial, 15-60 cm tall. Caudex often thickened; stems many from 1 crown. Leaves basal, persistent to anthesis; petiole wide; leaf blade obovate-oblong, oblong-lanceolate, or lanceolate, base attenuate, apex usually obtuse to acute. Inflorescences 3-5 from the same leaf rosette, ascending to erect, corymbose or paniculate, main axis 4-angular and 4-sulcate; spikes with 2-6 spikelets, axis 2-angular; spikelets 2-or 3-flowered, bracts ovate, calyx funnelform, limb white, corolla yellow.

Flower and Fruit: Fl. Apr. -Dec.

Distribution: Japan, Vietnam. Coastal areas of China. Main island (Guandiao, Sanpanwei), occasionally.

Habitat: Wet sandy, salty shales adjacent to the ocean.

A283. 蓼科
Polygonaceae

何首乌
***Fallopia multiflora*（*Polygonum multiflorum*）**
何首乌属 ***Fallopia***

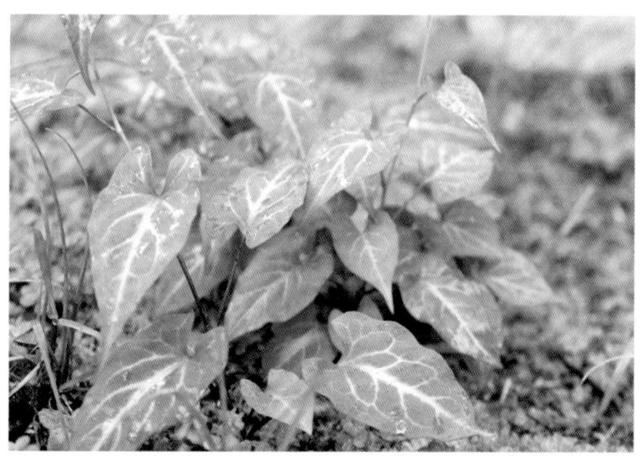

形态特征：多年生草本。块根肥厚，长椭圆形，黑褐色。茎缠绕，多分枝，具纵棱，下部木质化。叶互生，卵形或长卵形，顶端渐尖，基部心形或近心形，两面无毛或沿脉具乳突，全缘。圆锥花序顶生或腋生，分枝开展，花序梗被小突起；苞片三角状卵形，具小突起；花梗下部具关节。瘦果卵形，具3棱，黑褐色，有光泽。

花果期：花期6～10月，果期7～11月。

分布：日本。我国西北南部至长江以南各地。南麂主岛（百亩山、国姓岙、后隆、门屿尾、三盘尾），常见。

生境：山谷灌丛、山坡林下及沟边石隙。

Description: Herbs perennial. Root tuber hypertrophic, narrowly elliptic, black-brown. Stems twining, much branched, striate, ligneous at base. Leaves alternate, leaf blade ovate or narrowly ovate, apex acuminate, both surfaces glabrous or abaxially papillate along veins, margin entire. Inflorescence terminal or axillary, paniculate, spreading, ridged, peduncle minutely papillate; bracts triangular-ovate, papillate, pedicel articulate at base. Achenes ovoid, trigonous, black-brown, shiny.

Flower and Fruit: Fl. Jun. -Oct. , fr. Jul. -Nov.

Distribution: Japan. Southern Northwest China to the south areas of Yangtze River. Main island (Baimushan, Guoxingao, Houlong, Menyuwei, Sanpanwei), commonly.

Habitat: Mountain slopes, rock crevices, thickets in valleys.

A283. 蓼科
Polygonaceae

萹蓄
Polygonum aviculare
萹蓄属 *Polygonum*

形态特征：一年生草本。茎平卧、上升或直立，自基部多分枝，具纵棱。叶椭圆形，狭椭圆形或披针形，先端圆钝或急尖，基部楔形，全缘，两面无毛；叶柄短或近无柄，基部具关节；托叶鞘膜质，撕裂脉明显。花单生或 2 ~ 5 枚簇生叶腋；苞片薄膜质；花梗顶部具关节。瘦果卵形，黑褐色，密被由小点组成的细条纹，无光泽。

花果期：花期 5 ~ 7 月，果期 7 ~ 8 月。

分布：北温带广布。分布于全国各地。南麂主岛各处常见。

生境：田边、路边、荒地。

Description: Herbs annual. Stems prostrate, ascending, or erect, much branched from base, with longitudinal ridge. Leaf blade lanceolate or narrowly elliptic, apex acute or nearly obtuse, base cuneate, margin entire, both surfaces glabrous; petiole short or nearly absent, articulate at base; ochrea membranous, veined, apex lacerate. Flowers solitary or 2-5 fascicled at axils; bracts thinly membranous; articulate at apex. Achenes black-brown, ovoid, minutely granular striate.

Flower and Fruit: Fl. May-Jul., fr. Jul.-Aug.

Distribution: Widely distributed in north temperate zone. Throughout China. Main island of Nanji, commonly.

Habitat: Near fields, roadsides, waste places.

A283. 蓼科
Polygonaceae

火炭母
Polygonum chinense
萹蓄属 *Polygonum*

形态特征：多年生草本，基部近木质。茎直立，具纵棱，多分枝。叶卵形或长卵形，顶端短渐尖，基部截形或宽心形，全缘，两面无毛，下部叶具柄，基部常具叶耳，上部叶近无柄或抱茎；托叶鞘膜质，无毛，具脉纹。花序头状，数枚排成圆锥状，顶生或腋生，花序梗被腺毛。瘦果宽卵形，黑色，无光泽。

花果期：花期 7 ~ 9 月，果期 8 ~ 10 月。

分布：亚洲东部至南部。我国西北南部至长江以南各地。南麂各岛屿常见。

生境：山谷湿地、草坡，山坡灌丛。

Description: Herbs perennial, ligneous at base. Stems erect, striate, much branched. Leaf blade ovate, or elliptic; apex shortly acuminate; base truncate or broadly cordate, margin entire, both surfaces glabrous or hispid, lower leaves petiolate, usually auriculate at base, upper leaves subsessile; ocrea tubular; glabrous, much veined. Inflorescence terminal or axillary, capitate, usually several capitula aggregated and panicle-like; peduncle densely glandular hairy. Achenes black, opaque, broadly ovoid.

Flower and Fruit: Fl. Jul. -Sep. , fr. Aug. -Oct.

Distribution: Eastern to southern Asia. Southern Northwest China to the south areas of Yangtze River. Throughout Nanji Islands, commonly.

Habitat: Wet valleys, grassy slopes, thickets in valleys, mountain slopes.

A283. 蓼科
Polygonaceae

稀花蓼
Polygonum dissitiflorum
萹蓄属 *Polygonum*

形态特征：一年生草本。茎直立或下部平卧，分枝，具稀疏的倒生皮刺，常疏生星状毛。叶互生，卵状椭圆形，具短缘毛，上面疏生星状毛及刺毛，下面疏生星状毛，沿中脉具倒生皮刺；叶柄具星状毛及倒生皮刺。花序圆锥状，稀疏顶生或腋生；花序梗密被紫红色腺毛；苞片漏斗状，包围花序轴，具缘毛。瘦果近球形，暗褐色。

花果期：花期 6 ~ 8 月，果期 7 ~ 9 月。

分布：朝鲜、俄罗斯远东地区。分布于我国南北各地。南麂主岛（百亩山、后隆、三盘尾），偶见。

生境：山谷、丘陵草地，溪岸。

Description: Herbs annual. Stems erect or prostrate at base, branched, sparsely retrorsely prickly, usually with few stellate hairs. Leaves alternate, leaf blade ovate-elliptic, abaxially sparsely stellate hairy and retrorsely prickly along midvein, adaxially sparsely stellate hairy and bristly, base cordate or hastate; petiole stellate hairy, often retrorsely prickly. Inflorescence terminal or axillary, paniculate; peduncles reddish purple glandular hairy; bracts funnel-shaped, around rachis; margin strongly ciliate. Achenes globose, dark brown.

Flower and Fruit: Fl. Jun. -Aug. , fr. Jul. -Sep.

Distribution: Korea and Far East Russia. Throughout the north and south areas of China. Main island (Baimushan, Houlong, Sanpanwei), occasionally.

Habitat: Valleys, hilly grasslands, stream banks.

A283. 蓼科
Polygonaceae

蚕茧草
Polygonum japonicum
萹蓄属 *Polygonum*

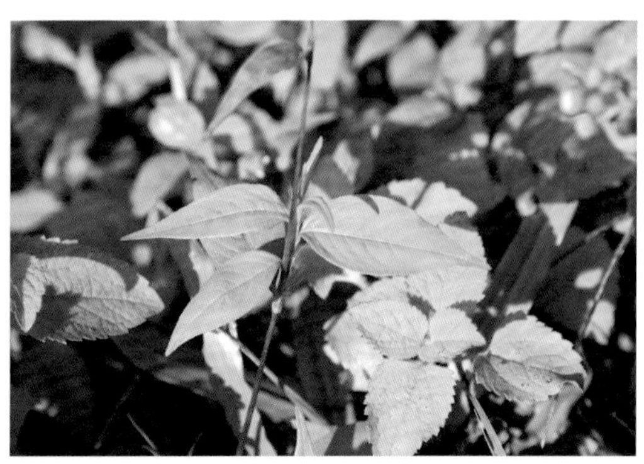

形态特征：多年生草本。根状茎横走。茎直立，无毛，有时疏生短硬毛。叶柄短或近无；叶片披针形，薄革质，具微小点，两面具稀疏贴伏的短硬毛，基部楔形，边缘全缘，具短硬毛，先端渐尖；托叶鞘管状，膜质，具硬伏毛，先端截形。花序顶生，穗状；苞片绿色，漏斗状，具缘毛；雌雄异株，雌花花被白色或者带粉红色，5深裂。瘦果黑色，具3棱或双凸镜状。

花果期：花期8～10月，果期9～11月。

分布：日本，朝鲜。广布我国黄河流域以南地区。南麂主岛（百亩山、国姓岙、后隆、门屿尾、三盘尾），常见。

生境：沼泽地，沟渠，溪边，河岸。

Description: Herbs perennial. Rhizomes horizontal. Stems erect, glabrous, sometimes sparsely hispidulous. Petiole short or nearly absent; leaf blade lanceolate, thinly leathery, densely minutely punctate, both surfaces sparsely appressed hispidulous, base cuneate, margin entire, hispidulous, apex acuminate; ocrea tubular, membranous, appressed hirsute, apex truncate. Inflorescences terminal, spicate; bracts green, funnel-shaped, ciliate; flowers diecious, female flowers perianth white or pinkish, 5-parted. Achenes black, trigonous or biconvex.

Flower and Fruit: Fl. Aug. -Oct. , fr. Sep. -Nov.

Distribution: Japan, Korea. Widely distributed to south of the Yellow River Basin. Main island (Baimushan, Guoxingao, Houlong, Menyuwei, Sanpanwei), commonly.

Habitat: Marshy areas, ditches, streamsides, riverbanks.

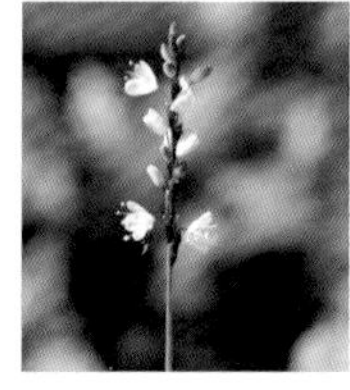

A283. 蓼科
Polygonaceae

绵毛酸模叶蓼
Polygonum lapathifolium var. *salicifolium*
扁蓄属 *Polygonum*

形态特征：一年生草本。茎直立，具分枝，无毛，节部膨大。叶披针形或宽披针形，下面密生白色绵毛，上面具黑点；叶柄短，具短硬伏毛；托叶鞘筒状，膜质，淡褐色，顶端截形。圆锥花序顶生或腋生，穗状；花序梗具腺体，苞片漏斗状，边缘具稀疏短毛；花淡红色或白色，4～5 深裂。瘦果包于宿存花被内，宽卵形，黑褐色。

花果期：花期 5～7 月，果期 6～10 月。

分布：东亚和东南亚。分布于我国南北各地。南麂主岛各处常见。

生境：路边，水沟边，田野边，水边。

Description: Herbs annual. Stems erect, branched, glabrous, swollen at nodes. Leaf blade lanceolate or broadly lanceolate, densely lanose abaxially, with large blackish spot adaxially; petiole short, appressed hispidulous; ocrea brownish, tubular, membranous, hazel, glabrous, apex truncate. Inflorescence terminal or axillary, spicate; peduncles glandular; bracts funnel-shaped, margin sparsely shortly ciliate, perianth pink or white, 4(or 5)-parted. Achenes included in persistent perianth, broadly ovoid, black-brown.

Flower and Fruit: Fl. May-Jul. , fr. Jun. -Oct.

Distribution: East and Southeast Asia. Throughout the north and south areas of China. Main island of Nanji, commonly.

Habitat: Roadsides, along ditches, field margins, watersides.

A283. 蓼科 Polygonaceae

长鬃蓼
Polygonum longisetum
萹蓄属 *Polygonum*

形态特征：一年生草本。茎直立、上升或基部近平卧，自基部分枝，节部稍膨大。叶互生，披针形或宽披针形，顶端急尖或狭尖，基部楔形；上面近无毛，下面沿叶脉具短伏毛，具缘毛；托叶鞘筒状，疏生柔毛。总状花序呈穗状，顶生或腋生，下部间断；苞片漏斗状，无毛，边缘具长缘毛；花梗与苞片近等长。瘦果宽卵形，黑色，有光泽。

花果期：花期 5 ~ 6 月，果期 6 ~ 8 月。

分布：东亚和东南亚。分布于我国南北各地。南麂主岛（关帝岙、门屿尾、三盘尾），偶见。

生境：潮湿的山谷，沿河岸，沟渠、水边背阴处。

Description: Herbs annual. Stems erect, ascending or prostrate at base, branched from base, swollen at nodes. Leaves alternate, leaf blade lanceolate or broadly lanceolate, apex acute or acuminate, base cuneate to rounded; nearly glabrous above, abaxially appressed hispidulous along veins below, ciliolate; ocrea tubular, sparsely pubescent. Inflorescence terminal or axillary, spicate, erect, interrupted; bracts funnel-shaped, glabrous, margin long ciliate; pedicels equaling bracts. Achenes broadly ovoid, trigonous, black, shiny.

Flower and Fruit: Fl. May-Jun. , fr. Jun. -Aug.

Distribution: East and Southeast Asia. Throughout the north and south areas of China. Main island (Guandiao, Menyuwei, Sanpanwei), occasionally.

Habitat: Moist valleys, along stream banks, shaded places along ditches, water sides.

A283. 蓼科 Polygonaceae

尼泊尔蓼
Polygonum nepalense
蓝蓄属 *Polygonum*

形态特征：一年生草本。茎匍匐或上升，基部多分枝。下部叶具叶柄，叶柄具翅；叶片卵形或三角状卵形，两面稀疏具刚毛的或无毛，疏生黄色透明腺点；托叶鞘管状，膜质，顶端斜截形。花序头状，顶生或腋生，基部常具 1 叶状总苞片；花被通常 4 裂，紫红色或白色，长圆形，先端钝。瘦果在宿存花被内藏，黑色，不透明，宽卵形，两面凸，密生洼点。

花果期：花期 5 ~ 8 月，果期 6 ~ 10 月。

分布：东亚和东南亚。除新疆外广布我国南北各地。南麂主岛（百亩山、国姓岙、门屿尾），偶见。

生境：潮湿的山谷，山坡。

Description: Herbs annual. Stems decumbent or ascending, much branched at base. Lower leaves petiolate, petiole winged; leaf blade ovate or triangular-ovate, both surfaces sparsely setose or glabrous, sparsely yellow pellucid glandular punctate; ocrea brownish, tubular, membranous, apex obliquely truncate. Inflorescence capitate, terminal or axillary, included by an involucral leaf; perianth purplish red or white, usually 4-parted, tepals oblong, apex obtuse. Achenes included in persistent perianth, black, opaque, broadly ovoid, biconvex, densely pitted.

Flower and Fruit: Fl. May-Aug. , fr. Jun. -Oct.

Distribution: East and Southeast Asia. Throughout the north and south areas of China, except Xingjiang. Main island (Baimushan, Guoxingao, Menyuwei), occasionally.

Habitat: Mountain slopes, moist valleys.

A283. 蓼科
Polygonaceae

杠板归
Polygonum perfoliatum
萹蓄属 *Polygonum*

形态特征：一年生草本。茎攀援，红棕色，多分枝，具棱，沿棱倒生皮刺。叶柄具倒生皮刺；叶盾三角形，下面沿叶脉疏生皮刺，上面无毛，基部截形或微心形，顶端微尖。穗状花序顶生或腋生；苞片卵圆形；花被白色或淡红色，5深裂，花被片椭圆形，果时增大，深蓝色，肉质。瘦果包于宿存花被内，黑色，有光泽，球形。

花果期：花期6～8月，果期7～10月。

分布：亚洲东部。分布于我国南北各地。南麂主岛各处常见。

生境：田间和路边，潮湿的山谷。

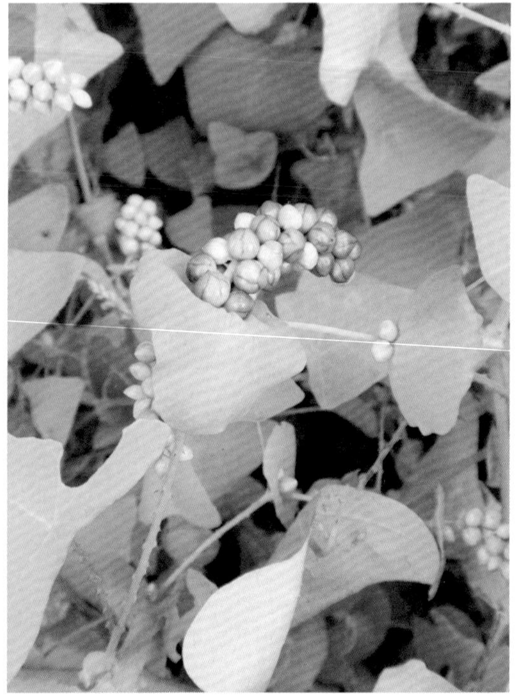

Description: Herbs annual. Stems trailing, red-brown, branched, angulate, with retrorse prickles along angles. Petiole sparsely retrorsely prickly; leaf blade triangular-peltate, abaxially usually sparsely retrorsely prickly along veins, adaxially glabrous, base truncate or subcordate, apex subacute. Inflorescence terminal or axillary, spicate; bracts ovate-orbicular; perianth white or pinkish, 5-parted, tepals elliptic, in fruit dark blue, accrescent, fleshy. Achenes included in persistent perianth, black, shiny, globose.

Flower and Fruit: Fl. Jun. -Aug. , fr. Jul. -Oct.

Distribution: Eastern areas of Asia. Throughout the north and south areas of China. Main island of Nanji, commonly.

Habitat: Near fields and roads, wet valleys.

A283. 蓼科 Polygonaceae

酸模
Rumex acetosa
酸模属 *Rumex*

形态特征：多年生草本，根为须根。茎直立，具深沟槽，无毛，常不分枝。基生叶和茎下部叶箭形，顶端急尖或圆钝，基部裂片急尖，全缘或微波状；上部叶较小，短柄或无柄；托叶鞘膜质，易破裂。花单性，雌雄异株；花序狭圆锥状，松散顶生，分枝稀疏。瘦果椭圆形，具3锐棱，黑褐色，有光泽。

花果期：花期5～7月，果期6～8月。

分布：亚洲、欧洲、北美洲。分布于我国南北各地。南麂主岛各处常见。

生境：山坡、林缘、沟边及路旁。

Description: Herbs perennial, fibrous root. Stems erect, grooved, glabrous, usually simple. Basal leaves ovate-lanceolate to lanceolate, base sagittate, apex acute, basal lobes acute at apices, margin entire or undulate; cauline leaves small, petiole short or nearly absent; ocrea membranous, fugacious. Flowers unisexual, dioecian; inflorescence terminal, paniculate, lax; branches simple or with a few secondary branches. Achenes trigonous, blackish brown, shiny.

Flower and Fruit: Fl. May-Jul. , fr. Jun. -Aug.

Distribution: Asia, Europe and North America. Throughout the north and south areas of China. Main island of Nanji, commonly.

Habitat: Mountain slopes, forest margins, moist valleys.

A283. 蓼科
Polygonaceae

羊蹄
Rumex japonicus
酸模属 *Rumex*

形态特征：多年生草本。茎直立，上部分枝，具沟槽，无毛。基生叶长圆形，边缘微波状，下面沿叶脉具小突起；茎上部叶狭长圆形；托叶鞘膜质，易破裂。花两性，多花轮生；花序圆锥状，花梗中下部具关节；外花被片椭圆形，内花被片宽心形，具不整齐的小齿，全部具长卵形小瘤。瘦果宽卵形，具3锐棱，暗褐色，有光泽。

花果期：花期5～6月，果期6～7月。

分布：东亚。分布于我国南北各地。南麂主岛各处常见。

生境：田边路旁、河滩及沟边湿地。

Description: Herbs perennial. Stems erect, branched above, grooved, glabrous. Basal leaves oblong, margin undulate, abaxially minutely papillate along veins; cauline leaves narrowly oblong; ocrea membranous, fugacious. Flowers bisexual, inflorescence paniculate, many flowers whorled, articulate below middle, outer tepals oval; inner tepals valves broadly cordate, margin irregularly denticulate, all valves with narrowly ovate tubercles. Achenes broadly ovoid, sharply trigonous, dark brown, shiny.

Flower and Fruit: Fl. May-Jun. , fr. Jun. -Jul.

Distribution: East Asia. Throughout the north and south areas of China. Main island of Nanji, commonly.

Habitat: Field margins, stream banks, wet valleys.

A295. 石竹科
Caryophyllaceae

无心菜
Arenaria serpyllifolia
无心菜属 *Arenaria*

形态特征：一年生或二年生草本。茎丛生，直立或铺散，密被白色长柔毛。叶无柄，叶片卵形，两面无毛或疏生长柔毛，背面3脉，基部渐狭，边缘具缘毛，先端锐尖。聚伞花序具多花；苞片卵形，草质，密被长柔毛，萼片5，披针形，背面长柔毛；花瓣5，白色，倒卵形。蒴果卵形，与宿存萼等长。种子淡褐色，肾形，小。

花果期：花期6～8月，果期8～9月。

分布：世界广布。分布于全国各地。南麂主岛各处常见。

生境：山坡草地，沙地或多石的荒地，田野。

Description: Herbs annual or biennial. Stems caespitose, erect or diffuse, densely white villous. Leaves sessile; leaf blade ovate, both surfaces glabrous or sparsely villous, 3-veined abaxially, base attenuate, margin ciliate, apex acute. Cymes many flowered; bracts ovate, herbaceous, usually densely villous,sepals 5, lanceolate, villous abaxially; petals 5, white, obovate. Capsule ovoid, equaling persistent sepals. Seeds pale brown, reniform, small.

Flower and Fruit: Fl. Jun. -Aug. , fr. Aug. -Sep.

Distribution: Cosmopolitan. Throughout China. Main island of Nanji, commonly.

Habitat: Mountain grassland slopes, sandy or stony barrens, fields.

A295. 石竹科
Caryophyllaceae

球序卷耳
Cerastium glomeratum
卷耳属 *Cerastium*

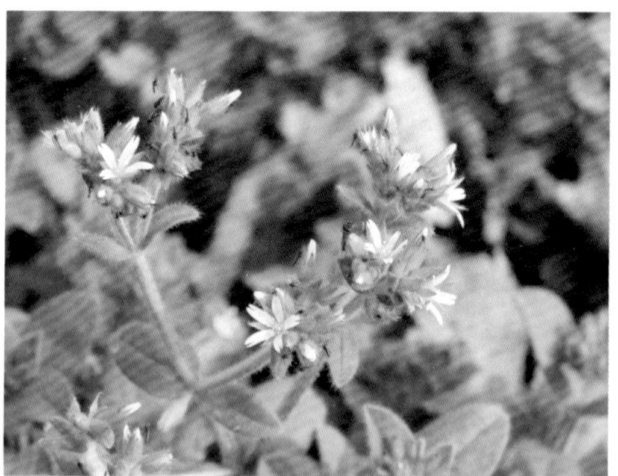

形态特征：一年生草本，高 10 ~ 20 cm。茎直立，单生或丛生，密被长柔毛，上部混生腺毛。茎下部叶匙形，顶端钝，基部渐狭成柄状；上部叶倒卵状椭圆形，顶端急尖，基部渐狭成短柄状，中脉明显。聚伞花序簇生或呈头状；花序轴密被腺柔毛；花瓣白色，线状长圆形，与萼片近等长或微长。蒴果长圆柱形，顶端 10 齿裂。种子扁三角形，具疣状凸起。

花果期：花期 4 月，果期 5 月。

分布：世界广布。我国南北各地广布。南麂主岛各处常见。

生境：林缘，山坡草地或河边沙土。

Description: Herbs annual, 10-20 cm tall. Stems erect, solitary or cespitose, densely villous, distally mixed glandular hairy. Lower leaves spatulate, apex obtuse, base attenuate petiolate; upper leaves obovate-elliptic, apex acute, base attenuate into short stalked, midvein conspicuous. Cymes clustered or capitate; rachis densely glandular pilose; petals white, linear-oblong, subequal or slightly longer than sepals. Capsule long-terete, apical 10 teeth-lobed. Seeds oblate-triangular, with verrucous protuberances.

Flower and Fruit: Fl. Apr. , fr. May.

Distribution: Cosmopolitan. Throughout the north and south areas of China. Main island of Nanji, commonly.

Habitat: Forest margins, mountain slope grasslands, sandy riversides.

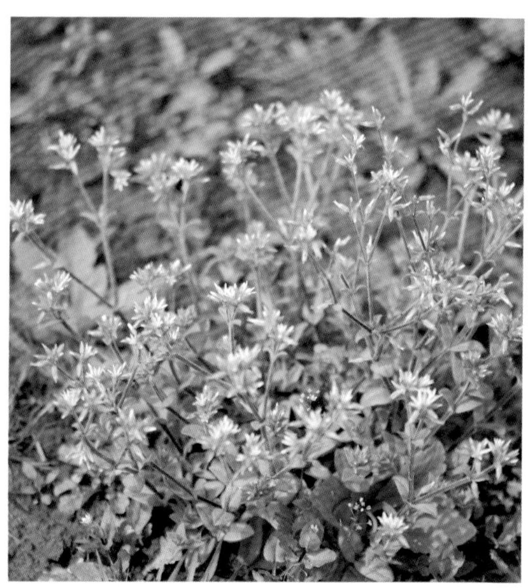

A295. 石竹科
Caryophyllaceae

石竹
Dianthus chinensis
石竹属 *Dianthus*

形态特征：多年生草本，高 30 ~ 50 cm，无毛。茎松散丛生，直立，上部分枝。叶线状披针形，全缘或有细小齿，先端渐尖。花单生的或数花集成聚伞花序；苞片 4，具缘毛，花萼筒状，有条纹，花瓣亮红色，紫红色，粉红色或白色，倒卵状三角形，喉部有斑纹，疏生髯毛，顶缘不整齐齿裂。蒴果圆筒形，包于宿存萼内，顶端 4 裂。

花果期：花期 5 ~ 6 月，果期 7 ~ 9 月。

分布：亚洲、欧洲。我国长江以北地区广泛分布，南方有归化。南麂主岛（关帝岙、门屿尾、三盘尾）、大檑山屿，常见。

生境：林缘沙地，森林草地，灌丛，山谷等。

Description: Herbs perennial, 30-50 cm tall, glabrous. Stems laxly caespitose, erect, distally branched. Leaves linear-lanceolate, margin entire or denticulate, apex acuminate. Flowers solitary or several in cymes; bracts 4, margin ciliate, calyx cylindric, striate, petals limb bright red, purple-red, pink, or white, obovate-triangular, throat spotted and laxly bearded, apex irregularly toothed. Capsule cylindric, surrounded by calyx, apex 4-toothed.

Flower and Fruit: Fl. May-Jun. , fr. Jul. -Sep.

Distribution: Asia, Europe. Widely distributed in the north of Yangtze River, naturalized in southern China. Main island (Guandiao, Menyuwei, Sanpanwei), commonly.

Habitat: Sandy forest margins, forest grasslands, scrubs, valleys, etc.

A295. 石竹科
Caryophyllaceae

瞿麦
Dianthus superbus
石竹属 ***Dianthus***

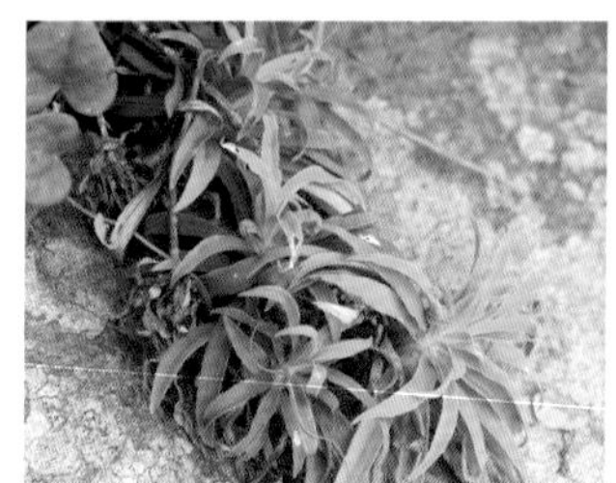

形态特征：多年生草本。茎直立，丛生，上部分枝。叶片线状披针形，中脉特明显，基部合生成鞘状。花1～2枚顶生，有时腋生；苞片倒卵形，顶端长尖，花萼圆筒形，常染紫红色晕，萼齿披针形，花瓣宽倒卵形，边缘缝裂至中部或中部以上，淡红色或带紫色，稀白色，喉部具丝毛状鳞片。蒴果圆筒形，顶端4裂。

花果期：花期6～9月，果期8～10月。

分布：东亚、欧洲。我国大部分地区有分布。南麂主岛（关帝岙、门屿尾、三盘尾）、大檑山屿，常见。

生境：丘陵，林窗和林缘，山坡，草地，山谷溪流及河岸。

Description: Herbs perennial. Stems erect, cespitose, branched distally. Leaf blade linear-lanceolate, midvein especially prominent, base connate into sheaths. Flowers 1-2 terminal, sometimes axillary; bracts obovate, apex long pointed, calyx cylindric, often purple-red, calyx teeth lanceolate, petals broadly obovate, margin cleft to middle or above, reddish or purplish, sparse white, throat with trichomes. Capsule cylindric, tip 4-lobed.

Flower and Fruit: Fl. Jun. -Sep. , fr. Aug. -Oct.

Distribution: East Asia, Europe. Most areas of China. Main island (Guandiao, Menyuwei, Sanpanwei), commonly.

Habitatat: Wooded hills, forest openings and margins, grassy hillsides, meadows, mountain valley streams, river banks.

A295. 石竹科
Caryophyllaceae

漆姑草
Sagina japonica
漆姑草属 *Sagina*

形态特征：一年生小草本，高 5 ~ 20 cm。茎丛生，稍铺散，上部被稀疏腺柔毛。叶对生，线形，顶端急尖。花单生枝端；花梗被稀疏短柔毛；萼片卵状椭圆形，外面疏生短腺柔毛，边缘膜质；花瓣白色，狭卵形，顶端圆钝，全缘。蒴果卵圆形，微长于宿存萼，5 瓣裂。

花果期：花期 4 ~ 5 月，果期 5 ~ 6 月。

分布：亚洲和欧洲。我国大部分地区有分布。南麂主岛各处常见。

生境：河岸沙质地、撂荒地、路旁草地。

Description: Small Herbs annual, 5-20 cm tall. Stems cespitose, slightly diffuse, distally sparsely glandular pilose. Leaves opposite, linear, apex acute. Flowers solitary; pedicel sparsely pubescent; sepals ovate-elliptic, outside sparsely pubescent, margin membranous; petals white, narrowly ovate, apex rounded-obtuse, margin entire. Capsule ovoid, slightly longer than persistent calyx, 5-valved.

Flower and Fruit: Fl. Apr. -May, fr. May-Jun.

Distribution: Asia and Europe. Most areas of China. Main island of Nanji, commonly.

Habitat: Riparian sandy, abandoned, roadside grasslands

A295. 石竹科
Caryophyllaceae

女娄菜
Silene aprica
蝇子草属 *Silene*

形态特征：一年生或二年生草本，全株密被灰色短柔毛。茎直立，单生或数个。基生叶倒披针形或狭匙形，基部渐狭成长柄状；茎生叶倒披针形、披针形或线状披针形。圆锥花序；花萼卵状钟形，萼齿三角状披针形；花瓣白色或淡红色，爪倒披针形，具缘毛，瓣片倒卵形，2 裂；副花冠片舌状。蒴果卵形。种子球状肾形，灰棕色。

花果期：花期 5 ~ 7 月，果期 6 ~ 8 月。

分布：东亚。我国大部分地区有分布。南麂主岛各处常见。

生境：山坡草地、灌丛中、林下、河岸或田埂。

Description: Herbs annual or biennial, densely gray pubescent throughout. Stems erect, solitary or several. Basal leaves oblanceolate or narrowly spatulate, base attenuate into long stalked; cauline leaves oblanceolate, lanceolate, or linear-lanceolate. Panicle; calyx ovate-campanulate, calyx teeth triangular-lanceolate; petals white or pale red, claws oblanceolate, ciliate, limb obovate, 2-lobed; corona segments ligulate. Capsule ovoid. Seeds orbicular-reniform, gray-brown.

Flower and Fruit: Fl. May-Jul. , fr. Jun. -Aug.

Distribution: East Asia. East. Most areas of China. Main island of Nanji, commonly.

Habitat: Hillside grassland, irrigation, subforests, river bank or ridges

A295. 石竹科
Caryophyllaceae

蝇子草（西欧蝇子草）
Silene gallica
蝇子草属 *Silene*

形态特征：一年生草本，全株被柔毛。茎单生，直立或上升，被短柔毛和腺毛。叶片长圆状匙形或披针形，顶端圆或钝，两面被柔毛和腺毛。总状花序单一；苞片披针形，草质；花萼卵形，被稀疏长柔毛和腺毛，萼齿线状披针形，顶端急尖，被腺毛；花瓣淡红色至白色，爪倒披针形，无毛，无耳，瓣片露出花萼，卵形或倒卵形，全缘。蒴果卵形。

花果期：花期 5 ~ 6 月，果期 6 ~ 7 月。

分布：原产西欧。我国各地有栽培，偶见逸生。南麂主岛（关帝岙、门屿尾、三盘尾）、大檑山屿，常见。

生境：干扰生境，路边、田野和草地。

Description: Herbs annual, pilose throughout. Stems solitary, erect or ascending, pubescent and glandular hairy. Leaf blade oblong-spatulate or lanceolate, apex rounded or obtuse, pilose and glandular hairs on both sides. Single raceme; bracts lanceolate, grassy; calyx ovate, sparsely villous and glandular hairy, calyx teeth linear-lanceolate, apex acute, glandular hairy; petals pale red to white, claws oblanceolate, glabrous, earless, petal segments exposing calyx, ovate or obovate, entire. Capsule ovate.

Flower and Fruit: Fl. May-Jun. , fr. Jun. -Jul.

Distribution: Native to Western Europe. Cultivated in most areas of China, occasionally escaped as wild. Main island (Guandiao, Menyuwei, Sanpanwei), commonly.

Habitat: Disturbed places, roadsides, fields and meadows.

A295. 石竹科
Caryophyllaceae

繁缕
Stellaria media
繁缕属 ***Stellaria***

形态特征：一年生或二年生草本，高 10 ～ 30 cm。茎俯仰或上升，基部分枝，常带淡紫红色。叶片宽卵形或卵形，顶端渐尖或急尖，基部渐狭或近心形，全缘。疏聚伞花序顶生；萼片 5，卵状披针形，外面被短腺毛；花瓣白色，长椭圆形，深 2 裂达基部，裂片近线形。蒴果卵形，顶端 6 裂，具多数种子。

花果期：花期 6 ～ 7 月，果期 7 ～ 8 月。

分布：亚洲东部，欧洲。我国大部分地区有分布。南麂主岛（关帝岙、三盘尾）、大檑山屿，常见。

生境：田野。

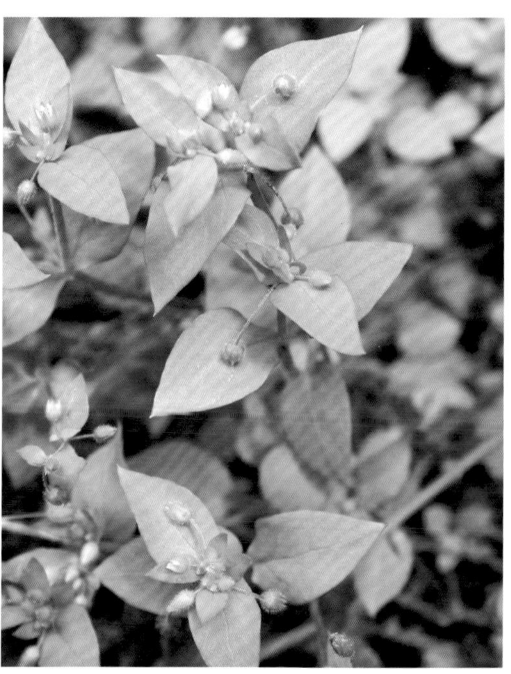

Description:Herbs annual or biennial, 10-30 cm tall. Stem pitching or ascending, base branching, often lavender red. Leaf blade broadly ovate or ovate, apex acuminate or acute, base tapered or subcordate, entire. Cymes terminal; sepals 5, ovate-lanceolate, outside covered with short glandular hairs; petals white, oblong, deeply 2-lobed to base, lobes sublinear. Capsule ovate, slightly longer than persistent calyx, apical 6-lobed, with many seeds.

Flower and Fruit: Fl. Jun. -Jul. , fr. Jul. -Aug.

Distribution: East Asia, Europe. Most areas of China. Main island (Guandiao, Sanpanwei), commonly.

Habitat: Fields.

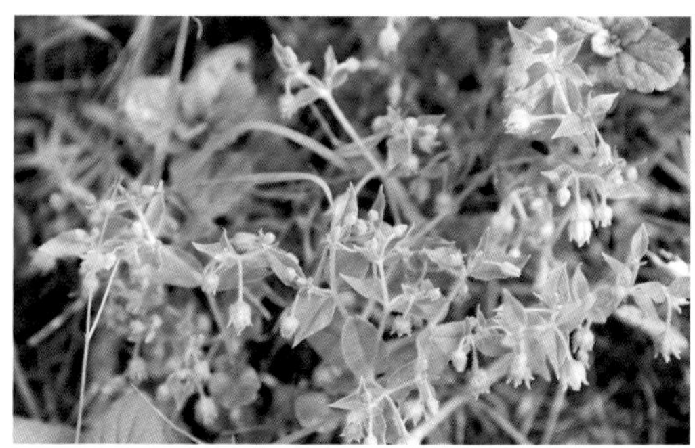

A297. 苋科
Amaranthaceae

牛膝
Achyranthes bidentata
牛膝属 *Achyranthes*

形态特征：多年生草本，高 70 ～ 120 cm。根圆柱形，茎有棱角或四方形；分枝对生。叶片椭圆形或椭圆披针形，两面有贴生或开展柔毛；叶柄具柔毛。穗状花序顶生及腋生，花后反折，总花梗具白色柔毛；苞片宽卵形；小苞片刺状，基部两侧各有 1 卵形膜质小裂片；花被片披针形，有 1 中脉。胞果矩圆形，光滑。

花果期：花期 7 ～ 9 月，果期 9 ～ 11 月。

分布：东亚至东南亚。除东北外广布我国。南麂主岛各处常见。

生境：山边、路旁。

Description: Herbs perennial, 70-120 cm tall. Roots terete. Stems angular or quadrate; branches opposite. Leaf blade elliptic or elliptic-lanceolate, both surfaces adnate or spreading pilose; petiole pilose. Spikes terminal and axillary, reflexed after anthesis; total pedicel with white pilose; bracts broadly ovate; bracteoles spiny, base with 1 ovate membranous lobule on each side; tepals lanceolate, with 1 midvein. Utricles roundish, smooth.

Flower and Fruit: Fl. Jul. -Sep. , fr. Sep. -Nov.

Distribution: East Asia to Southeast Asia. Throughout China except Northeast China. Main island of Nanji, commonly.

Habitat: Hillsides, roadsides.

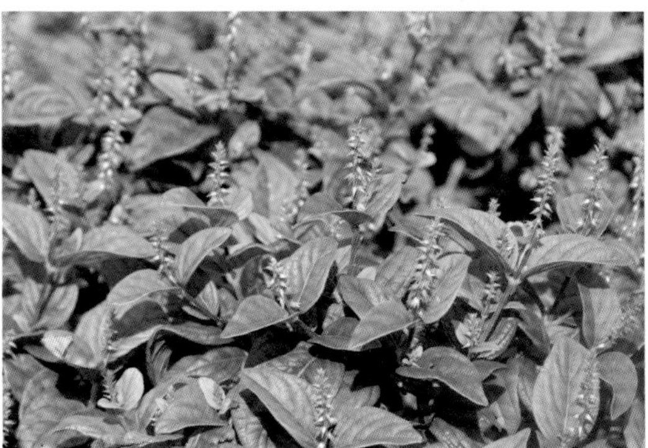

A297. 苋科
Amaranthaceae

喜旱莲子草（空心莲子草、水花生）
Alternanthera philoxeroides
莲子草属 *Alternanthera*

形态特征：多年生草本。茎匍匐，管状，上部上升，长达 120 cm，具分枝，幼茎及叶腋被白或锈色柔毛，老时无毛。叶长圆形或倒卵状披针形，先端尖或圆钝，具短尖，基部渐窄，全缘，下面具颗粒状突起。头状花序具花序梗，单生叶腋；苞片白色，卵形，具 1 脉，花被片长圆形，白色，光亮，无毛。

花果期：花期 5 ~ 10 月。

分布：原产巴西，我国引种栽培后逸为野生。南麂主岛各处常见。

生境：耕地、荒地。

Description: Herbs perennial. Stem ascending from a creeping base, tubular, to 120 cm, branched, young stem and leaf axil white hairy, old ones glabrous. Leaf blade oblong, or ovate-lanceolate, apex acute or obtuse, mucronate, base attenuate, margin entire, granular protuberance below. Heads with a peduncle, solitary at leaf axil; bracts white, ovate, 1-veined, tepals oblong, white, shiny, glabrous.

Flower and Fruit: Fl. May-Oct.

Distribution: Native to Brazil, naturalized in China after introduction and cultivation. Main island of Nanji, commonly.

Habitat: Cultivated, waste places.

A297. 苋科
Amaranthaceae

凹头苋
***Amaranthus blitum*（*Amaranthus lividus*）**
苋属 *Amaranthus*

形态特征：一年生草本。茎伏卧而上升，基部分枝，淡绿色或紫红色。叶片卵形或菱状卵形，顶端凹缺，具芒尖，基部宽楔形，全缘或微波状。花成束腋生，于茎端和枝端成直立穗状花序或圆锥花序；花被片矩圆形或披针形，背部有 1 隆起中脉。胞果扁卵形，微皱缩而近平滑。种子环形，边缘具环状边。

花果期：花期 7 ～ 8 月，果期 8 ～ 9 月。

分布：东亚、欧洲、非洲北部及南美。我国除内蒙古、宁夏、青海、西藏外，均有分布。南麂主岛（国姓岙、后隆、门屿尾、兴岙），偶见。

生境：田野，荒地。

Description: Herbs annual. Stems procumbent and ascending, branched from base, pale green or purplish red. Leaf blade ovate or ovate-rhombic, apex notched, with a mucro, base broadly cuneate, entire or slightly undulate. Flower clusters axillary, those of terminal clusters erect spikes or panicle; perianth segments rectangular-orbicular or lanceolate, with a raised midvein at back. Utricles compressed-ovoid, slightly shrunken and nearly smooth. Seeds annular, margin with annular margin.

Flower and Fruit: Fl. Jul. -Aug. , fr. Aug. -Sep.

Distribution: Native to East Asia, Europe, Northern Africa, and South America. Widely distributed in China, except Inner Mongolia, Ningxia, Qinghai, Xizang. Main island (Guoxingao, Houlong, Menyuwei, Xingao), occasionally.

Habitat: Fields, waste places.

A297. 苋科
Amaranthaceae

绿穗苋
Amaranthus hybridus
苋属 *Amaranthus*

形态特征：一年生草本，高 30 ～ 50 cm。茎直立，分枝，上部近弯曲，有开展柔毛。叶片卵形或菱状卵形，顶端急尖或微凹，基部楔形，边缘波状或有不明显锯齿，微粗糙，叶柄有柔毛。圆锥花序顶生，细长，有分枝；苞片及小苞片钻状披针形，花被片矩圆状披针形。胞果卵形，环状横裂。

花果期：花期 7 ～ 8 月，果期 9 ～ 10 月。

分布：我国大部分区域均有分布。南麂主岛(后隆、兴岙)，偶见。

生境：农场，荒地，山坡。

Description:Herbs annual, 30-50 cm tall. Stem erect, branched, distally subcurved, with spreading pilose. Leaf blade ovate or rhomboid ovate, apex acute or slightly concave, base cuneate, margin undulate or inconspicuously serrate, petiole pilose. Panicle terminal, slender, ascending slightly curved, branched, from spikes; bracts and bracteoles subulate lanceolate; tepals rectangular oblong-lanceolate. Utricle ovate, annular transverse lobed.

Flower and Fruit: Fl. Jul. -Aug. , fr. Sep. -Oct.

Distribution: Throughout China. Main island (Houlong, Xingao), occasionally.

Habitat: Farms, waste places, hillsides.

A297. 苋科
Amaranthaceae

刺苋
Amaranthus spinosus
苋属 *Amaranthus*

形态特征：一年生草本。茎直立，分枝，上部近弯曲，有开展柔毛。叶片卵形或菱状卵形，顶端急尖或微凹，基部楔形，边缘波状或有不明显锯齿，微粗糙，上面近无毛，下面疏生柔毛；叶柄有柔毛。圆锥花序顶生，细长，上升稍弯曲，有分枝，由穗状花序而成；苞片及小苞片钻状披针形，中脉坚硬，绿色；花被片矩圆状披针形，顶端锐尖，具凸尖。胞果卵形，环状横裂，超出宿存花被片。种子近球形，黑色。

花果期：花期 7 ~ 8 月，果期 9 ~ 10 月。

分布：广布全球暖温带和热带地区。我国广布。南麂主岛各处常见。

生境：荒地、花园。

Description: Herbs annual, Stem erect, branched, distally subcurved, with spreading pilose. Leaf blade ovate or rhomboid ovate, apex acute or slightly concave, base cuneate, margin undulate or inconspicuously serrate, slightly scabrous, subglabrous above, sparsely pilose below; petiole pilose. Panicle terminal, slender, ascending slightly curved, branched, composed of spikes; bracts and bracteoles subulate lanceolate, midvein hard, green; tepals rectangular oblong-lanceolate, apex acute, convex. Utricle ovate, annular transverse lobed, beyond persistent tepals. Seeds subglobose, black.

Flower and Fruit: Fl. Jul. -Aug. , fr. Sep. -Oct.

Distribution: Cosmopolitan in warm-temperate and tropical regions. Throughout China. Main island of Nanji, commonly.

Habitat: Waste places, gardens.

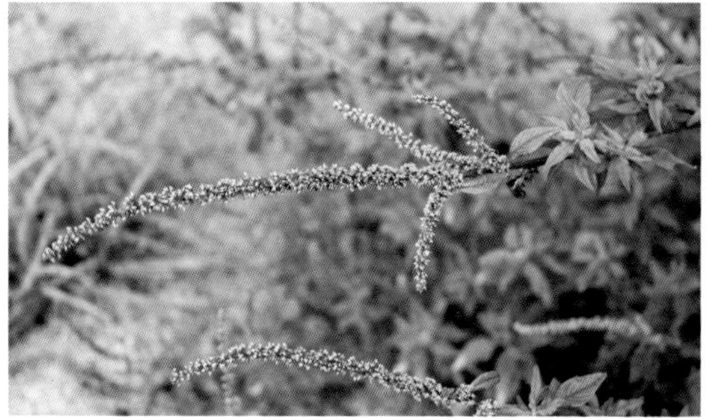

A297. 苋科
Amaranthaceae

皱果苋
Amaranthus viridis
苋属 *Amaranthus*

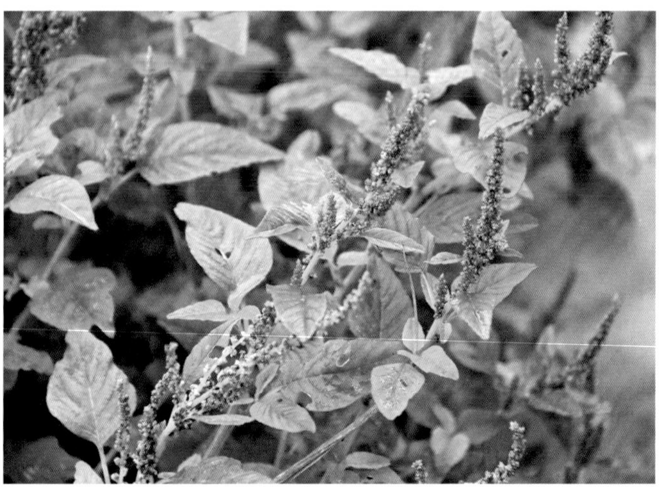

形态特征：一年生草本，高 40 ～ 80 cm。茎直立，不显明棱，稍分枝，绿色或带紫色。叶片卵形、卵状椭圆形，顶端凹，基部宽楔形或近截形，全缘或微呈波状缘。圆锥花序顶生，有分枝，由圆柱形穗状花序形成，直立；苞片及小苞片披针形，顶端具凸尖；花瓣矩圆形或宽倒披针形，顶端急尖。胞果扁球形，绿色，不裂。

花果期：花期 6 ～ 8 月，果期 8 ～ 10 月。

分布：泛热带地区。除西北和西藏外我国广布。南麂主岛各处常见。

生境：田野，荒地。

Description: Herbs annual, 40-80 cm tall. Stem erect, unangular, slightly branched, green or purplish. Leaf blade ovate, ovate-elliptic, apical apiculate concave or concave, base broadly cuneate or subtruncate, margin entire or slightly undulate. Panicle terminal, branched, composed of cylindrical spikes, erect; bracts and bracteoles lanceolate, apex convex; tepals oblong or broadly oblanceolate, apex acute. Utricle oblate, green, indehiscent.

Flower and Fruit: Fl. Jun. -Aug. , fr. Aug. -Oct.

Distribution: Pantropical. Throughout China except for Northwest China and Xizang. Main island of Nanji, commonly.

Habitat: Fields, waste places.

A297. 苋科
Amaranthaceae

青葙
Celosia argentea
青葙属 *Celosia*

形态特征：一年生草本，高 0.3 ~ 1 m，全体无毛。茎直立，有分枝，绿色或红色，具显明条纹。叶片矩圆状披针形，或披针状条形，绿色常带红色，顶端具小芒尖，基部渐狭。花多数，密生，在茎端或枝端成单一、无分枝的塔形穗状花序；苞片及小苞片披针形，白色；花瓣矩圆状披针形，初白色，顶端带红，或全粉红，后白色。胞果卵形，包裹在宿存花被片内。

花果期：花期 5 ~ 8 月，果期 6 ~ 10 月。

分布：东亚至东南亚，热带非洲。分布几遍我国。南鹿主岛（三盘尾），常见。

生境：山坡，田野边缘。

Description: Herbs annual, 0.3 -1 m tall, glabrous. Stem erect, branched, green or red, striate conspicuously. Leaf blade rectangular lanceolate, or lanceolate, green often reddish, apex with small awned apiculate, base tapering. Flowers numerous, dense, at stem or branch ends into a single, unbranched tower or terete spike; bracts and bracteoles lanceolate, white; tepals oblong-lanceolate, white at first with red tip, or all pink, later white. Utricle ovate, enclosed in persistent tepals.

Flower and Fruit: Fl. May-Aug. , fr. Jun. -Oct.

Distribution: East Asia, Southeast Asia, tropical Africa. Almost throughout China. Main island (Sanpanwei), commonly.

Habitat: Hillsides, field margins.

A297. 苋科
Amaranthaceae

狭叶尖头叶藜
Chenopodium acuminatum subsp. *virgatum*
藜属 *Chenopodium*

形态特征：一年生草本。茎直立，具绿色条棱，多分枝。叶较狭小，狭卵形、矩圆形乃至披针形，长显著大于宽。花两性，团伞花序于枝上部排列成穗状或穗状圆锥状花序，花序轴具圆柱状毛束；花被片扁球形，5深裂，有红色或黄色粉粒。胞果顶基扁，圆形或卵形。

花果期：花期6～7月，果期8～9月。

分布：日本，越南东北部。我国东部沿海各省区。南麂主岛各处常见。

生境：湖边、海滨、荒地。

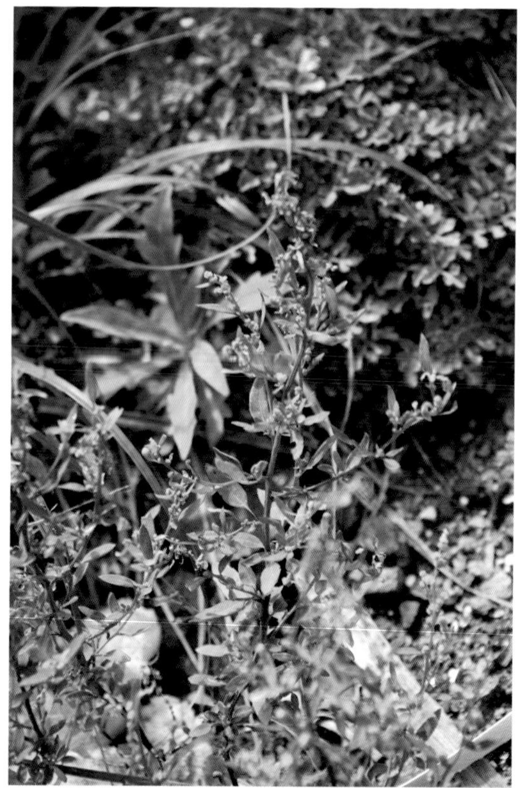

Description: Herbs annual. Stem erect, ribbed and green, much branched. Leaves narrow, narrowly ovate, rectangular or lanceolate, significantly longer than wide. Flowers bisexual, cymes arranged into spikelet or spikelet panicle on the upper part of the branch, inflorescence axis with terete hair fascicles; perianth oblate, 5-lobed, with reddish or yellowish farinose. Utricle apically flat, orbicular or ovate.

Flower and Fruit: Fl. Jun. -Jul. , fr. Aug. -Sep.

Distribution: Japan, Northwest Vietnam. Coastal areas of eastern China. Main island of Nanji, commonly.

Habitat: Lake shores, beaches, wastelands.

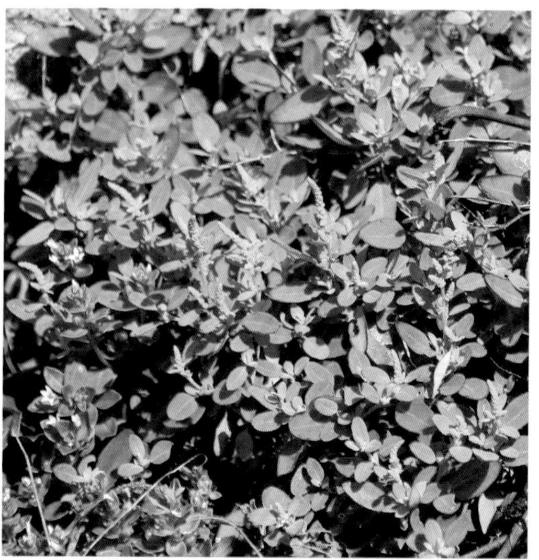

A297. 苋科
Amaranthaceae

灰绿藜
Chenopodium album
藜属 *Chenopodium*

形态特征：一年生草本。茎直立，多分枝，具条棱及绿色或紫红色色条。叶片菱状卵形至宽披针形，边缘具不整齐锯齿。花两性，花簇于枝上部排列成或大或小的穗状圆锥状或圆锥状花序；花被裂片 5，宽卵形至椭圆形，背面具纵隆脊，有粉。果皮与种子贴生。

花果期：花果期 5 ～ 10 月。

分布：东亚和东南亚。我国南北各地广布。南麂主岛各处常见。

生境：田野，荒地和路边。

Description: Herbs annual. Stem erect, much branched, with green or purple-red striate. Leaf blade rhombic-ovate to broadly lanceolate, margin irregularly serrate. Flowers bisexual, Glomerules arranged into large or small panicles or spikelike panicles on upper part of branches; perianth segments 5, broadly ovate to elliptic, abaxially longitudinally keeled, farinose. Pericarp adnate to seed.

Flower and Fruit: Fl. and fr. May-Oct.

Distribution: East and Southeast Asia. Throughout the north and south areas of China. Main island of Nanji, commonly.

Habitat: Fields, waste places, roadsides.

A297. 苋科
Amaranthaceae

土荆芥
Dysphania ambrosioides (*Chenopodium ambrosioides*)
腺毛藜属 *Dysphania*

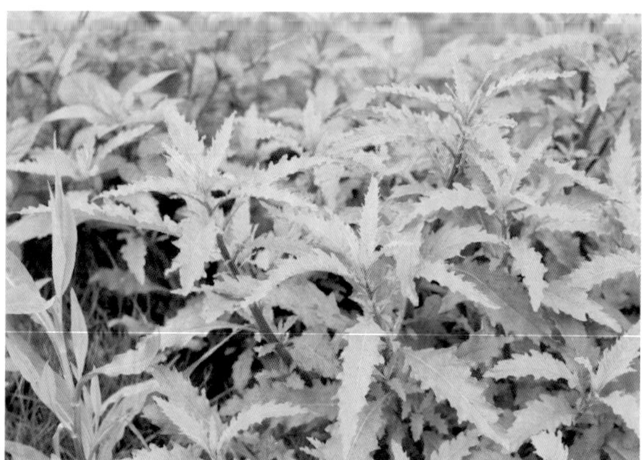

形态特征：一年生或多年生草本。茎直立，多分枝，有色条及钝条棱。叶片矩圆状披针形至披针形，先端急尖或渐尖，边缘具稀疏不整齐的大锯齿，有强烈香味。花两性及雌性，通常 3 ~ 5 个朵簇生于上部叶腋；花被片 5，稀 3，卵形，绿色。胞果扁球形，完全包于花被内。种子横生或斜生，黑色或暗红色，平滑，有光泽。

花果期：花果期长。

分布：广布世界热带及温带地区。在我国北方常作药物栽培。南麂主岛各处常见。

生境：喜生于村旁、路边、河岸等处。

Description: Herbs annual or perennial. Stem erect, much branched, striate, obtusely ribbed. Leaf blade oblong-lanceolate to lanceolate, apex acute or acuminate, margin sparsely and irregularly coarsely serrate, with strong odor. Flowers bisexual and female, usually 3-5 per glomerule borne in upper leaf axils; perianth segments (3 or)5, ovate, green. Utricle enclosed by perianth, depressed globose. Seed horizontal or oblique, black or dark red, smooth, glossy.

Flower and Fruit: Fl. and fl. over a lengthy period.

Distribution: Tropical and temperate regions of the world. Naturalized; often cultivated for medicine in North China.Main island of Nanji, commonly.

Habitat: Village side, road side and river bank.

A297. 苋科
Amaranthaceae

地肤
Kochia scoparia f. *trichophylla*
地肤属 *Kochia*

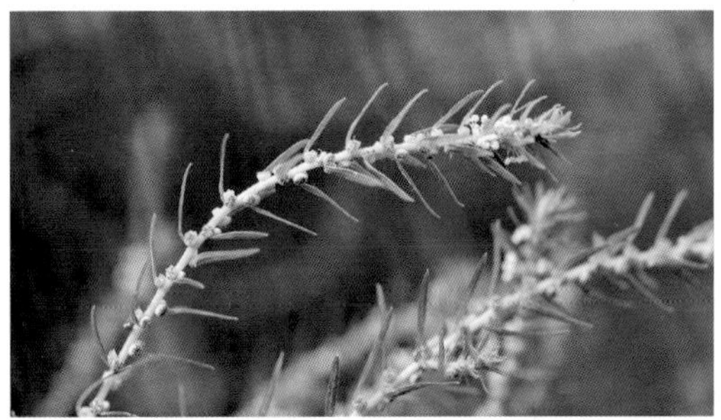

形态特征：一年生草本。茎直立，多分枝，整个植株外形卵球形。叶互生，披针形，具 3 条主脉；茎部叶小，具 1 脉。花常 1 ~ 3 枚簇生于叶腋，构成穗状圆锥花序；花被近球形，淡绿色，裂片三角形。胞果扁球形，果皮膜质，与种子离生。种子黑色，具光泽。

花果期：花期 6 ~ 9 月，果期 7 ~ 10 月。

分布：亚洲、欧洲。分布于我国各地。南麂主岛各处常见。

生境：山谷、河岸、海滩、荒地、田边、路旁。

Description: Herbs annual. Stems erect, much branched, entire plant shaped ovoid. Leaves alternate, lanceolate, with 3 main veins; stem leaves small, with 1 vein. Flowers often 1-3 clustered in leaf axils, forming spikes of panicles; perianth subglobose, greenish, lobes triangular. Utricles oblate, pericarp membranous, free from seeds. Seeds black, lustrous.

Flower and Fruit: Fl. Jun. -Sep. , fr. Jul. -Oct.

Distribution: Europe and Asia. Throughout China. Main island of Nanji, commonly.

Habitat: Valleys, river banks, beaches, wastelands, field margins, roadsides.

A304. 番杏科
Aizoaceae

番杏
Tetragonia tetragonioides
番杏属 *Tetragonia*

形态特征：一年生肉质草本。表皮细胞内有针状结晶体，呈颗粒状凸起。茎粗壮，初直立，后平卧上升，从基部分枝。叶片卵状菱形或卵状三角形，边缘波状。花单生或 2 ～ 3 枚簇生叶腋，近无梗；萼片开展，内面黄绿色。坚果陀螺形，具钝棱，有 4 ～ 5 角，附有宿存花被。

花果期：花果期 8 ～ 10 月。

分布：亚洲南部、大洋洲和南美洲。我国华东和广东、云南等地。南麂主岛（关帝岙、门屿尾、三盘尾）、大檑山屿，常见。

生境：海滨沙地，常栽培。

Description: Annual fleshy herbs. Epidermal cells with needle-like crystals, grainy bumps like. Stem stout, erect when young, then decumbent, branched at base. Leaf blade rhomboid-ovate or deltoid-ovate, margin undulate. Flowers solitary or 2-3 fascicled at axils, subsessile; sepals spreading, yellowish green inside. Fruit turbinate, with blunt ridge, 4-or 5-corniculate, attached with persistent perianth.

Flower and Fruit: Fl. and fr. Aug. -Oct.

Distribution: Southern Asia, Oceania, and South America. East China, Guangdong and Yunnan. Main island (Guandiao, Menyuwei, Sanpanwei), commonly.

Habitat: Sandy shores, usually cultivated.

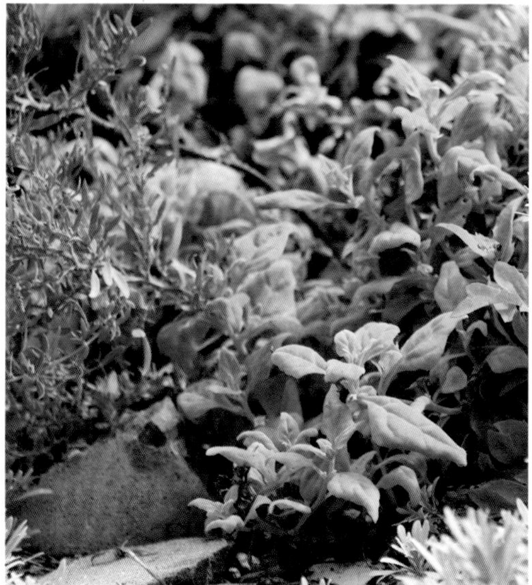

A308. 紫茉莉科 Nyctaginaceae

叶子花（三角梅、簕杜鹃）
Bougainvillea spectabilis
叶子花属 *Bougainvillea*

形态特征：藤状灌木。枝、叶密生柔毛；刺腋生、下弯。叶片椭圆形或卵形，基部圆形，有柄。花序腋生或顶生；苞片椭圆状卵形，基部圆形至心形，暗红色或淡紫红色，花被管狭筒形，绿色，密被柔毛，顶端5～6裂，裂片开展，黄色，子房具柄。果实密生毛。

花果期：花期冬春季。

分布：原产热带美洲。我国南方广泛栽培。南麂主岛（大沙岙、美龄居、镇政府），栽培。

生境：温暖湿润气候，光照充足的环境。

Description: Shrubs hederiform. Branches and leaves densely pubescent; spines axillary, recurved. Leaves petiolate; leaf blade elliptic or ovate, base rotund. Inflorescence axillary or terminal; bracts dark red or light purple-red, elliptic-ovate, base rotund to cordate, perianth tube green, narrowly tubular, rounded, densely pubescent, apex 5-6-lobed, lobes spreading, yellow, hairs copious, spreading, ovary stipitate. Fruit densely hairy.

Flower and Fruit: Fl. winter-spring.

Distribution: Native to tropical America. Widely cultivated in southern China. Main island (Dashaao, Meilingju, Zhenzhengfu), cultivated.

Habitat: Warm and humid climate with plenty of light.

A308. 紫茉莉科
Nyctaginaceae

紫茉莉（胭脂花）
Mirabilis jalapa (Nyctago jalapa)
紫茉莉属 *Mirabilis*

形态特征：一年生草本，高可达1米。根肥粗，倒圆锥形，黑色或黑褐色。茎直立，圆柱形，多分枝，无毛或疏生细柔毛，节稍膨大。叶片卵形或卵状三角形，顶端渐尖，基部截形或心形，全缘。花常数朵簇生枝端；总苞钟形，5裂，裂片三角状卵形，顶端渐尖，具脉纹，果时宿存；花被紫红色、黄色、白色或杂色，高脚碟状，5浅裂；花午后开放，次日午前凋萎。瘦果球形，革质，黑色，表面具皱纹。种子胚乳白粉质。

花果期：花期6～10月，果期8～11月。

分布：原产热带美洲。我国南北各地栽培，常逸生为杂草。南麂主岛偶见，栽培或归化。

生境：温暖湿润，通风良好环境。

Description: Herbs annual, to 1 m tall. Roots fat, conical, black or dark brown. Stem erect, terete, much branched, glabrous or sparsely pilose, nodes slightly inflated. Leaf blade ovate or ovate-triangular, apex acuminate, base truncate or cordate, entire. Flowers constant fascicled branch apex; involucre campanulate, 5-lobed, lobes triangular-ovate, apex acuminate, veined, persistent when fruit; perianth purplish red, yellow, white or variegated, high saucer, 5-lobed; openning in late afternoon, closing next morning. Achene globose, leathery, black, surface rugose. Seed endosperm silky.

Flower and Fruit: Fl. Jun. -Oct. , fr. Aug. -Nov.

Distribution: Native to tropical America. Cultivated in most areas of China, usually as a ruderal weed. Occasionally on main island of Nanji, cultivated or naturalized.

Habitat: Warm and humid, well-ventilated environment.

A312. 落葵科
Basellaceae

落葵薯
Anredera cordifolia
落葵薯属 *Anredera*

形态特征：缠绕藤本，根茎粗壮。叶卵形或近圆形，先端尖，基部圆或心形，稍肉质，腋生珠芽。总状花序具多花，花序轴纤细，下垂；苞片窄，宿存，花托杯状，花常由此脱落，花被片白色，渐变黑，卵形、长圆形或椭圆形，雄蕊白色，花丝顶端在芽内反折，开花时伸出，花柱白色，3叉裂，柱头棍棒状或宽椭圆形。

花果期：花期6～10月。

分布：原产南美洲。我国华东和华南有栽培或归化。南麂主岛各处常见。

生境：沟谷边、河岸岩石上、村旁墙垣、荒地或灌丛中。

Description: Vines twining, stout rhizome. Leaves ovate or suborbicular, apex acute, base rounded or cordate, thinly fleshy, axillary bulbule. Racemes many flowered, rachis slender, pendulous; bracts narrow, persistent, receptacle goblet, flowers fallen from here, perianth white, gradually black, ovate, oblong or elliptic, stamens white, filaments reflexed at apex in bud, spreading in anthesis, style white, split to 3 stigmatic arms, each with 1 club-shaped or broadly elliptic stigma.

Flower and Fruit: Fl. Jun. -Oct.

Distribution: Native to South America. Cultivated or naturalized in East China and South China. Main island of Nanji, commonly.

Habitat: Valley, on rocks of river banks, walls, waste places or thickets.

A315. 马齿苋科
Portulacaceae

马齿苋
Portulaca oleracea
马齿苋属 *Portulaca*

形态特征：一年生草本，高 10 ～ 15 cm。茎平卧或斜倚，多分枝，圆柱形。叶互生或近对生，倒卵形，似马齿状，顶端圆钝或平截，基部楔形，全缘，中脉微隆起。花无梗，常簇生枝端；苞片近轮生，萼片对生，盔形，背部具龙骨状凸起，花瓣黄色，倒卵形。蒴果卵球形，盖裂。

花果期：花期 6 ～ 8 月，果期 7 ～ 9 月。

分布：广布世界温带和热带地区。分布于我国各地。南麂主岛各处常见。

生境：菜园、农田及路旁。

Description: Herbs annual, 10-15 cm tall. Stems prostrate or reclining, much branched, terete. Leaves alternate or subopposite, obovate, horse-toothed, apex obtuse or truncate, base cuneate, entire, midrib slightly elevated. Flowers sessile, often clustered; bracts subwhorled, sepals opposite, galeate, back keeled, petals yellow, obovate. Capsule ovoid, dehiscent.

Flower and Fruit: Fl. Jun. -Aug. , fr. Jul. -Sep.

Distribution: Tropical and temperate regions worldwide. Throughout China. Main island of Nanji, commonly.

Habitat: In gardens, farmlands and roadsides.

A317. 仙人掌科
Cactaceae

单刺仙人掌
Opuntia monacantha
仙人掌属 *Opuntia*

形态特征：灌木或乔木状，高 1.3 ~ 4m。树干圆柱状，顶生分枝亮绿色，倒卵形，基部狭窄，先端边缘波状。分枝上刺稀疏，每小窠 1 或 2(或者 3) 刺，直立或开展，灰色；钩毛褐色。叶圆锥形，脱落。萼状花被片具红色中脉和黄边，倒卵形；瓣状花被片平展，黄色至橙色，边近全缘；花丝绿色，花药浅黄色。果实红紫色，倒卵球形。种子光褐色，不规则椭圆形。

花果期：花期 5 ~ 6 月，果期 11 ~ 12 月。

分布：原产南美洲，热带、亚热带地区广泛栽培和归化。我国南方地区有归化。南麂主岛（关帝岙、门屿尾、三盘尾）、大檑山屿常见。

生境：滨海岩质山坡，悬崖石隙。

Description: Shrubs or treelike, 1.3-4 m tall. Trunk terete, terminal joints glossy green, obovate, narrowed basally, margin undulate toward apex. Spines sparse on joint 1 or 2 (or 3) per areole, erect or spreading, grayish; glochids brownish. Leaves conic, deciduous. Sepaloids with red midrib and yellow margin, obovate; petaloids spreading, yellow to orange, margin subentire; filaments greenish, anthers pale yellow. Fruit reddish purple, obovoid. Seeds light tan, irregularly elliptic.

Flower and Fruit: Fl. May-Jun. , fr. Nov. -Dec.

Distribution: Native to South America, widely introduced and naturalized in tropical and subtropical regions. Naturalized in southern China. Main island (Guandiao, Menyuwei, Sanpanwei), commonly.

Habitat: Coastal rocky slopes, cliff crevices.

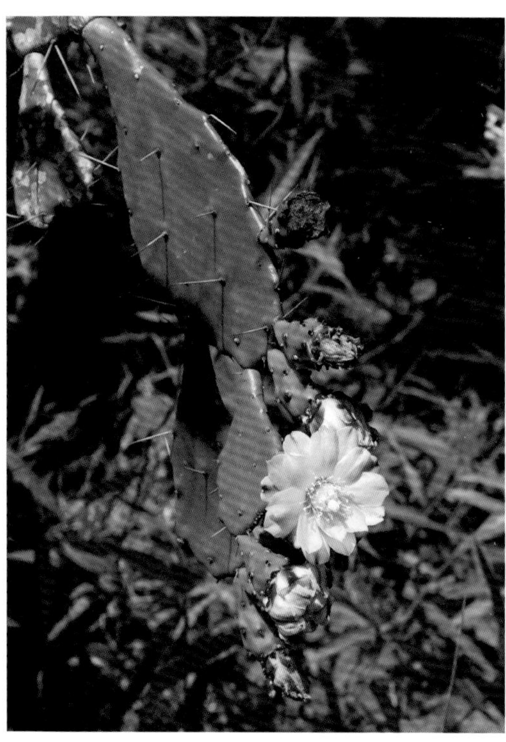

A318. 蓝果树科
Nyssaceae

喜树
Camptotheca acuminata
喜树属 *Camptotheca*

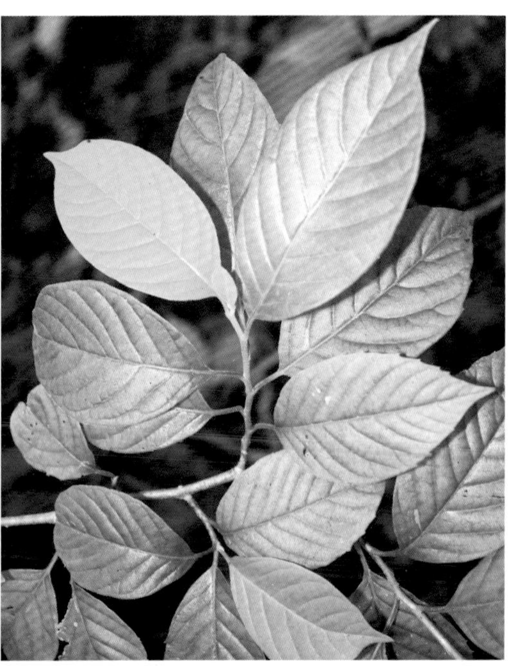

形态特征：落叶乔木，高达 20 m；树皮浅灰色，具深沟；小枝紫色，具长柔毛，老枝无毛。叶片背面绿色透明，长圆状卵形、长圆状椭圆形或圆形，纸质，稍短柔毛，基部近圆形，边缘全缘。头状花序顶生或腋生，常 2 ~ 9 花，苞片 3，三角形，两面被毛，花萼杯状，花瓣 5，亮绿色。果具薄翅，灰褐色，干燥时光滑透明。

花果期：花期 5 ~ 7 月，果期 9 月。

分布：我国长江流域以南地区。南麂主岛偶见栽培。

生境：林缘，溪边，路旁常见栽培。

Description: Trees deciduous, to 20 m high; bark light gray, deeply furrowed; young branchlets purplish, villous, old branchlets glabrous. Leaf blade abaxially greenish and lucid, oblong-ovate, oblong-elliptic, or orbicular, papery, slightly pubescent, base subrounded, margin entire. Head terminal or axillary, often 2-9-flowered, bracts 3, triangular, both surfaces pubescent, calyx cup-shaped, petals 5, light green. Fruit thinly winged, gray-brown, smooth and lucid when dry.

Flower and Fruit: Fl. May-Jul. , fr. Sep.

Distribution: To the South of Yangtze River Basin. Cultivated on main island of Nanji, occasionally.

Habitat: Forest margins, by streams, commonly cultivated along roadsides.

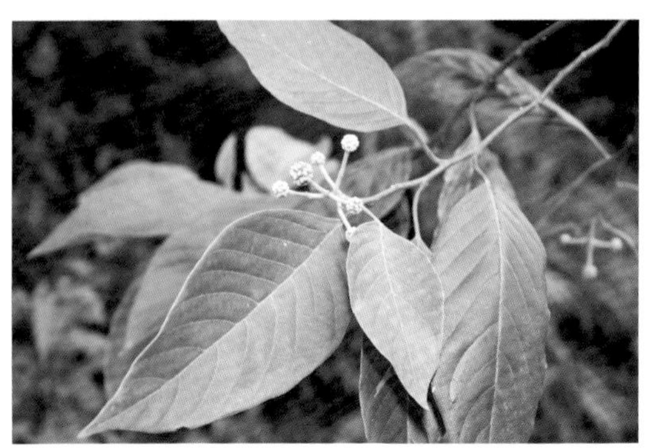

A320. 绣球科
Hydrangeaceae

绣球（八仙花）
Hydrangea macrophylla
绣球属 *Hydrangea*

形态特征：灌木，高 1 ~ 4 m。枝圆柱形，粗壮，紫灰色，具少数长皮孔。叶纸质或近革质，倒卵形或阔椭圆形，两面无毛或仅下面中脉两侧被毛；叶柄粗壮，无毛。伞房状聚伞花序近球形，具短总梗，分枝粗壮，紧贴短柔毛；花密集，多数不育；不育花萼片 4，卵形或阔卵形，粉红色、淡蓝色或白色；可育花极少，萼筒倒圆锥状，花瓣长圆形。蒴果长陀螺状。

花果期：花期 6 ~ 8 月。

分布：日本、朝鲜。我国华东至华南及沿海岛屿，常栽培。南麂主岛（大沙岙、美龄居、镇政府），栽培。

生境：山谷溪旁或山顶疏林。

Description: Shrubs, 1-4 m tall. Branches cylindrical, stout, purplish gray, glabrous, with a few long lenticels. Leaves papery or subleathery, obovate or broadly elliptic, glabrous on both surfaces or sparsely curled pubescent on both sides of lower midrib only; petiole stout, glabrous. Cymes subglobose, with short pedicels, branches stout, densely appressed pubescent, flowers dense, mostly sterile; sterile sepals 4, ovate, or broadly ovate, pink, pale blue, or white; fertile flowers few, calyx tube oboconical, petals oblong. Capsule long turbinate.

Flower and Fruit: Fl. Jun. -Aug.

Distribution: Japan, Korea. East China to South China and coastal islands, usually cultivated. Main island (Dashaao, Meilingju, Zhenzhengfu), cultivated.

Habitat: Valley creek, mountain sparse forest.

A332. 五列木科
Pentaphylacaceae

滨柃
Eurya emarginata
柃木属 *Eurya*

形态特征：常绿灌木，高 1 ～ 2 m。嫩枝圆柱形，密被黄褐色短柔毛；顶芽长锥形，被短柔毛或几无毛。叶互生，厚革质，倒卵形或倒卵状披针形，顶端圆而有微凹，基部楔形，具细微锯齿，稍反卷，上面中脉凹下；叶柄无毛。花 1 ～ 2 枚生于叶腋；萼片圆形；雄蕊约 20 枚，柱头 3 裂。果实圆球形。

花果期：花期 11 ～ 12 月，果期 5 ～ 8 月。

分布：东亚。我国华东地区。南麂各岛屿常见。

生境：滨海山坡或岩石缝灌丛。

Description: Evergreen shrubs, 1-2 m tall. Shoots terete, densely tawny pubescent; terminal buds long conic, pubescent or few glabrous. Leaves alternate, thickly leathery, obovate or obovate-lanceolate, apex rounded and retuse, base cuneate, minutely serrate, slightly revolute, upper midvein concave below; petiole glabrous. Flowers 1-2 in leaf axils; sepals rounded; stamens ca. 20, style tip 3-lobed. Fruit globose.

Flower and Fruit: Fl. Nov. -Dec. , fr. May-Aug.

Distribution: Native to East Asia. East China. Throughout Nanji Islands, commonly.

Habitat: Thickets on mountain slopes, rock crevices along seacoasts.

A332. 五列木科
Pentaphylacaceae

柃木
Eurya japonica
柃木属 *Eurya*

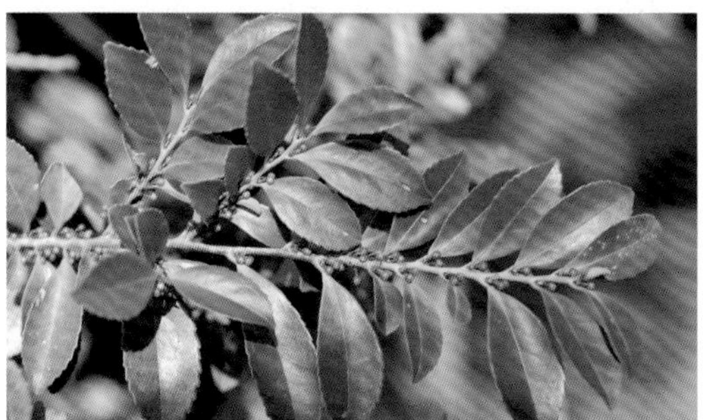

形态特征：灌木，高 1 ～ 3.5 m，全株无毛。嫩枝黄绿色或淡褐色，具 2 棱；顶芽披针形。叶互生，厚革质或革质，倒卵形、倒卵状椭圆形至长圆状椭圆形，顶端钝或近圆形，微凹，基部楔形，具疏钝齿，上面中脉凹下。花 1 ～ 3 枚腋生；萼片卵形；雄蕊 12 ～ 15 枚，子房圆球形，3 室，花柱顶端 3 浅裂。果实圆球形。

花果期：花期 2 ～ 3 月，果期 9 ～ 10 月。

分布：东亚。我国华东地区。南麂各岛屿常见。

生境：山坡或山谷灌丛。

Description: Shrubs, 1-3.5 m tall, glabrous throughout. Shoots yellowish green or pale brown, with 2 ribs; terminal buds lanceolate. Leaves alternate, thickly leathery or leathery, obovate, obovate-elliptic to oblong-elliptic, apex obtuse or suborbicular, retuse, base cuneate, with sparsely coarsely obtuse teeth, upper midrib concave below. Flowers 1-3 axillary; sepals ovate; stamens 12-15, anthers undivided; ovary orbicular, 3-loculed, apically 3-lobed. Fruit globose.

Flower and Fruit: Fl. Feb. -Mar. , fr. Sep. -Oct.

Distribution: East Asia. East China. Throughout Nanji Islands, commonly.

Habitat: Thickets on mountain slopes or in valleys.

A332. 五列木科
Pentaphylacaceae

细枝柃
Eurya loquaiana
柃木属 *Eurya*

形态特征：灌木或小乔木，高 2 ～ 10 m。嫩枝圆，密被微毛；顶芽狭披针形，密被微毛。叶互生，薄革质，长圆状披针形，下面干后常变为红褐色，上面中脉下凹。花 1 ～ 4 枚簇生于叶腋；雄蕊 10 ～ 15，子房卵圆形，花柱顶端 3 裂。果实圆球形，熟时黑色。

花果期：花期 10 ～ 12 月，果期翌年 7 ～ 9 月。

分布：我国长江以南各地。南麂各岛屿常见。

生境：山坡沟谷的林中或灌丛。

Description: Shrubs or small trees, 2-10m tall. Shoots rounded, densely hirsute; terminal buds narrowly lanceolate, densely pubescent. Leaves alternate, thinly leathery, oblong-lanceolate, often reddish brown below when dry, concave above midrib. Flowers 1-4 clustered in leaf axils; stamens 10-15, ovary ovoid, apically 3-lobed. Fruit black when mature, globose.

Flower and Fruit: Fl. Oct. -Dec. , fr. Jul. -Sep. of 2nd year.

Distribution: To the south of Yangtze River. Throughout Nanji Islands, commonly.

Habitat: Forests or thickets on mountain slopes or in valleys.

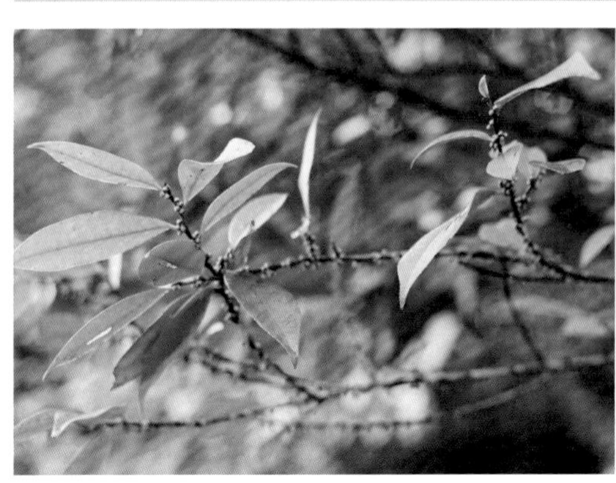

A332. 五列木科
Pentaphylacaceae

窄基红褐柃
Eurya rubiginosa var. *attenuata*
柃木属 *Eurya*

形态特征：灌木至小乔木，高达5 m。幼枝具2棱，无毛；顶芽无毛。叶革质，互生，长圆状披针形，基部楔形到宽楔形，具细锯齿，略反卷。花1～3簇生叶腋；萼片卵圆形，近革质，先端圆；雄蕊约15，花瓣长圆状披针形；子房球形，无毛；花柱先端3裂，几乎离生。果圆球形。

花果期：花期10～11月，果期翌年5～8月。

分布：我国华东至华南地区。南麂各岛屿常见。

生境：林中、灌丛。

Description: Shrubs to small trees, up to 5 m tall. Young shoots with 2 ribs, glabrous; terminal bud glabrous. Leaves leathery, alternate, oblong-lanceolate, base cuneate to broadly cuneate, serrulate, slightly revolute. Flowers 1-3 clustered leaf axils; sepals ovoid, subleathery, apex rounded; stamens ca. 15, anthers undivided; petals oblong-lanceolate; ovary globose, glabrous; style apically 3-parted to almost distinct. Fruit globose.

Flower and Fruit: Fl. Oct. -Nov. , fr. May-Aug. of 2nd year.

Distribution: East China to South China. Throughout Nanji Islands, commonly.

Habitat: Forests, thickets.

A335. 报春花科
Primulaceae

蓝花琉璃繁缕
Anagallis arvensis f. *coerulea*
琉璃繁缕属 *Anagallis*

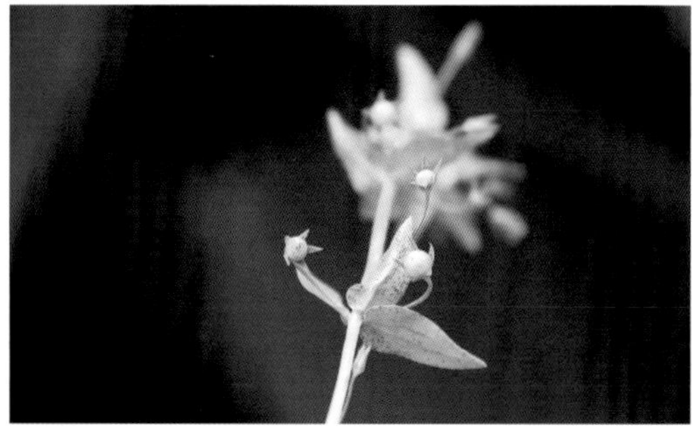

形态特征：一年生或二年生草本，高 10 ~ 30 cm。茎四棱，棱边狭翅状。叶交互对生或有时 3 枚轮生，卵圆形至狭卵形。单花腋生，花萼深裂，裂片线状披针形；花冠辐状，浅蓝色，分裂近达基部。蒴果球形。

花果期：花期 3 ~ 4 月，果期 5 ~ 7 月。

分布：世界温带和热带地区。我国浙江、福建、广东、台湾。南麂主岛（百亩山、关帝岙、三盘尾），常见。

生境：海岸沙丘或田野荒地中。

Description: Herbs annual or biennial, 10-30 cm tall. Stem quadrangular, short winged on ridges. Leaves opposite, occasionally in whorls of 3, ovate to narrowly ovate. Flower axillary, calyx parted, lobes linear-lanceolate; corolla rotate, wathet, parted nearly to base. Capsule globose.

Flower and Fruit: Fl. Mar. -Apr. , fr. May -Jul.

Distribution: Widely in temperate and tropical areas. Zhejiang, Fujian, Guangdong, Taiwan,China. Main island (Baimushan, Guandiao, Sanpanwei), commonly.

Habitat: Coastal dunes or field wastelands.

A335. 报春花科
Primulaceae

朱砂根
Ardisia crenata
紫金牛属 *Ardisia*

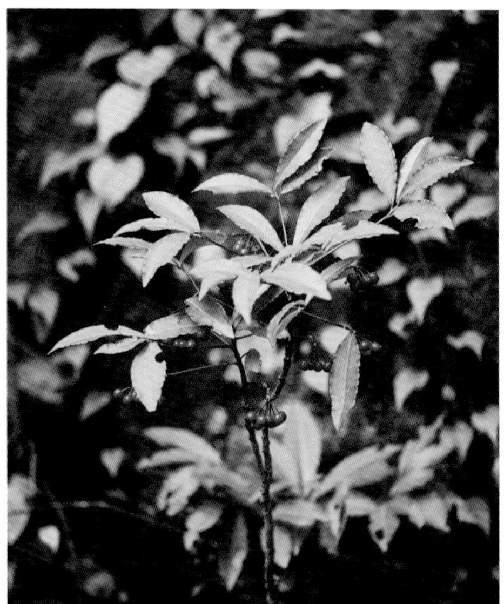

形态特征：灌木，高 1 ~ 2 m。茎粗壮无毛。叶革质，椭圆形，边缘具皱波状或波状齿，具明显的边缘腺点，两面无毛。伞形花序或聚伞花序，侧枝顶端；萼片长圆状卵形，花瓣白色，盛开时反卷。果鲜红色。

花果期：花期 5 ~ 6 月，果期 10 ~ 12 月。

分布：亚洲东部。我国长江流域以南地区。南麂主岛（百亩山、国姓岙、三盘尾），偶见。

生境：森林，山腰，山谷，灌木状的地区，阴湿处。

Description: Shrubs 1-2 m tall. Stem stout glabrous. Leaf blade leathery, elliptic, margin crenate, or undulate, with distinct marginal glandular spots, glabrous on both surfaces. Inflorescences umbellate or cymose, terminal; sepals oblong-ovate, petals white, inversely curled when blooming. Fruit red.

Flower and Fruit: Fl. May-Jun. , fr. Oct. -Dec.

Distribution: Eastern Asia. To the south of Yangtze River Basin. Main island (Baimushan, Guoxingao, Sanpanwei), occasionally.

Habitat: Forests, hillsides, valleys, shrubby areas, dark damp places.

A335. 报春花科
Primulaceae

山血丹（沿海紫金牛）
Ardisia lindleyana
紫金牛属 *Ardisia*

形态特征：灌木，高 1 ~ 2 m。小枝、叶柄、叶背和花序轴密被红色微柔毛。叶片长圆形或倒披针形，革质或纸质，无毛，边缘近全缘或波状，有斑点。伞形花序在侧枝弯曲的末端；花纸质，白色，萼片长圆形或狭卵形，被柔毛，花瓣近离生，椭圆形或卵形，正面在基部被黄色腺体。果深红色，球形，疏生斑点。

花果期：花期 5 ~ 7 月，果期 10 ~ 12 月。

分布：越南北部。我国长江流域以南地区。南麂主岛（国姓岙、门屿尾、三盘尾），偶见。

生境：茂密的混交林、丘陵、山谷、沿溪阴湿处。

Description: Shrubs 1-2 m tall. densely and minutely reddish puberulent on branchlets, petioles, abaxial leaf surface, and inflorescence rachis. Leaf blade oblong or oblanceolate, leathery or papery, glabrous, margin subentire or undulate, prominently punctate. Umbels on curved ends of specialized lateral branches; Flowers papery, white; sepals oblong or narrowly ovate, puberulent,petals nearly free, elliptic or ovate, punctate, yellow glandular granulose adaxially at base. Fruit dark red, globose, sparsely punctate.

Flower and Fruit: Fl. May-Jul. , fr. Oct. -Dec.

Distribution: North Vietnam. To the south of Yangtze River Basin. Main island (Guoxingao, Menyuwei, Sanpanwei), occasionally.

Habitat: Dense mixed forests, hills, valleys, along streams, dark damp places.

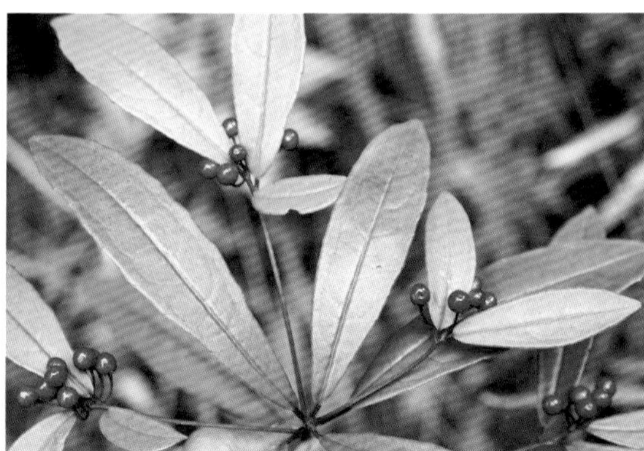

A335. 报春花科
Primulaceae

多枝紫金牛
Ardisia sieboldii
紫金牛属 *Ardisia*

形态特征：灌木，稀小乔木，高可达 10 m。树皮带银灰色，分枝多，小枝粗壮。叶片纸质或革质，倒卵形或椭圆状卵形，全缘，两面无毛。复亚伞形花序或复聚伞花序，腋生枝顶；多花，被鳞片和柔毛，花萼卵形，具腺点，花瓣白色，广卵形。果球形，红色至黑色，略肉质。

花果期：花期 3 ～ 6 月，果期 1 ～ 4 月。

分布：日本。我国浙江、福建、台湾。南麂各岛屿常见。

生境：海岸山坡、林缘、灌丛。

Description: Shrubs or rarely small trees, to 10 m tall. Bark belt silver-gray, branching much, branchlets stout. Leaf blade papery or leathery, obovate to elliptic ovate, margin entire, glabrous on both surfaces. Inflorescences branches compound subumbellate or compound cymose, axillary at branch apex; many flowers, rust-colored scales and puberulent, calyx ovoid, glandular spot, petals white, broadly ovate. Fruit globose, red to blackish, somewhat fleshy.

Flower and Fruit: Fl. Mar. -Jun. , fr. Jan. -Apr.

Distribution: Japan. Zhejiang, Fujian, Taiwan,China. Throughout Nanji Islands, commonly.

Habitat: Costal slopes, forest edge, thickets.

A335. 报春花科
Primulaceae

滨海珍珠菜
Lysimachia mauritiana
珍珠菜属 *Lysimachia*

形态特征：二年生草本，高 10 ~ 50 cm。茎簇生，直立，圆柱形，上部分枝。叶互生，匙形或倒卵形，先端钝圆，两面散生黑色粒状腺点。总状花序顶生，苞片匙形，花冠白色，裂片舌状长圆形。蒴果梨形。

花果期：花期 5 ~ 6 月，果期 6 ~ 8 月。

分布：东亚、东南亚及印度洋和太平洋岛屿。我国沿海各地。南麂各岛屿常见。

生境：海滨的沙滩、石缝。

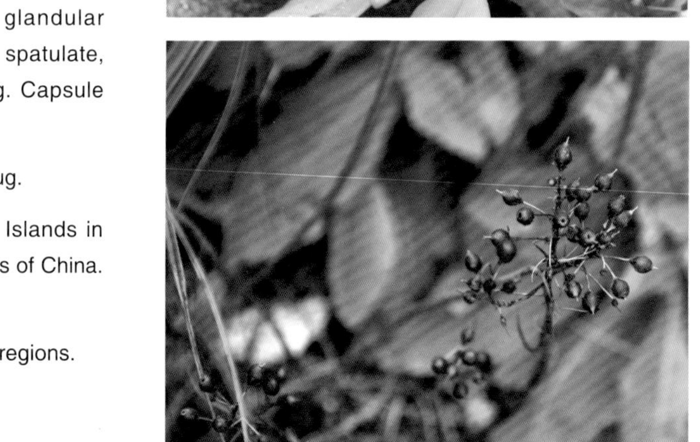

Description: Herbs biennial, 10-50 cm tall.Stems often many, erect, terete, usually branched in upper part.Leaves alternate, spatulate or obovate, apex obtuse to subrounded, sparsely black glandular punctate. Raceme terminal, lower bracts spatulate, corolla white, lobes erect, ligulate-oblong. Capsule pyriform.

Flower and Fruit: Fl. May-Jun. , fr. Jun. -Aug.

Distribution: East Asia, Southeast Asia, Islands in the Indian and Pacific oceans. Coastal areas of China. Throughout Nanji Islands, commonly.

Habitat: Rock crevices, beaches in coastal regions.

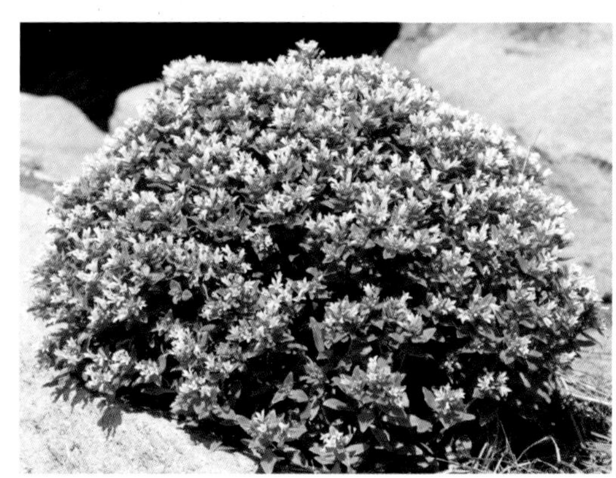

A335. 报春花科
Primulaceae

杜茎山
Maesa japonica
杜茎山属 *Maesa*

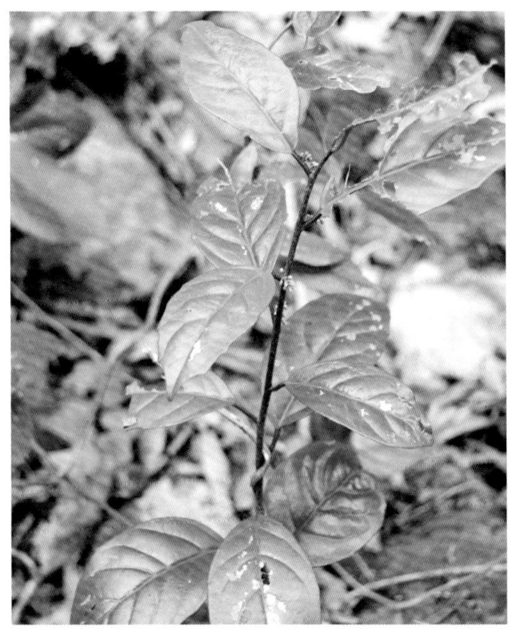

形态特征：灌木，高 1 ～ 3 m。小枝无毛，具细条纹，疏生皮孔。叶革质，椭圆形、倒卵形或披针形。总状或圆锥花序，腋生；苞片卵形，花冠白色，长钟形，具明显脉状腺纹，裂片边缘具细齿。果球形，肉质，具脉状腺条纹，宿存萼包果顶端，常宿存花柱。

花果期：花期 1 ～ 3 月，果期 10 月至翌年 5 月。

分布：日本、越南北部。我国长江流域以南地区。南麂主岛各处常见。

生境：混交林、山坡、石灰岩山地。

Description: Shrubs, 1-3 m tall. Branchlets glabrous, with fine stripes, sparsely lenticellate. Leaf blade leathery, elliptic, obovate or lanceolate. Inflorescences racemose or paniculate, axillary; bracteoles ovate, corolla white, long bell-shaped, with distinct veined glandular stripes, lobes margin finely serrulate. Fruit globose, fleshy, veined glandular striate, calyx persistent apical pericarp, style often persistent.

Flower and Fruit: Fl. Jan. -Mar. , fr. Oct. -May.

Distribution: Japan, North Vietnam. To the south of Yangtze River Basin. Main island of Nanji, commonly.

Habitat: Mixed forests, hillsides, limestone mountains.

A335. 报春花科
Primulaceae

密花树
Myrsine seguinii
密花树属 *Rapanea*

形态特征：灌木或乔木，高 2 ~ 7 m。小枝无毛，具皱纹。叶片革质，长圆状倒披针形至倒披针形，全缘，两面无毛。伞形花序或花簇生，白色或淡绿色，有时为紫红色。果球形或近卵形，灰绿色或紫黑色，有时具纵肋。

花果期：花期 4 ~ 5 月，果期 10 ~ 12 月。

分布：日本、缅甸、越南。我国西南各省至台湾。南麂主岛（后隆、三盘尾），偶见。

生境：林缘、灌丛、路旁。

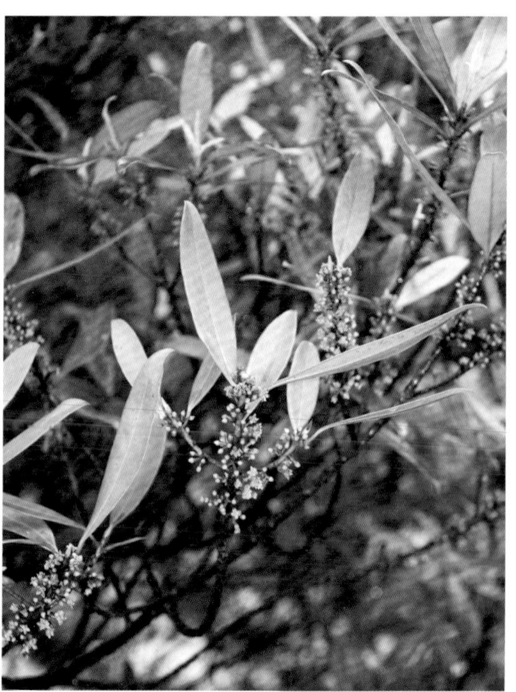

Description: Shrubs or trees, 2-7 m tall. Branchlets glabrous, rugose. Leaf blade leathery, oblong-oblanceolate to oblanceolate, margin entire, glabrous on both surfaces. Umbels or flowers clustered, white or pale green, sometimes purplish red. Fruit globose or subovate, grayish green or purplish black, sometimes longitudinally ribbed.

Flower and Fruit: Fl. Apr. -May , fr. Oct. -Dec.

Distribution: Japan, Burma, Vietnam. Southwest China to Taiwan, China. Main island (Houlong, Sanpanwei), occasionally.

Habitat: Mixed forests, shrubby areas, roadsides.

A337. 山矾科
Symplocaceae

白檀
Symplocos paniculata
山矾属 *Symplocos*

形态特征：落叶灌木或小乔木，高达 5 m。嫩枝有灰白色柔毛，老枝无毛。叶膜质或薄纸质，阔倒卵形、椭圆状倒卵形或卵形，边缘有细尖锯齿。圆锥花序，花萼淡黄色，花冠白色，5 深裂。核果熟时蓝色，卵状球形，稍偏斜，顶端宿萼裂片直立。

花果期：花期 4 ~ 6 月，果期 9 ~ 11 月。

分布：东亚、东南亚。我国东北、华北及长江以南各地。南麂各岛屿常见。

生境：山坡、路边、疏林或密林中。

Description:Shrubs or small trees, deciduous,5 m tall. Young branchlets grayish white pilose, old branchlets glabrous. Leaf blade membranous, or thinly papery, broadly obovate, elliptic-obovate, or ovate, margin sharply glandular dentate. Panicles, calyx pale yellow, corolla white, 5-lobed. Drupes bluish, ovate-globose, slightly oblique, apex persistent calyx lobes erect.

Flower and Fruit:Fl. Apr. -Jun. , fr. Sep. -Nov.

Distribution:East Asia, Southeast Asia. Northeast, North China, and the south of Yangtze River. Throughout Nanji Islands, commonly.

Habitat:Slopes, roadsides, open or dense forests.

A345. 杜鹃花科
Ericaceae

锦绣杜鹃
Rhododendron* × *pulchrum
杜鹃属 ***Rhododendron***

形态特征：半常绿灌木，高达 2 ~ 5 m。幼枝密被淡棕色扁平糙伏毛。叶椭圆形或椭圆披针形，先端钝尖，基部楔形，被毛。顶生伞形花序 1 ~ 5 花，花梗被红棕色扁平糙毛；花萼 5 裂，裂片披针形，被糙毛；花冠漏斗形，玫瑰色，有深紫红色斑点，5 裂。蒴果长圆状卵圆形，被糙毛，有宿萼。

花果期：花期 4 ~ 5 月，果期 9 ~ 10 月。

分布：我国长江流域以南广泛分布、栽培。南麂主岛（大沙岙、美龄居、镇政府），栽培。

生境：温暖、半阴、凉爽、湿润、通风的环境。

Description: Semi-evergreen shrubs, 2-5m tall. Cladium densely pale brown flattened strigose. Leaves elliptic or elliptic lanceolate, apex obtuse, base cuneate, pubescent. Terminal umbel with 1-5 flowers. Pedicels red-brown flattened strigose; calyx 5-lobed, lobes lanceolate, strigose; corolla funnelform, rose, with deep purplish -red spots, 5-lobed. Capsule oblong ovoid, strigose, with persistent calyx.

Flower and Fruit: Fl. Apr. -May, fr. Sep. -Oct.

Distribution: Widely to the south of Yangtze River Basin, cultivated. Main island (Dashaao, Meilingju, Zhenzhengfu), cultivated.

Habitat: Warm, semi-shaded, cool, moist and ventilated environment.

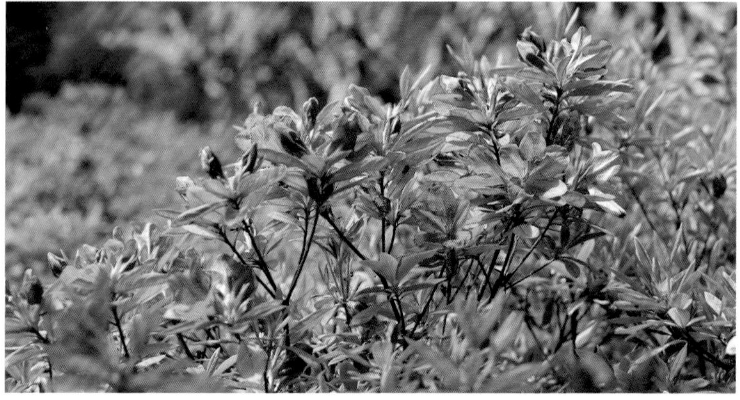

A345. 杜鹃花科
Ericaceae

南烛（乌饭树）
Vaccinium bracteatum
越橘属 *Vaccinium*

形态特征：常绿灌木或小乔木，高 2 ～ 6 m，多分枝。小枝不明显具棱，被短柔毛或无毛。叶薄革质，椭圆形或披针形，先端尖。总状花序顶生和腋生，多花，苞片叶状，披针形，边缘有锯齿，花冠白色，筒状或坛状。浆果紫黑色。

花果期：花期 6 ～ 7 月，果期 8 ～ 10 月。

分布：东亚、东南亚。广布我国长江以南各地。南麂主岛各处常见。

生境：林下、灌丛、路边草地。

Description: Shrubs or small trees, evergreen, 2-6 m tall, much branched. Twigs inconspicuously angled, pubescent or glabrous. Leaf blade thinly leathery, elliptic or lanceolate, apex acute. Racemose terminal or axillary, many flowered, bracts leaflike, lanceolate, margin serrate, corolla white, tubular or altar shaped. Berry dark purple.

Flower and Fruit: Fl. Jun. -Jul. , fr. Aug. -Oct.

Distribution: East Asia, Southeast Asia. Widely to the south of Yangtze River. Main island of Nanji, commonly.

Habitat: Forests, thickets, grassy places at roadsides.

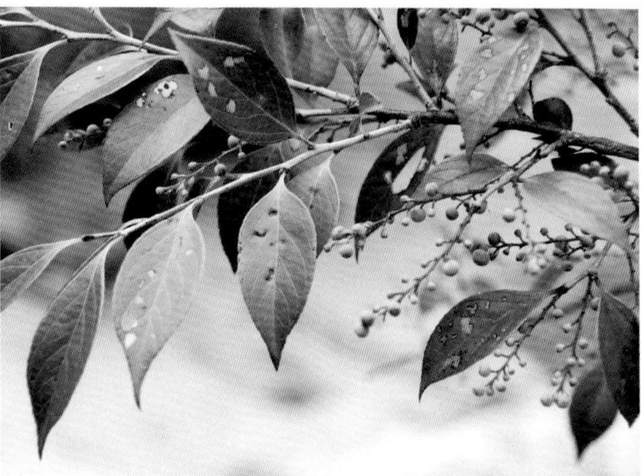

A352. 茜草科
Rubiaceae

山东丰花草
Diodia teres (*Borreria shandongensis*)
双角草属 *Diodia*

形态特征：一年生草本，分枝多，直立或斜升，高10～30 cm。枝微呈四棱形，被短毛。叶纸质，无柄，线状披针形，顶端渐尖，两面粗糙，干后边缘微背卷。花单生叶腋，无梗；花冠粉红色，近漏斗形，顶部4裂，裂片长圆形。蒴果倒卵形，被疏柔毛。种子2，长圆形。

花果期：花果期8～9月。

分布：原产安的列斯，南美洲和北美洲。我国福建（金门），山东（青岛）。南麂主岛（门屿尾），常见。

生境：受干扰，退化的开阔地带。

Description: Herbs annual, many branches, erect or ascending, 10-30 cm tall. Branches slightly quadrangular, covered with short hairs. Leaves papery, sessile, linear-lanceolate, tapered at the top, rough on both sides, and slightly rolled back after drying. Flowers solitary in leaf axils, sessile; corolla pink, nearly funnel-shaped, 4-lobed at the top, lobes oblong. Capsule obovate, sparsely pilose. Seeds 2, oblong.

Flower and Fruit: Fl. and fr. Aug. -Sep.

Distribution: Native to Antilles and North and South America. Fujian (Jinmen), Shandong (Qingdao). Main island (Menyuwei), commonly.

Habitat: Disturbed, often degraded open ground.

A352. 茜草科
Rubiaceae

四叶葎
Galium bungei
拉拉藤属 *Galium*

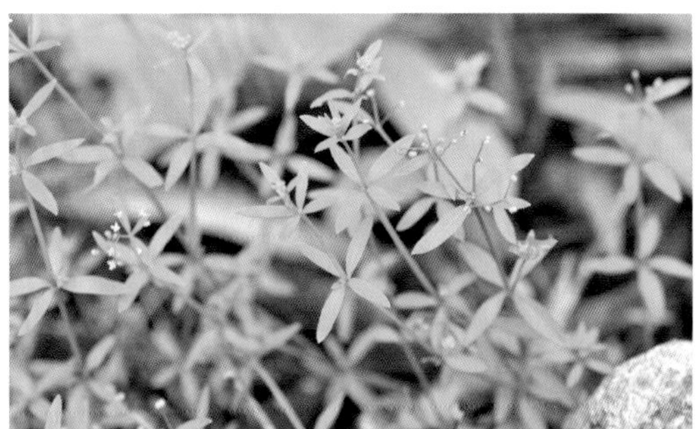

形态特征：多年生丛生草本，高 5 ~ 50 cm。茎纤细，具 4 棱。叶 4 片轮生，中部以上茎生叶片线状披针形至椭圆形，先端急尖，基部楔形。聚伞花序顶生或腋生，通常 3 ~ 10 朵；花冠淡黄绿色，4 裂。分果 2，半球形，表面具鳞片状突起。

花果期：花期 4 ~ 9 月，果期 5 月至翌年 1 月。

分布：日本、朝鲜。广布我国各地，长江流域中下游和华北较常见。南麂主岛（百亩山、国姓岙、门屿尾、三盘尾），常见

生境：山地森林、灌丛或草地，开阔的田野、河沟、海滩或溪边。

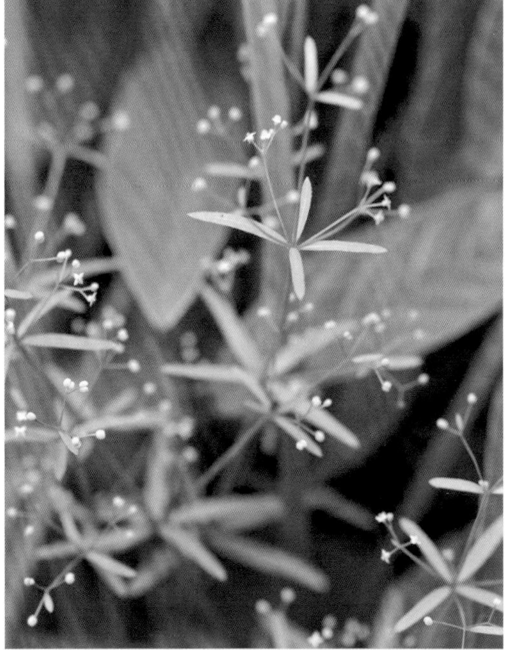

Description: Herbs perennial, 5-50 cm tall, tufted. Stems slender, 4-angled. Leaves in whorls of 4, stem leaves above middle linear lanceolate to elliptic, apex acute, base cuneate. Cymes terminal or axillary, usually 3-10; corolla yellowish green, 4-lobed. Schizocarp 2, hemispheric, surface with scaly projections.

Flower and Fruit: Fl. Apr. -Sep. , fr. May-Jan of 2nd year.

Distribution: Japan, Korea. Throughout China, usually in middle and lower Yangtze River Basin and North China. Main island (Baimushan, Guoxingao, Menyuwei, Sanpanwei), commonly.

Habitat: Forests, thickets, or meadows on mountains, hills, open fields, farmlands, ditch sides, riversides and beaches, streamsides.

A352. 茜草科
Rubiaceae

猪殃殃
Galium spurium (Galium aparine var. *tenerum)*
拉拉藤属 *Galium*

形态特征：一年生草本，平卧或攀援，高 30 ~ 50 cm。茎 4 棱，基部分枝，棱上倒生锐刺。茎叶 6 ~ 8 轮生，近无柄；叶片干纸质，狭倒披针形，上面具柔毛或短刚毛。花序顶生和腋生，聚伞花序有 2 到多花；苞片叶状或无；花冠黄绿色或白色，旋转，浅裂，裂片 4，三角形到卵形，锐尖。果实近球形到宽肾形，无毛或密被钩状毛。

花果期：花期 3 ~ 7 月，果期 4 ~ 11 月。

分布：非洲、欧亚大陆和地中海。除了海南和南海诸岛遍布我国。南麂主岛（百亩山、国姓岙、门屿尾、三盘尾），常见

生境：开阔的田野，河边，农田，山坡。

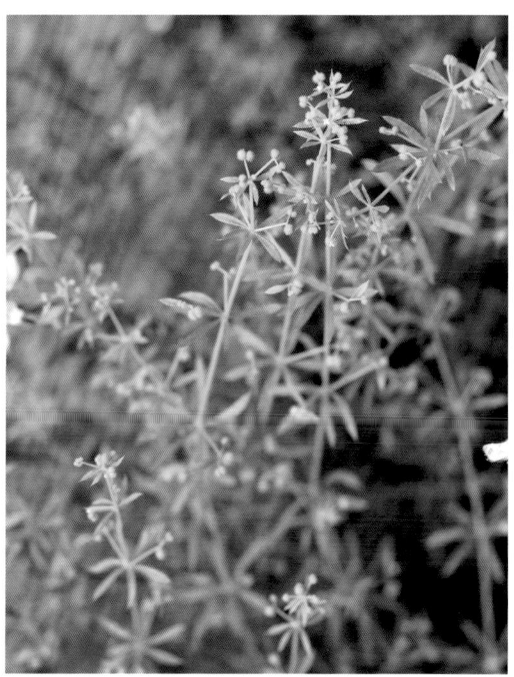

Description: Herbs annual, procumbent or climbing, 30-50 cm tall. Stems 4-angled, branched from base, retrorsely aculeate on angles. Leaves at middle stem region in whorls of 6-8, subsessile; blade drying papery, narrowly oblanceolate, usually pilosulous or hispidulous adaxially. Inflorescences terminal and axillary, cymes 2-to several flowered; bracts leaflike or none; Corolla yellowish green or white, rotate, lobed, lobes 4, triangular to ovate, acute. Mericarps subglobose to broadly kidney-shaped, glabrous or often densely covered with uncinate trichomes.

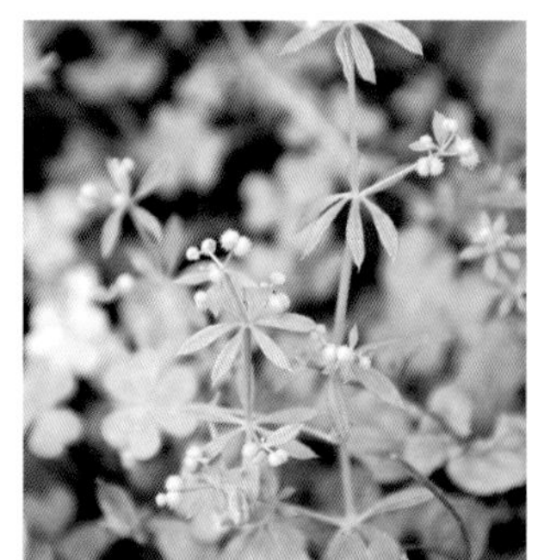

Flower and Fruit: Fl. Mar. -Jul. , fr. Apr. -Nov.

Distribution: Africa, Eurasia, and the Mediterranean. Throughout China except Hainan and Nanhai Zhudao. Main island (Baimushan, Guoxingao, Menyuwei, Sanpanwei), commonly.

Habitat: Open fields, riversides, farmlands, mountain slopes.

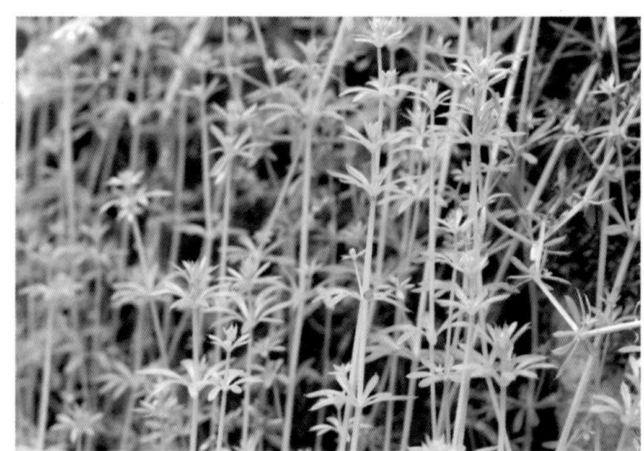

A352. 茜草科
Rubiaceae

栀子
Gardenia jasminoides
栀子属 *Gardenia*

形态特征：常绿直立灌木，高0.3～3 m。叶对生或3叶轮生，叶片革质，倒卵状椭圆形至长椭圆形；托叶膜质。花单生枝顶，花冠白色或乳黄色，芳香，高脚碟状，顶端常6裂，倒卵形或勺形；花药线形，伸出；花柱粗厚，柱头纺锤形，伸出。果黄色至橙红色，卵球形，有5～8纵棱。

花果期：花期3～7月，果期5月至翌年2月。

分布：原产亚洲，世界各地栽培。我国华北以南各地。南麂各岛屿常见。

生境：丘陵山坡、山谷溪边或旷野灌丛中。

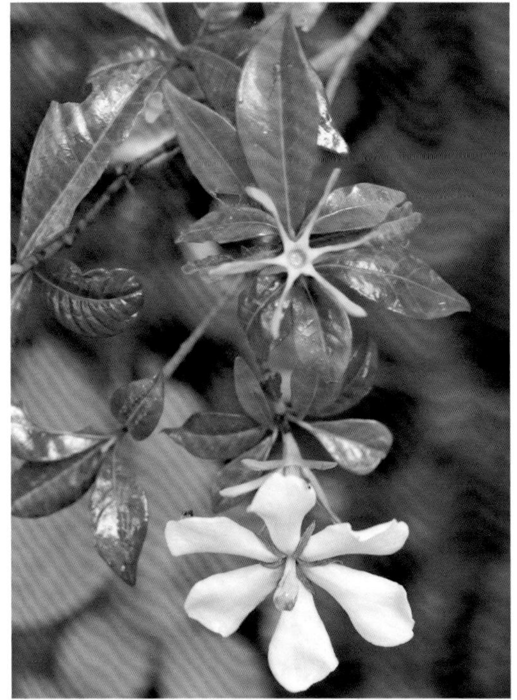

Description: Evergreen shrubs, erect, 0.3-3 m tall. Leaves opposite or ternate, obovate-elliptic to long elliptic; stipules membranous. Flower solitary, terminal, corolla white or pale yellow, fragrant, high-legged saucer, apex usually 6-lobed, obovate or ladle-shaped; anthers linear, extended; style thick, stigma fusiform, extended. Fruit yellow to orange-red, ovoid, with 5-8 longitudinal ribs.

Flower and Fruit: Fl. Mar. -Jul. , fr. May-Feb.

Distribution: Native in Asia, widely cultivated. To the south of North China. Throughout Nanji Islands, commonly.

Habitat: Thickets and forests at streamsides, on mountain slopes or hills, or in valleys or fields.

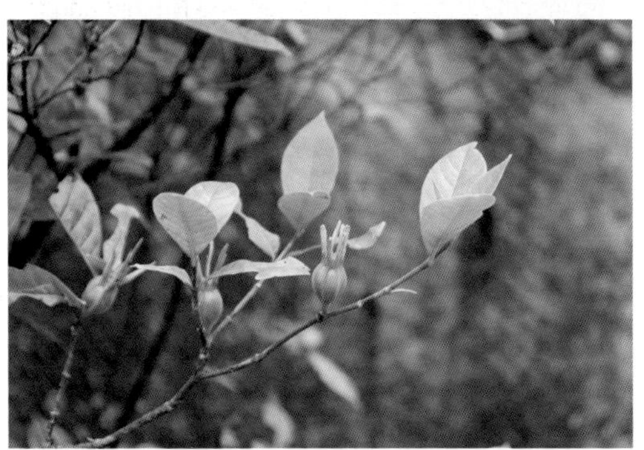

A352. 茜草科
Rubiaceae

白花蛇舌草
Hedyotis diffusa
耳草属 *Hedyotis*

形态特征：一年生无毛纤细披散草本，高 20 ~ 50 cm。茎稍扁，基部分枝。叶对生，无柄，膜质，线形，顶端短尖，正面光滑，反面面有时粗糙。花 4 数，单生或双叶腋；花梗略粗壮；萼管球形；花冠白色，管形，喉部无毛，花冠裂片卵状长圆形，顶端钝。蒴果膜质，扁球形，成熟时顶部室背开裂。种子具棱，干后深褐色，有深而粗的窝孔。

花果期：花果期 5 ~ 10 月。

分布：日本、东南亚至南亚。我国长江流域以南地区。南麂主岛（关帝岙、三盘尾），偶见。

生境：稻田，田埂，潮湿的开阔地。

Description: Herbs annual, glabrous, slender scattered, 20-50 cm tall. Stem slightly flattened, branching from base. Leaves opposite, sessile, membranous, linear, apically mucronate, adaxially smooth, sometimes rough abaxially. Flowers 4-numbered, solitary or twin in leaf axils; pedicels slightly stout; calyx tube globose; corolla white, tubular, throat glabrous, corolla lobes ovate-oblong, apex blunt. Capsule membranous, oblate, apical compartment dehiscent at maturity. Seeds angulate, dry after dark brown, with deep and thick pit holes.

Flower and Fruit: Fl. and fr. May-Oct.

Distribution: Japan, Southeast Asia to South Asia. To the south of Yangtze River Basin. Main island (Guandiao, Sanpanwei), occasionally.

Habitat: Paddy fields, ridges of farmlands, humid open fields.

A352. 茜草科
Rubiaceae

肉叶耳草
Hedyotis strigulosa（*Hedyotis coreana*）
耳草属 *Hedyotis*

形态特征：一年生(或多年生)无毛肉质草本，高 10 ~ 20 cm。多分枝，近丛生状。叶肉质有光泽，对生，无柄，长圆状倒卵形或长圆形。聚伞花序或有时排成短圆锥花序式，有花 3 ~ 8 朵，顶生或腋生；花 4 数，具纤细的花梗。蒴果扁陀螺形。种子细小，黑褐色。

花果期：花果期 12 月至翌年 4 月。

分布：日本、朝鲜、密克罗尼西亚。我国浙江、台湾、广东。南麂各岛屿常见。

生境：海边的沙滩、泥滩、荒地及峭壁。

Description: Herbs annual or perennial, glabrous, fleshy. Branches much, subfasciculate, 10-20 cm tall. Leaves fleshy, glossy, opposite, sessile, oblong-obovate or oblong. Cymes sometimes arranged in short panicle type, with flowers of 3-8, terminal or axillary; flowers 4-numbered, with slender pedicels. Capsule compressed turbinate. Seeds small, dark brown.

Flower and Fruit: Fl. and fr. Dec. -Apr.

Distribution: Japan, Korea, Micronesia. Zhejiang, Taiwan, Guangdong,China. Throughout Nanji Islands, commonly.

Habitat: Sandy or muddy beaches, wastelands, on rocks near sea.

A352. 茜草科
Rubiaceae

纤花耳草
***Hedyotis tenelliflora*（*Hedyotis angustifolia*）**
耳草属 ***Hedyotis***

形态特征：一年生或多年生草本，高 15 ~ 50 cm。茎直立，多分枝。小枝具 4 棱。叶片薄革质，线形或线状披针形，中脉在上面凹陷，下面隆起。花 2 ~ 3 朵簇生叶腋，无梗；花冠白色，漏斗状，顶端 4 裂，花药突出。蒴果卵形，具宿存的萼裂片，成熟时，顶端开裂。

花果期：花果期 4 ~ 12 月。

分布：热带亚洲。我国东南至西南地区。南麂主岛各处常见。

生境：山谷坡地、田埂。

Description: Herbs annual or perennial, 15-50 cm tall. Stems erect, diffusely branched. Branchlets 4-angulate. Leaves thinly leathery, linear, linear-lanceolate, midrib concave above, raised below. Flowers 2 to 3 fascicled leaf axils, sessile; corolla white, funnelform, apical 4-lobed, anthers prominent. Capsule ovate, with persistent calyx lobes, apex dehiscent at maturity.

Flower and Fruit: Fl. and fr. Apr. -Dec.

Distribution: Tropical Asia. Southeast to southwest of China. Main island of Nanji, commonly.

Habitat: Slopes in valleys, ridges of fields.

A352. 茜草科
Rubiaceae

羊角藤
Morinda umbellata subsp. *obovata*
巴戟天属 *Morinda*

形态特征：常绿攀援灌木，有时呈披散灌木状。叶对生；叶片薄革质或纸质革质，卵形、倒卵状披针形或长圆形；全缘；托叶干膜质。花序顶生，4 ~ 10 个伞形排列；花冠白色，稍呈钟状。聚花核果成熟时红色，近球形或扁球形。

花果期：花期 6 ~ 7 月，果期 10 ~ 11 月。

分布：热带亚洲。我国长江以南各地。南麂主岛各处常见。

生境：山地林缘或阴密灌丛。

Description: Evergreen climbing shrubs, sometimes scattered shrubby. Leaves opposite; leaf blade thinly leathery or papery leathery, ovate, obovate-lanceolate or oblong; entire; stipules dry membranous. Inflorescences terminal, 4-10 umbellate; corolla white, slightly campanulate. Polyfloral drupe red when ripe, subglobose or oblate.

Flower and Fruit: Fl. Jun. -Jul. , fr. Oct. -Nov.

Distribution: Tropical Asia. To the south of Yangtze River.Main island of Nanji, commonly.

Habitat: Slope, forest edge, dense thickets.

A352. 茜草科
Rubiaceae

鸡矢藤（鸡屎藤）
Paederia foetida（*Paederia scandens*）
鸡矢藤属 *Paederia*

形态特征：藤状灌木，无毛或被柔毛。叶对生，卵形或披针形，基部浑圆，叶上面无毛，下面脉上被微毛；侧脉每边 4 ～ 5 条，在上面平展，在下面突起；托叶卵状披针形，顶部 2 裂。圆锥花序腋生或顶生，扩展；花有小梗，三歧蝎尾状聚伞花序；花萼钟形，萼檐裂片钝齿形；花冠紫蓝色，通常被绒毛，裂片短。果球形，无毛。

花果期：花期 5 ～ 10 月，果期 7 ～ 12 月。

分布：亚洲东部和南部。我国大部分地区有分布。南麂各岛屿常见。

生境：森林、林缘，峡谷和山坡的灌丛。

Description: Shrubs hederiform, glabrous or pilose. Leaves opposite, ovate or lanceolate, base rounded, leaves glabrous above, glabrous on veins below; lateral veins 4-5 per side, spreading above, protruding below; stipules ovate-lanceolate, apically 2-lobed. Panicles axillary or terminal, spreading; flowers sessile, borne on cymes of scorpioid tricidum; calyx campanulate, calyx limb lobes obtuse-toothed; corolla violet-blue, usually tomentose, lobes short. Fruit globose, glabrescent.

Flower and Fruit: Fl. May-Oct. , fr. Jul. -Dec.

Distribution: Eastern and southern Asia. Most areas of China. Throughout Nanji Islands, commonly.

Habitat: Forests, forest margins, thickets in ravines and mountain slopes.

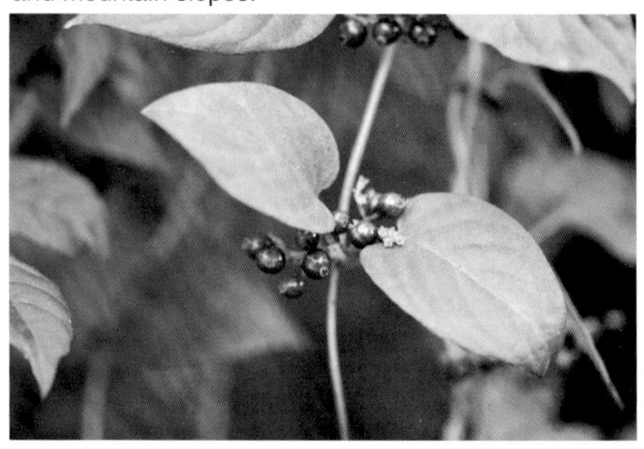

A352. 茜草科
Rubiaceae

九节
***Psychotria asiatica*（*Psychotria rubra*）**
九节属 ***Psychotria***

形态特征：常绿灌木或小乔木，高 0.5 ～ 5 m。叶对生，纸质或革质，长圆形或披针状长圆形；叶脉在上面凹下，下面凸起。聚伞花序顶生，多花；花冠白色，漏斗状，喉部被白色长柔毛，裂片近三角形，开放时反折。核果红色，近球形或宽椭圆形；小核 3 ～ 5 浅棱。

花果期：花果期全年。

分布：日本、中南半岛和印度。广布我国西南、华南，东至台湾。南麂各岛屿常见。

生境：峡谷、山坡或村庄边缘的灌丛或森林。

Description: Evergreen shrubs or small trees, 0.5-5 m tall. Leaves opposite, papery or leathery, oblong or lanceolate-oblong; midrib and lateral veins concave above, convex below. Cymose terminal, many flowered; corolla white, funnelform, in throat white villous, lobes triangular, reflexed when open. Drupes red, subglobose to broadly ellipsoid; pyrenes shallowly 3-5-ribbed.

Flower and Fruit: Fl. and fr. year-round.

Distribution: Japan, Indochina and India. Widely in South and Southwest China, east to Taiwan,China. Throughout Nanji Islands, commonly.

Habitat: Thickets or forests in ravines, on hill slopes, or at village margins.

A352. 茜草科
Rubiaceae

蔓九节（匍匐九节）
Psychotria serpens
九节属 *Psychotria*

形态特征：常绿攀援藤本，常以气根攀附在岩石或树上。叶对生，干后纸质至革质，叶形多变，幼枝上卵形或倒卵形，生殖枝上椭圆形、披针形或倒卵状长圆形，边全缘或稍反卷。聚伞花序顶生，少数到多花，3 ~ 5 分枝；花冠白色，漏斗状，花冠管喉部被密毛，裂片舌状长圆形。核果白色，近球形或椭圆形，小核具 4 ~ 5 浅棱。

花果期：花期 5 ~ 7 月，果期 6 ~ 12 月。

分布：东亚和中南半岛。我国华东、华南各省沿海山地。南麂各岛屿常见。

生境：平地、丘陵、山地、平地的灌丛或林中。

Description: Evergreen climbing vines, often clinging to rocks or trees by air roots. Leaves opposite, leaf blade drying papery to leathery, ovate or obovate on juvenile stems and elliptic, lanceolate, or obovate-oblong on reproductive stems, margins plane or sometimes thinly revolute. Cymes terminal, few to many flowered, branched for 3-5 orders, pedunculate; corolla white, funnelform, tube in throat densely villous, lobes ligulate-oblong. Drupe white, subglobose or ellipsoid, when ripe, pyrenes shallowly 4-or 5-ribbed.

Flower and Fruit: Fl. May-Jul. , fr. Jun. -Dec.

Distribution: East Asia and Indochina peninsula. Coastal mountainous areas of East China, South China. Throughout Nanji Islands, commonly.

Habitat: Thickets or forests in ravines, mountains, hills, flat lands.

A352. 茜草科
Rubiaceae

墨苜蓿
Richardia scabra
墨苜蓿属 *Richardia*

形态特征：一年生草本，匍匐或近直立。茎扁平到近圆柱状，具长硬毛。叶片干膜质至厚纸质，卵形，椭圆形，两面粗糙至无毛。花萼具子房部分倒卵球形，具小乳突到具短硬毛，裂片 6，披针形；具缘毛；花冠白色，裂片 6，三角形。果具分果爿 3，椭圆形到倒卵球形，背面密被小乳突到具短硬毛，腹面有一条狭沟槽。

花果期：花果期 2 ～ 11 月。

分布：原产安的列斯群岛和南美洲、北美洲。我国华南和台湾。南麂主岛（门屿尾），偶见。

生境：荒地。

Description: Herbs, annual, decumbent or suberect. Stems flattened to subterete, hirsute. Leaf blade drying membranous to thickly papery, ovate, elliptic, or lanceolate, both surfaces scabrous to glabrescent. Calyx with ovary portion obovoid, papillose to hispidulous, lobes 6, lanceolate, margins ciliate; corolla white, lobes 6, triangular. Fruit with mericarps 3, ellipsoid to obovoid, dorsally densely papillose to hispidulous, ventrally with 1 narrow groove along length of face.

Flower and Fruit: Fl. and fr. Feb. -Nov.

Distribution: Native to Antilles and North and South America. South China, Taiwan,China. Main island (Menyuwei), occasionally.

Habitat: Wastelands.

A352. 茜草科
Rubiaceae

阔叶丰花草
***Spermacoce alata*（*Borreria latifolia*）**
钮扣草属 ***Spermacoce***

形态特征：披散、粗壮草本，被毛，茎和枝均为明显的四棱柱形。叶片椭圆形或卵状长圆形，长度变化大，边缘波浪形，鲜时黄绿色，叶面平滑；叶柄扁平；托叶膜质。花数朵丛托叶鞘内，无梗，小苞片长于花萼，萼管圆筒形，花冠漏斗形，浅紫色，裂片线形。蒴果椭圆形，种子近椭圆形，两端钝。

花果期：花期 5 ～ 11 月。

分布：原产新热带地区。我国华东、华南地区有归化。南麂主岛（门屿尾），偶见。

生境：废墟和荒地。

Description: Herbs, scattered, stout, hirsute, stem and branches are distinct quadrilateral. Leaf blade elliptic or ovate-oblong, variable in length, margin wavy, yellowish green when fresh, smooth; petiole flattened; stipules membranous. Several flowers clustered in stipules sheath, sessile; bracteoles longer than calyx, calyx tube cylindrical, corolla funnelform, light purple, lobes linear. Capsule elliptic, seed subelliptic, obtuse at both ends.

Flower and Fruit: Fl. and fr. May-Nov.

Distribution: Native to the Neotropics. Naturalized in East China and South China. Main island (Menyuwei), occasionally.

Habitat: Disturbed ground and wastelands.

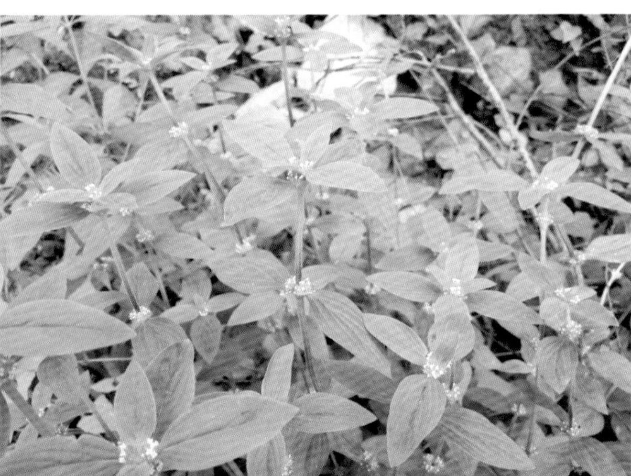

A356. 夹竹桃科
Apocynaceae

匙羹藤
Gymnema sylvestre
匙羹藤属 *Gymnema*

形态特征：木质藤本。茎皮灰褐色，疏具皮孔，幼枝被微毛，老渐无毛。叶倒卵形或卵状长圆形，厚纸质，侧脉每边 4 ~ 5 条，弯拱上升；叶柄被短柔毛，顶端具丛生腺体。聚伞花序短于叶；萼片卵形，具缘毛，花冠绿白色，裂片卵形，无毛，附属物外露。蓇葖果常单生，宽披针形，无毛，喙状渐尖。种子卵圆形，种缨白色绢状。

花果期：花期 6 ~ 8 月；果期 9 ~ 10 月。

分布：热带亚洲、非洲。我国华东至华南地区。南麂各岛屿常见。

生境：开阔的林地、灌丛。

Description: Woody vines. Stem grayish brown, sparsely lenticellate, young branchlets pubescent, glabrescent. Leaf blades obovate or ovate-oblong, thick papery, lateral veins 4 or 5 pairs, convergent; petiole sparsely pubescent, apex tufted glands. Cymes much shorter than leaves; sepals ovate, ciliate, corolla greenish white, lobes ovate, glabrous, appendages exserted. Follicles mostly solitary, broadly lanceolate in outline, glabrous, beak acuminate. Seeds ovate, coma silky white.

Flower and Fruit: Fl. Jun. -Aug. , fr. Sep. -Oct.

Distribution: Tropical Asia, Africa. East China, to South China. Throughout Nanji Islands, commonly.

Habitat: Open woods, bushland.

A356. 夹竹桃科
Apocynaceae

球兰
Hoya carnosa
球兰属 *Hoya*

形态特征：攀援灌木，附生，除花序外无毛。叶对生，肉质，宽卵状心形至卵圆状长圆形或椭圆形，顶端钝，基部圆形。假伞形花序腋外生，球形约30朵花，花白色，裂片三角形，内面多乳头状突起，边缘下弯，先端反折；副花冠星状，外角急尖，中脊隆起，边缘反折而成1孔隙，内角急尖。蓇葖果线状披针形。

花果期：花期4～6月，果期7～8月。

分布：日本、马来西亚、越南。我国华东、华南地区。南麂主岛（百亩山、国姓岙、门屿尾、三盘尾），常见。

生境：海岛的山地阴湿处或岩石上。

Description: Shrubs epiphytic, climbing, glabrous except inflorescences. Leaves opposite, fleshy, blades broadly ovate-cordate to ovate-oblong or elliptic. Pseudumbels extra-axillary, globose, ca. 30-flowered, pubescent; corolla white, lobes triangular, densely papillate inside, margin recurved, apex reflexed; corona lobes stellate spreading, outer angle acute, middle ridge prominent, margin strongly reflexed and enclosing a hollow space at base, inner angle acute. Follicles linear-lanceolate.

Flower and Fruit: Fl. Apr. -Jun. , fr. Jul. -Aug.

Distribution: Japan, Malaysia, Vietnam. East China, South China. Main island (Baimushan, Guoxingao, Menyuwei, Sanpanwei), commonly.

Habitat: Mountain damp places or rocks of coastal islands.

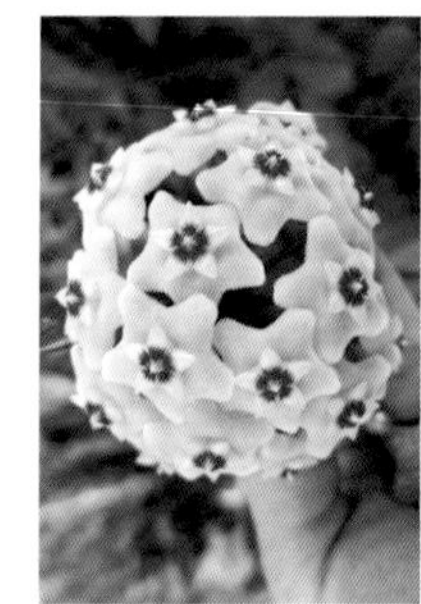

A356. 夹竹桃科
Apocynaceae

夹竹桃（红花夹竹桃、欧洲夹竹桃）
Nerium oleander（*Nerium indicum*）
夹竹桃属 *Nerium*

形态特征：常绿小乔木或灌木状，高达6米，枝条含水液。叶窄椭圆形，革质，基部楔形或下延，先端渐尖或尖。伞房状聚伞花序顶生；花芳香，花萼裂片窄三角形或窄卵形；花冠漏斗状，紫红、粉红、橙红、黄或白色；副花冠裂片5，花瓣状，流苏状撕裂。蓇葖果2，离生，圆柱形。种子长圆形，基部较窄，顶端钝、褐色，种皮被锈色短柔毛。

花果期：花期春秋季，果期冬春季。

分布：亚洲、欧洲、北美洲。热带、亚热带、温带地区广泛栽培归化。南麂主岛（大沙岙、美龄居、镇政府），栽培。

Description: Evergreen shrubs or small trees, to 6 m tall, juice watery. Leaves very narrowly elliptic, leathery, base cuneate or decurrent on petiole, apex acuminate or acute. Cymes forming terminal corymbose; flowers fragrant, sepals narrowly triangular to narrowly ovate; funnel-shaped corolla purplish red, pink, salmon, yellow, or white; corona lobes 5, petalike, tasselike lacerate. Follicles 2, distinct, cylindric. Seeds oblong, base narrower, apex blunt, brown, seed coat rust-colored pubescent.

Flower and Fruit: Fl. spring-autumn, fr. winter-spring.

Distribution: Asia, Europe, North America. Widely cultivated and naturalized in tropical, sub-tropical, and temperate parts. Main island (Dashaao, Meilingju, Zhenzhengfu), cultivated.

A356. 夹竹桃科
Apocynaceae

络石
Trachelospermum jasminoides
络石属 *Trachelospermum*

形态特征：常绿木质藤本，长达 10 m。茎赤褐色，有皮孔。叶革质或近革质，椭圆形至卵状椭圆形或宽倒卵形。圆锥状聚伞花序腋生或顶生；花白色，芳香；苞片及小苞片狭披针形，花萼 5 深裂，顶部反卷。蓇葖果线形，双生，叉开，无毛。种子多褐色，顶端具种毛。

花果期：花期 4 ～ 6 月；果期 8 ～ 10 月。

分布：日本、朝鲜、越南。我国除东北、西北地区外，其他各省区广泛分布。南麂各岛屿常见。

生境：向阳的林缘、灌丛。

Description: Evergreen lianas woody, to 10 m. Stem brownish, lenticellate. Leaves leathery or subleathery, elliptic to ovate-elliptic or broadly obovate. Cymes paniculate, terminal and axillary; flowers white, fragrant; bracts and bracteoles narrowly lanceolate, calyx 5-parted, apically recurved. Follicles linear, twining, divaricate, glabrous. Seeds much brown, coma.

Flower and Fruit: Fl. Apr. -Jun, fr. Aug. -Oct.

Distribution: Japan, Korea, Vietnam. Most areas except Northeast and Northwest China. Throughout Nanji Islands, commonly.

Habitat: Sunny edges of forests, brushwoods.

A359. 旋花科
Convolvulaceae

肾叶打碗花（滨旋花）
Calystegia soldanella
打碗花属 *Calystegia*

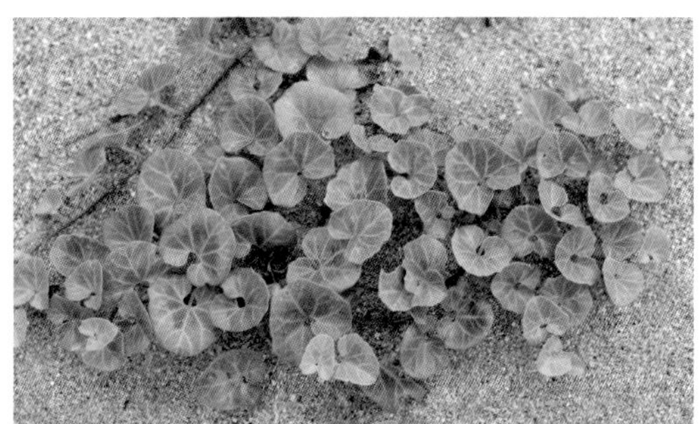

形态特征：多年生草本，无毛。茎匍匐，具棱或有时具狭翅。叶互生，肾形，顶端圆或凹，具小短尖头，全缘或浅波状；叶柄长于叶片。花单生叶腋，苞片 2，宽卵形，外萼片长圆形，内萼片卵形；花冠淡红色，钟状，冠檐微裂；雄蕊花丝基部扩大，无毛；子房无毛，柱头 2 裂，扁平。蒴果卵球形。种子黑色，平滑。

花果期：花期 5 ～ 7 月，果期 7 ～ 9 月。

分布：广泛分布于欧、亚温带及大洋洲海滨。我国沿海地区。南麂主岛（大沙岙、关帝岙、三盘尾），常见。

生境：海滨沙地、沿海沙丘。

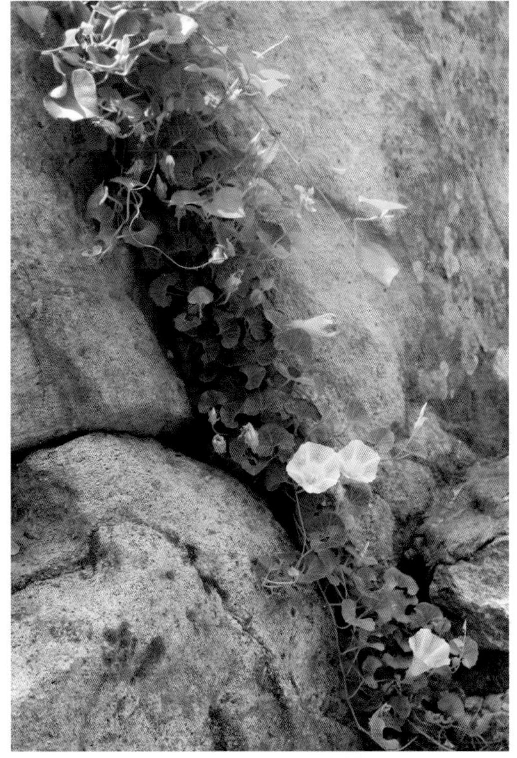

Description: Herbs perennial, glabrous. Stem prostrate, angulate or narrowly winged. Leaf alternate, reniform, apex round or concave, mucronate, margin entire or shallow sinuate; petiole longer than blade. Flowers solitary leaf axils, bracts 2, broadly ovate, outer sepal oblong, inner sepals ovate; corolla pink, campanulate, crown eaves slightly cracked; stamens filaments base enlarged, glabrous; ovary glabrous, stigma 2-lobed, flat. Capsule ovoid. Seeds black, smooth.

Flower and Fruit: Fl. May-Jul. , fr. Jul. -Sep.

Distribution: Widespread in temperate areas of Eurasia and Oceania coast. Coastal areas of China. Main island (Dashaao, Guandiao, Sanpanwei), commonly.

Habitat: Sandy seashores, coastal dunes.

A359. 旋花科
Convolvulaceae

马蹄金
Dichondra micrantha（*Dichondra repens*）
马蹄金属 *Dichondra*

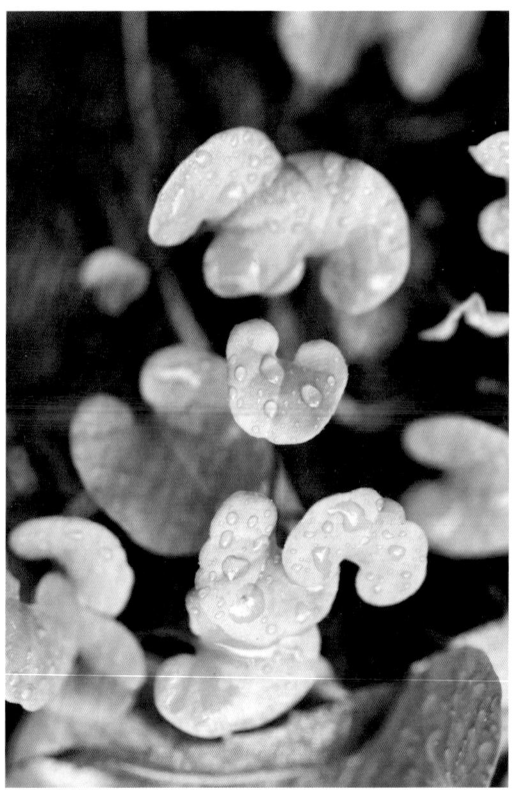

形态特征：多年生草本。茎匍匐地面，被细柔毛，节上生根。叶肾形至近圆形，背面具稀疏贴伏毛，正面无毛。花单生叶腋，花柄短于叶柄，花萼宽钟状，裂片倒卵状长圆形至匙形，边缘密生短柔毛；花冠钟状，黄色，深5裂，无毛。蒴果近球形，膜质，疏生短柔毛。种子1～2，深褐色，无毛。

花果期：花期4～5月；果期7～8月。

分布：广布两半球热带亚热带地区。我国长江流域以南和台湾。南麂主岛各处常见。

生境：山坡、路边的草地中。

Description: Herbs perennial. Stems prostrate, pubescent, rooting at nodes. Leaf blade reniform to nearly circular, abaxially sparsely appressed pilose, adaxially glabrous. Flowers solitary, axillary, pedicel shorter than petiole; calyx broadly campanulate, lobes obovate-oblong to spatulate, margin densely pubescent; corolla campanulate, yellow, 5-lobed to middle, glabrous. Capsule subglobose, membranous, sparsely pubescent. Seeds 1 -2, dark brown, glabrous.

Flower and Fruit: Fl. Apr. -May, fr. Jul. -Aug.

Distribution: Widespread in tropical and subtropical areas of both hemispheres. To the south of the Yangtze River Basin and Taiwan,China. Main island of Nanji, commonly.

Habitat: Grasslands on mountain slopes, roadsides.

A359. 旋花科
Convolvulaceae

毛牵牛（心萼薯）
Ipomoea biflora（*Aniseia biflora*）
虎掌藤属 *Ipomoea*

形态特征：攀援或缠绕草本。茎细长，有细棱，被灰白色倒向硬毛。叶心形或心状三角形，顶端渐尖，基部心形，两面被长硬毛。花序腋生，花常2朵；苞片小，线状披针形，被疏长硬毛；萼片5，结果时稍增大；花冠白色，狭钟状，雄蕊5，内藏，子房圆锥状，无毛。蒴果近球形，果瓣内面光亮。种子4，卵状三棱形，被微毛或短绒毛。

花果期：花期4～6月；果期8～10月。

分布：亚洲热带、东非及澳洲北部。我国华东至华南、西南地区。南麂主岛、大檑山屿，常见。

生境：山谷，山坡，路旁或林下，常见于干燥处。

Description: Herds, scandent or twining. Stem slender, ribbed, with grayish hirsute axial parts. Leaf blade cordate or deltate-cordate, apex acuminate, base cordate, hirsute-villous. Inflorescences axillary, often 2-flowered; bracts small and linear-lanceolate, sparse long bristles; pedicel thin, sepals 5, slightly enlarged in fruit; corolla white, narrowly campanulate, stamens 5, hide inner, ovary conical, glabrous. Capsule subglobose, inner surface bright. Seeds 4, ovoid-trigonous, uberulent to tomentellous.

Flower and Fruit: Fl. Apr. -Jun. , fr. Aug. -Oct.

Distribution: Tropical Asia, East Africa, North Australia. East China to South China, Southwest China. Main island, Daleishan Island, commonly.

Habitat: Valleys, mountain slopes, roadsides, forests, usually in dry places.

A359. 旋花科 Convolvulaceae

瘤梗番薯（瘤梗甘薯）
Ipomoea lacunosa
虎掌藤属 *Ipomoea*

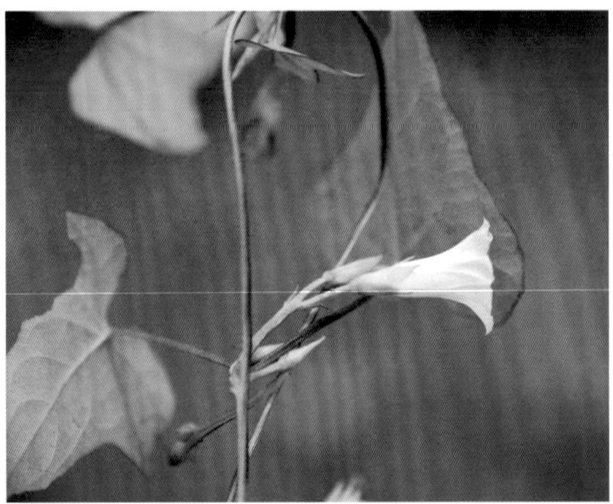

形态特征：茎缠绕，多分枝，被稀疏的疣基毛。叶互生，叶卵形至宽卵形，全缘，基部心形，先端具尾状尖。花序腋生，花序梗无毛但明显具棱，有瘤状凸起；花冠漏斗状，无毛，白色、淡红色或淡紫红色；雄蕊内藏，花丝基部有毛；子房近卵球形，被毛。蒴果近球形，中部以上被毛，先端 4 瓣裂。

花果期：花果期 7 ～ 11 月。

分布：原产热带美洲，我国华东地区有归化。南麂主岛（三盘尾），偶见。

生境：路旁、荒地。

Description: Stem twining, branched, with sparse verrucous hairs. Leaves alternate, blades ovate to broadly ovate, entire, base cordate, apex caudate. Inflorescences axillary, peduncle glabrous but distinctly angular, tuberculate convex; corolla funnelform, glabrous, white, reddish or lavender red; stamens hidden, filaments hairy base; ovary subovoid, hairy. Capsule subglobose, hairy above middle, apical 4-valved.

Flower and Fruit: Fl. and fr. Jul. -Nov.

Distribution: Native to Tropical America, naturalized in East China. Main island (Sanpanwei), occasionally.

Habitat: roadsides, wastelands.

A359. 旋花科
Convolvulaceae

牵牛
Ipomoea nil
虎掌藤属 *Ipomoea*

形态特征：一年生缠绕草本，茎上被倒向毛。叶宽卵形或近圆形，边全缘或 3 ～ 5 裂，基部心形，先端渐尖。花腋生，1 至数朵，苞片线形或叶状，被开展的微硬毛；萼片披针形，外被开展刚毛，花冠漏斗状，蓝紫色或紫红色，花冠管色淡。蒴果近球形，3 瓣裂。种子卵状三棱形，黑褐色或米黄色，被灰色短绒毛。

花果期：花期 5 ～ 6 月。

分布：原产热带美洲，现热带广布。我国大部分地区栽培或归化。南麂主岛各处常见。

生境：山坡灌丛、路边、田野，树篱。

Description: Herbs annual, twining, with retrorsely hirsute axial parts. Leaf blade broadly ovate margin entire or 3-5-lobed, apex acuminate. Inflorescences axillary, 1-to few flowered, bracts linear or filiform, spreading hirtellous; sepals lanceolate, abaxially spreading hirsute, corolla funnelform, bluish purple or purplish red, corolla tube pale. Capsule subglobose, 3-valved. Seeds black or straw, ovoid-trigonous, gray puberulent.

Flower and Fruit: Fl. May-Jun.

Distribution: Native to tropical America, now nearly circumtropical. Cultivated or naturalized in most areas of China. Main island of Nanji, commonly.

Habitat: Thickets on mountain slopes, waysides, fields, hedges.

A359. 旋花科
Convolvulaceae

厚藤（马鞍藤）
Ipomoea pes-caprae
虎掌藤属 *Ipomoea*

形态特征：多年生匍匐草本，全株无毛。茎紫红色，基部木质化。叶互生，厚纸质，宽椭圆形或肾形，顶端微缺或2裂，裂片圆，基部阔楔形、截平至浅心形，背面有2枚腺体。多歧聚伞花序腋生，1至多花；花序梗粗壮，苞片小，早落，萼片厚纸质，卵形，花冠紫色或深红色，漏斗状。蒴果球形，果皮革质，4瓣裂。种子密被褐色茸毛。

花果期：花果期5～10月。

分布：泛热带海滨。我国华东、华南等沿海地区。南麂主岛（大沙岙），常见。

生境：沿海沙地或海边的开阔地带。

Description: Herbs perennial, prostrate, glabrous. Stems purple and red, base lignified. Leaf alternate, thick papery, broadly elliptic or reniform, apex emarginate or deeply 2-lobed, round lobes, base broadly cuneate, truncate, to shallowly cordate; 2-glandular abaxially. Pleiochasium axillary, 1-to several flowered; peduncle thick, bracts small, deciduous, sepals thick papery, ovate, corolla purple or crimson, funnelform. Capsule globose, pericarp leathery, 4-lobed. Seeds densely brown pubescence.

Flower and Fruit: Fl. and fr. May-Oct.

Distribution: Pantropical littoral species. Coastal areas of East China and South China. Main island (Dashaao), commonly.

Habitat: Sandy seashores, open fields near seashores.

A359. 旋花科
Convolvulaceae

圆叶牵牛
Ipomoea purpurea
虎掌藤属 *Ipomoea*

形态特征：一年生缠绕草本，茎上被毛。叶圆卵形或宽卵形，被刚伏毛，基部心形，顶端锐尖。花序腋生，1 ～ 5 花，苞片线形，被开展的长硬毛；萼片近相等，外面 3 片长椭圆形，渐尖，内面 2 片线状披针形，外面基部均被开展的硬毛；花冠漏斗状，紫红色、红色或蓝紫色，花冠管白色。蒴果近球形，3 瓣裂。种子卵状三棱形，黑褐色或米黄色。

花果期：花期 6 ～ 9 月，果期 9 ～ 11 月。

分布：原产南北美洲，全世界广泛引种归化。我国大部分地区有分布。南麂主岛（门屿尾、三盘尾），偶见。

生境：路边，树篱，田野。

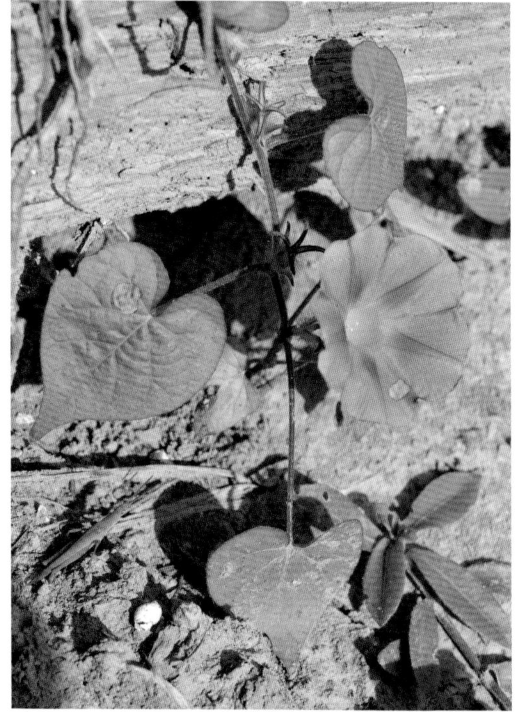

Description: Herbs annual, twining, axial parts pubescent. Leaf blade circular-ovate or broadly ovate, strigose, base cordate, apex acute. Inflorescences 1-5-flowered, axillary, bracts linear, spreading hirsute; sepals subequal, outer 3 oblong, apex acuminate, inner 2 linear-lanceolate, spreading hirsute abaxially in basal; corolla red, reddish purple, or blue-purple, with a fading to white center, funnelform. Capsule subglobose, 3-valved. Seeds black or straw colored, ovoid-trigonous.

Flower and Fruit: Fl. Jun. -Sep. , Fr. Sep. -Nov.

Distribution: Native of North and South America, introduced and naturalized worldwide. Most areas of China. Main island (Menyuwei, Sanpanwei), occasionally.

Habitat: Waysides, hedges, fields.

A359. 旋花科
Convolvulaceae

茑萝（茑萝松、羽叶茑萝）
Ipomoea quamoclit（*Quamoclit pennata*）
虎掌藤属 *Ipomoea*

形态特征：一年生缠绕草本，全株无毛。叶卵形或长圆形，羽状深裂至中脉，具 10 ~ 18 对线形细裂片。聚伞花序腋生，花少数，直立，花梗果时增厚成棒状；萼片绿色，椭圆形，先端钝，具小凸尖，花冠深红色，高脚碟状，雄蕊及柱头伸出。蒴果卵形。种子 4，卵状长圆形，黑褐色。

花果期：花期 6 ~ 10 月。

分布：原产热带美洲，现全球温带及热带地区广布。我国广泛栽培。南麂主岛偶见栽培。

Description: Herbs annual, twining glabrous. Leave blades ovate or oblong, pinnate deeply lobed to midvein, with 10-18 pairs of linear lobules. Inflorescences axillary, cymes composed of a few flowers; flowers erect, peduncle thickened into clavate at fruit; sepals green, elliptic, apex obtuse, mucroniculate, corolla deep red, salverform, stamens and stigma extend. Capsule ovate. Seeds 4, ovate-oblong, dark brown.

Flower and Fruit: Fl. Jun. -Oct.

Distribution: Native to tropical America; global temperate and tropical areas now. Widely cultivated in China.Cultivated on main island of Nanji, occasionally.

A359. 旋花科
Convolvulaceae

三裂叶薯
Ipomoea triloba
虎掌藤属 *Ipomoea*

形态特征：一年生草本。茎缠绕或平卧，无毛或茎节疏被柔毛。叶宽卵形或卵圆形，基部心形，全缘，具粗齿或3裂，无毛或疏被柔毛。伞形聚伞花序，1至数花，花序梗无毛；花梗无毛，被小瘤；萼片边缘流苏状，先端具小尖头；花冠淡红或淡紫色，漏斗状。蒴果近球形，被细刚毛，4瓣裂。

花果期：花期5～10月，果期8～11月。

分布：原产西印度群岛，现为热带广布杂草。我国华东、广东等地及其沿海岛屿。南麂主岛（门屿尾、三盘尾），偶见。

生境：田野或路边。

Description: Herbs annual. Stems twining or prostrate, glabrous or nodes sparsely pubescent. Leaf blade broadly ovate to circular, base cordate, margin entire, coarsely dentate to deeply 3-lobed, glabrous or sparsely pilose abaxially. Inflorescences dense umbellate cymes, 1-to several flowered, peduncle glabrous; pedicel glabrous, tuberculate, sepals margin fimbriate, apex mucronate; corolla pink or pale purple, funnelform, glabrous. Capsule globular, bristly pubescent, 4-valved.

Flower and Fruit: Fl. May-Oct. , fr. Aug. -Nov.

Distribution: Native to the West Indies, now a circumtropical weed. East China, Guangdong, and the coastal islands. Main island (Menyuwei, Sanpanwei), occasionally.

Habitat: Roadsides or fields.

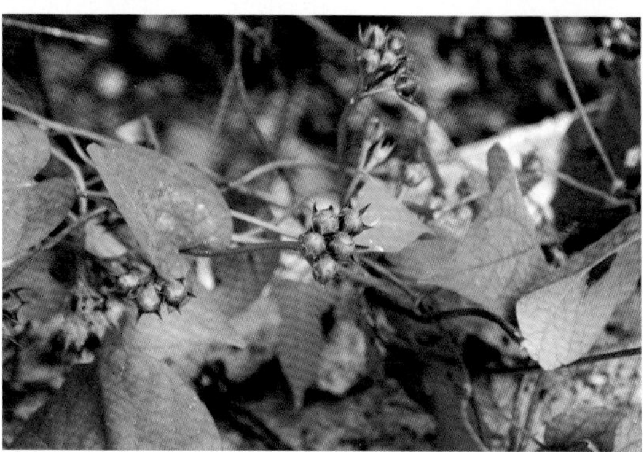

A360. 茄科
Solanaceae

红丝线
Lycianthes biflora
红丝线属 ***Lycianthes***

形态特征：亚灌木。小枝、叶背、叶柄、花梗及花萼上密被淡黄色的单毛及分枝绒毛。叶互生，二型，大小不等，大叶片椭圆状卵形，小叶片宽卵形，膜质，全缘。花序无柄，通常 2～3 朵花生于叶腋，偶见 5 朵；萼杯状，花冠淡紫色或白色，5 深裂，花冠筒隐于萼内。浆果球形，绯红色。种子多数，淡黄色，卵状三角形，具凸起网纹。

花果期：花期 5～8 月，果期 7～11 月。

分布：日本、印度和东南亚。我国华东、华南、西南地区。南麂主岛各处常见。

生境：荒野阴湿地、林下、路旁、水边及山谷中。

Description: Subshrubs. Branchlets, leaf adaxial surface, petiole, pedicel, and calyx densely covered with yellowish simple hairs and divergent tomentose. Leaves alternate, dimorphic; blade of major leaf elliptic-ovate, blade of minor leaf broadly ovate, both membranous, entire. Inflorescences sessile, usually 2-3 (5) flowered in leaf axils; calyx cup-shaped, corolla lavender or white, 5-lobed, tube concealed in calyx. Berry globose, scarlet. Seeds numerous, pale yellow, ovate triangular, with raised reticulate veins.

Flower and Fruit: Fl. May-Aug., fr. Jul. -Nov.

Distribution: Japan, India and Southeast Aisa. East China, South China, Southwest China. Main island of Nanji, commonly.

Habitat: Wet places of wastelands, forests, roadsides, by waters, valleys.

A360. 茄科
Solanaceae

构杞
Lycium chinense
枸杞属 *Lycium*

形态特征：多分枝灌木，高 0.5 ～ 2 m。枝条细弱，弯曲或俯垂，淡灰色，有纵条纹，具刺。叶纸质，单叶互生或 2 ～ 4 枚簇生，卵形、或线状披针形，顶端急尖，基部楔形。花序单生或双生叶腋，或在短枝上同叶簇生；花萼 3 ～ 5 裂，裂片有缘毛；花冠漏斗状，淡紫色，5 深裂，裂片边缘有缘毛。浆果红色，卵状。种子黄色。

花果期：花果期 6 ～ 11 月。

分布：亚洲、欧洲。我国南北各地。南麂各岛屿常见。

生境：山坡、荒地、盐碱地、路旁及村边宅旁。

Description: Multibranched shrubs, 0.5-2 m tall. Branches pale gray, slender, curved or pendulous, vertical stripes, thorns. Leaves papery, solitary or in clusters of 2-4, ovate, rhombic, or linear-lanceolate, apex acute, base cuneate. Inflorescences solitary or paired flowers on long shoots or fasciculate among leaves on short shoots; calyx campanulate, 3-5-divided to halfway, lobes densely ciliate; corolla pale purple, tube funnel-form, 5-parted, lobes pubescent at margin. Berry red, ovoid. Seeds yellow.

Flower and Fruit: Fl. and fr. Jun. -Nov.

Distribution: Asia, Europe. Throughout the south and north arcas of China. Throughout Nanji Islands, commonly.

Habitat: Slopes, wastelands, saline places, roadsides, near houses.

A360. 茄科
Solanaceae

假酸浆
Nicandra physalodes
假酸浆属 *Nicandra*

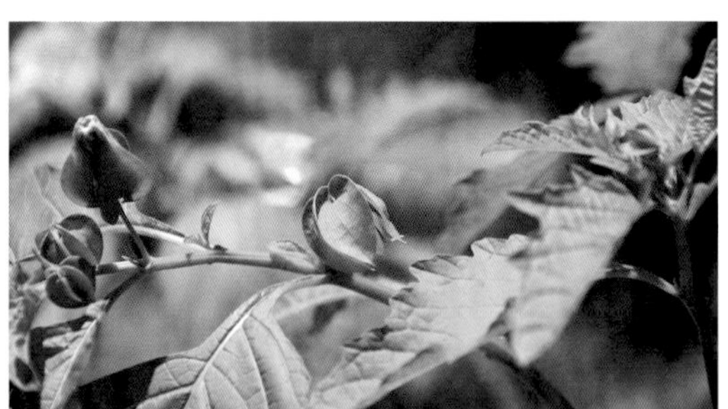

形态特征：草本，高 0.4 ～ 1.5 m。茎直立，有棱。叶卵形或椭圆形，草质，顶端急尖或短渐尖，基部楔形，边缘具圆粗齿或浅裂，两面有稀疏毛。花单生枝腋而与叶对生，花梗长于叶柄，俯垂；花萼深裂，裂片顶端尖锐，基部心脏状箭形，有 2 尖锐的耳片，果时包围果实；花冠钟状，浅蓝色，5 浅裂。浆果球状，黄色。种子淡褐色。

花果期：花期夏季，果期秋季。

分布：原产秘鲁，广泛分布。我国南北均有栽培，常归化。南麂主岛（关帝岙、三盘尾），偶见。

生境：田边、荒地、丘陵或房前屋后。

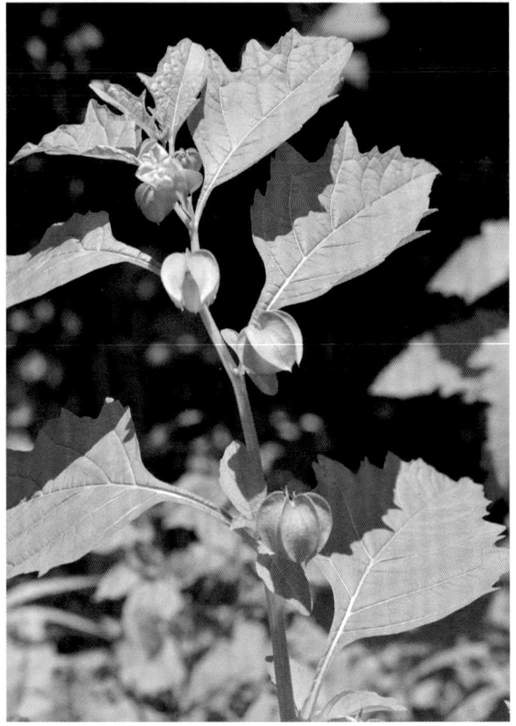

Description: Herbs, 0.4-1.5 m tall. Stems erect, angular. Leaf blade ovate or elliptic, papery, sparsely pubescent on both surfaces, base cuneate, margin lobed or coarsely sinuate-dentate, apex acute or short acuminate. Flowers solitary in axils of branches opposite leaves, pedicels usually longer than petioles, drooping; calyx deeply parted, lobes apex acute, base heart-shaped arrow-shaped, with 2 sharp auricles, enclosing fruit when fruit; corolla campanulate, light blue, 5-lobed. Berry yellow. Seeds pale brown.

Flower and Fruit: Fl. summer, fr. autumn.

Distribution: Native to Peru, widely distributed. Cultivated in north and south areas of China, usually naturalized. Main island (Guandiao, Sanpanwei), occasionally.

Habitat: Near fields, houses, and hills, wastelands.

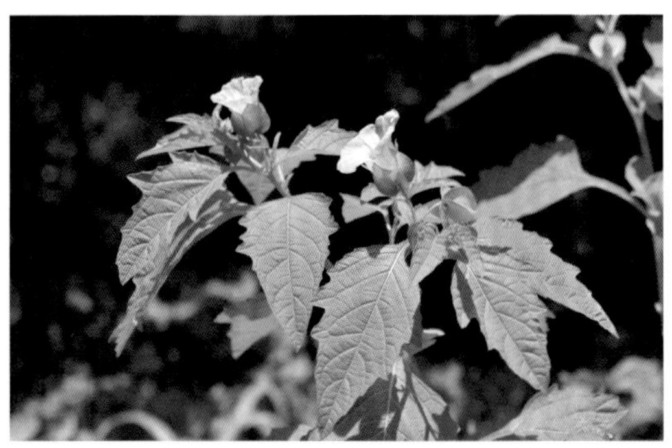

A360. 茄科
Solanaceae

烟草
Nicotiana tabacum
烟草属 *Nicotiana*

形态特征：一年生或短暂多年生草本，高 0.7 ~ 2 m，全体被腺毛。根粗壮，茎基部稍木质化。叶片卵形到椭圆形或披针形，膜质，具腺毛，顶端渐尖，基部渐狭至茎成耳状而半抱茎。花序顶生，圆锥状，多花；花萼筒状或筒状钟形，裂片三角形，渐尖，不等长；花冠漏斗状，淡红色，筒部色更淡。蒴果卵状或矩圆状。种子圆形，褐色。

花果期：花期夏季，果期秋季。

分布：原产南美洲。我国南北各省区广为栽培。大檑山屿，栽培或野生。

Description: Herbs annual or short-lived perennial, 0.7-2 m tall, glandular hairy overall. Roots stout, stem base slightly woody. Leaf blade ovate to elliptic or lanceolate, membranous, glandular hairy, base narrowed, nearly half clasping, apex acuminate. Inflorescences many-flowered, terminal, much-branched panicles; calyx tubular or tubular-campanulate, lobes deltate, acuminate, unequal;corolla funnelform, pink, tube color lighter. Capsules ellipsoid or ovoid. Seeds brown, rounded.

Flower and Fruit: Fl. summer, fr. autumn.

Distribution: Native to South America. Widely cultivated throughout China. Daleishan Island, cultivated or wild.

A360. 茄科
Solanaceae

苦蘵
Physalis angulata
灯笼果属 ***Physalis***

形态特征：一年生草本，高 30 ~ 50 cm。茎多分枝，分枝纤细。叶片卵形至卵状椭圆形，顶端渐尖或急尖，基部阔楔形或楔形，全缘或有不等大的牙齿，两面近无毛。花梗纤细和花萼一样生短柔毛，5 裂，裂片披针形，生缘毛；花冠淡黄色或白色，喉部常有紫色斑纹。果萼卵球状，薄纸质；浆果，种子圆盘状。

花果期：花期 5 ~ 7 月，果期 7 ~ 12 月。

分布：全球广布。我国长江流域以南地区。南麂主岛各处常见。

生境：受干扰区，森林、村庄、路旁。

Description: Herbs annual, 30-50 cm tall. Stems much branched, branches slender. Leaf blade ovate to obovate-elliptic, apex acuminate or acute, base cuneate or broadly cuneate, margin entire or dentate, both surfaces subglabrous. Pedicels slender and calyx pubescent, 5-lobed, lobes lanceolate, ciliate; corolla pale yellow or white, fspotted in throat. Fruiting calyx ovoid, papery; berry, seeds discoid.

Flower and Fruit: Fl. May-Jul. , fr. Jul. -Dec.

Distribution: Worldwide. To the south of Yangtze River Basin. Main island of Nanji, commonly.

Habitat: Disturbed sites, forests, villages, roadsides.

A360. 茄科
Solanaceae

小酸浆（毛苦蘵）
***Physalis minima*（*Physalis angulata* var. *villosa*）**
灯笼果属 *Physalis*

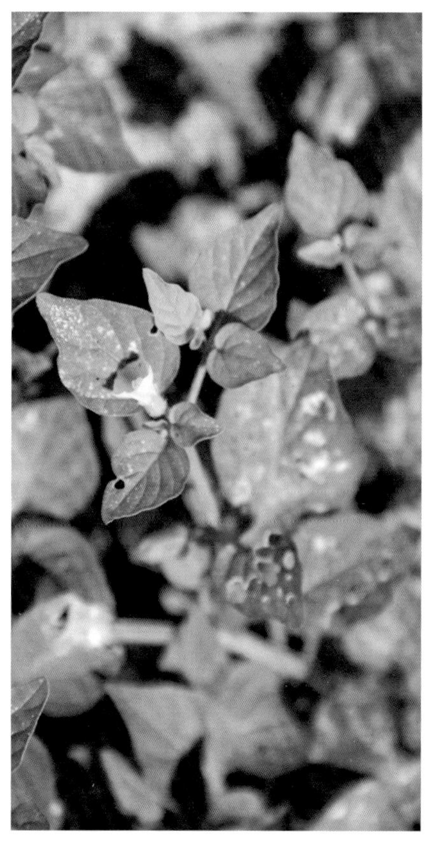

形态特征：一年生草本。茎匍匐或直立，被多细胞柔毛。叶柄细弱，叶片卵形或卵状披针形，顶端渐尖，基部歪斜楔形，全缘、波状或有少数粗齿，两面脉上有柔毛；花梗被短柔毛；花萼钟状，外生短柔毛，裂片三角形，花冠黄色。果萼近球状或卵球状，果实球状。

花果期：花期 4 ～ 6 月；果期 8 ～ 10 月。

分布：全球广布。我国华南、西南地区。南麂主岛（东方岙、门屿尾），偶见。

生境：山谷、林缘。

Description: Herbs annual. Stems prostrate or erect, pubescent with long many-celled hairs. Petiole thin, leaf blade ovate or ovate-lanceolate, apex acuminate, base cuneate, often oblique, margin entire and sinuate, or with a few coarse teeth, pubescent along veins; pedicel pubescent; calyx campanulate, pubescent, lobes deltate, corolla yellow. Fruiting calyx subglobose or ovoid, berry globose.

Flower and Fruit: Fl. Apr. -Jun. , fr. Aug. -Oct.

Distribution: Worldwide. South China, Southwest China.Main island (Dongfangao, Menyuwei), occasionally.

Habitat: Valleys, forest edge.

A360. 茄科
Solanaceae

毛酸浆
Physalis philadelphica (Physalis pubescens)
灯笼果属 ***Physalis***

形态特征：一年生草本。茎多分枝，被毛，后脱落。叶阔卵形，顶端急尖，基部斜心形，边缘常有不等大的尖牙齿，两面疏生毛，脉上毛较密；叶柄密生短柔毛。花单生叶腋，花梗密生短柔毛，花萼钟状，半裂，花冠淡黄色，喉部具紫斑。果萼卵状，具5棱角，基部微凹；浆果球状，黄色或有时带紫色。种子近圆盘状。

花果期：花果期5～8月。

分布：原产墨西哥，我国有广泛栽培和归化。南麂主岛（三盘尾），偶见。

生境：路边、荒地、田野及住宅旁。

Description: Herbs annual. Stems branched, glabrescent or sparingly pubescent. Leaf blade broadly ovate, apex acute, base cordate, often oblique, margin usually unequal dentate, sparsely hairy on both surfaces, densely hairy on veins; petiole densely pubescent. Flowers alone axillary, pedicels densely pubescent, calyx campanulate, divided to halfway, corolla pale yellow, spotted in throat. Fruiting calyx ovate, weakly 5-angled, slightly invaginated at base; berry globose, yellow or sometimes purple. Seeds discoid.

Flower and Fruit: Fl. and fr. May-Aug.

Distribution: Native to Mexico, widely cultivated and naturalized in China. Main island (Sanpanwei), occasionally.

Habitat: Roadsides, wastelands, fields and near houses.

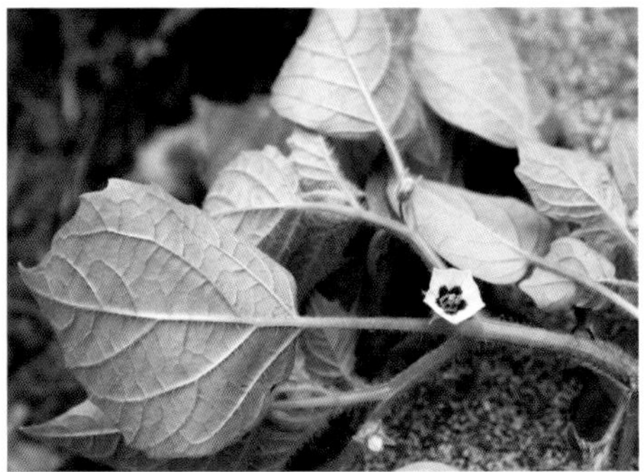

A360. 茄科
Solanaceae

少花龙葵
Solanum americanum
茄属 *Solanum*

形态特征：一年生草本或短暂多年生植物，高 25 ~ 100 cm。茎紫色或绿色，多直立。叶卵形至卵状长圆形，先端渐尖，基部楔形下延至叶柄而成翅，波状或有不规则粗齿。伞形花序腋生，花 3 ~ 6 朵；萼绿色，花冠白色，筒部隐于萼内，5 裂，裂片卵状披针形。浆果球状，幼时绿色，成熟后黑色。种子近卵形，两侧压扁。

花果期：花期 6 ~ 10 月，果期 7 月至翌年 1 月。

分布：热带温带地区广布。我国华东至华南地区。南麂各岛屿常见。

生境：荒地、路旁、田野。

Description: Herbs annual or short-lived perennial, 25-100 cm talls. Stems green or purple, mostly erect. Leaf blade ovate to oval oblong, apex acute, base cuneate and extends to petiole to form a wing, sinuate or sparingly dentate. Umbel axillary, 3-6-flowered; calyx green, corolla white, tube hide in calyx, 5-lobed, lobes ovate-lanceolate. Berry globose, green when young, ripening black. Seeds subovate, flatten on both sides.

Flower and Fruit: Fl. Jun. -Oct. , Fr. Jul. -Jan.

Distribution: Widespread in all tropical and temperate regions. East China to South China. Throughout Nanji Islands, commonly.

Habitat: Waste places, roadsides, fields.

A360. 茄科
Solanaceae

牛茄子
Solanum capsicoides
茄属 *Solanum*

形态特征：草本或亚灌木，直立或蔓生，高 30 ~ 100 cm。茎上白色皮孔明显，具淡黄色针状刺，直或微弯。叶阔卵形，基部心形，5 ~ 7 浅裂或半裂；叶柄粗壮，微具纤毛及较大的直刺。总状花序腋外生，花 1 ~ 4 朵，萼杯状，毛被同茎，花冠白色，裂片披针形。浆果扁球状，橙红色。种子浅黄色，盘状，具明显狭圆翅。

花果期：花期 6 ~ 8 月，果期 8 ~ 10 月。

分布：原产巴西，现温暖地区广布。我国大部分地区均有分布。南麂主岛（三盘尾、镇政府、兴岙），偶见。

生境：路旁荒地、疏林或灌木丛。

Description: Herbs or subshrubs, erect or sprawling, 30-100 cm tall. Stems conspicuously white lenticellate, copiously armed with pale yellow, needlelike prickles, straight to slightly recurved. Leaf blade broadly ovate, base cordate, margin 5-7-lobed to halfway; petiole stout, slightly ciliated and larger straight thorns. Inflorescences extra-axillary, racemose, 1-4-flowered, calyx cup-shaped, pubescent as on stems, corolla white, lobes lanceolate. Berry subglobose, orange-red. Seeds yellowish, discoid, with conspicuous thin orbicular wing.

Flower and Fruit: Fl. Jun-Aug. , fr. Aug.-Oct.

Distribution: Native to Brazil; now widespread of warm regions. Most areas of China. Main island (Sanpanwei, Zhenzhengfu, Xingao), occasionally.

Habitat: Wastelands, near roadsides, open forests, thickets.

A360. 茄科
Solanaceae

北美刺龙葵
Solanum carolinense
茄属 *Solanum*

形态特征：多年生草本植物，高可达100 cm。茎分枝松散，有淡黄色的刺和星状短绒毛。叶片卵形至长圆形，具不规则波状齿，叶两面均具有微黄色的星状短毛。聚伞圆锥花序，花梗有刺，萼裂片无刺；花冠紫色，偶有白色。浆果球状，熟时淡橙色或黄色，光滑无毛。种子倒卵形，扁平，浅橙色或黄色。

花果期：花期7～8月，果期9～11月。

分布：原产美洲加勒比地区，热带地区广布。南麂主岛（三盘尾、镇政府），偶见。

生境：荒地、路旁。

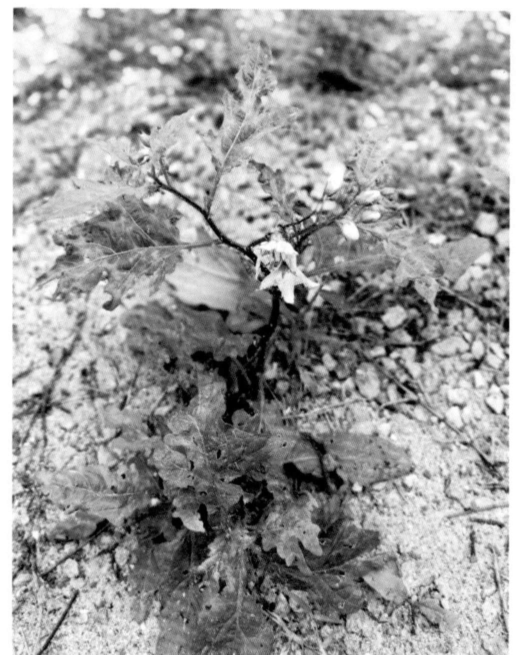

Description: Herbs perennial, up to 100 cm tall. Stems loosely branched, with yellowish spines and stellate tomentose. Leaf blade ovate to oblong, with irregular undulate teeth, both sides with yellowish stellate short hairs. Cymous panicle, pedicel thorns, sepals stingless; corolla purple, occasionally white. Berry globose at maturity, pale orange or yellow, smooth, glabrous. Seeds obovate, flattened, pale orange or yellow.

Flower and Fruit: Fl. Jul. -Aug. , fr. Sep. -Nov.

Distribution: Native to Caribbean, widely distributed in tropical areas. Main island (Sanpanwei, Zhenzhengfu), occasionally.

Habitat: Wastelands, roadsides.

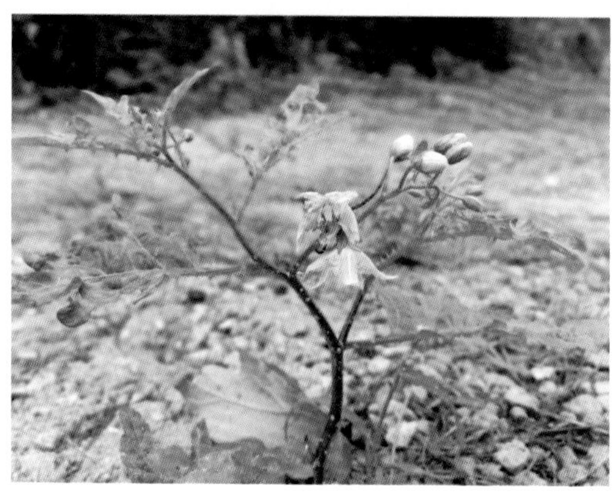

A360. 茄科
Solanaceae

白英
Solanum lyratum
茄属 *Solanum*

形态特征：多年生草质藤本。茎及小枝均密被具节长柔毛。叶互生，多数为琴形，基部常 3 ~ 5 深裂，裂片全缘，两面均被白色发亮的长柔毛，中脉明显。聚伞花序顶生或腋外生，疏花，花萼环状，无毛；花冠蓝紫色或白色。浆果球状，成熟时红黑色。种子近盘状，扁平。

花果期：花期 7 ~ 8 月；果期 10 ~ 11 月。

分布：东亚、东南亚。我国大部分地区有分布。南麂主岛各处常见。

生境：山谷草地、路旁、田边。

Description: Perennial vines herbaceous. Stems and twigs are densely nodular with many long-celled hairy. Leaf alternate, most piano shaped, the base often deep 3-5-parted, lobes margin entire, both sides are covered with shinny villous, obvious midrib. Cyme terminal or extra-axillary, sparsely flowered; corolla bluish violet or white. Berry globose. Seeds nearly discoid, flat.

Flower and Fruit: Fl. Jul. -Aug. , fr. Oct. -Nov.

Distribution: East Asia and Southeast Asia. Most areas of China. Main island of Nanji, commonly.

Habitat: Grasslands in valleys, near roads and fields.

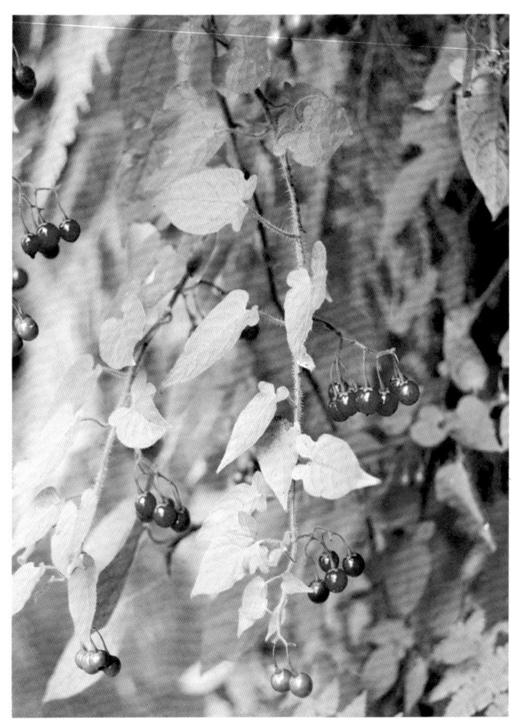

A360. 茄科
Solanaceae

龙葵
Solanum nigrum
茄属 *Solanum*

形态特征：一年生直立草本，高 25 ～ 100 cm，被短柔毛，无刺。茎有纵棱，疏生短柔毛。叶互生，卵形，先端短尖，基部楔形至阔楔形，近全缘，无毛或两面被稀疏短柔毛。蝎尾状花序腋外生，有花 4 ～ 10 朵，花萼小，浅杯状；花冠白色，5 深裂，裂片卵圆形。浆果球形，熟时黑色。种子盘状。

花果期：花期 6 ～ 9 月；果期 7 ～ 11 月。

分布：亚洲、欧洲。我国各地区均有分布。南麂主岛各处常见。

生境：田边，荒地及村庄附近。

Description: Herbs annual, erect, 25-100 cm, pubescent, unarmed. Stems often angular, sparsely pubescent. Leaf alternate, oval, apex acuminate, base cuneate to wide cuneate, nearly margin entire, glabrous or both sides sparse pubescent. Inflorescences extra-axillary umbels, 4-10 flowers, calyx small, shallow cup shaped; corolla white, the base deep 5-lobed, lobes oval. Berry dull black, globose. Seeds discoid.

Flower and Fruit: Fl. Jun. -Sep. , fr. Jul. -Nov.

Distribution: Asia, Europe. Throughout China. Main island of Nanji, commonly.

Habitat: Fields, wastelands, and near villages.

A366. 木樨科
Oleaceae

探春花
Jasminum floridum
素馨属 *Jasminum*

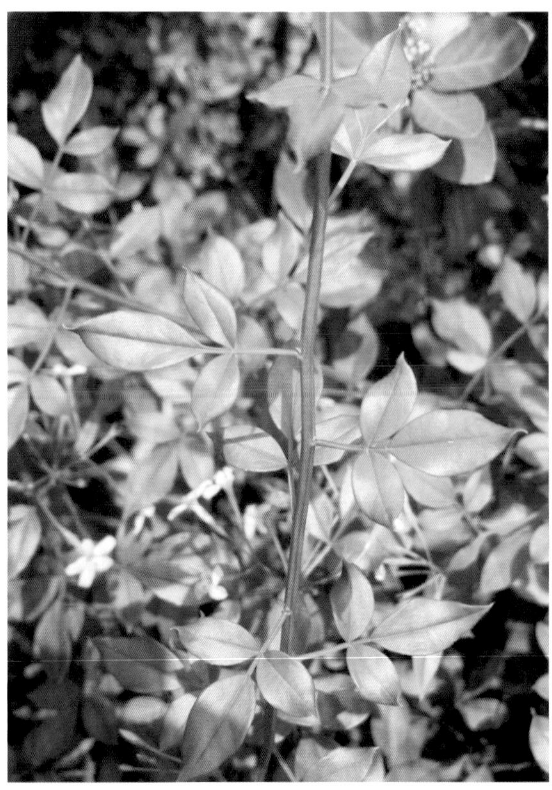

形态特征：直立或攀援灌木，高 0.4 ～ 3 m。小枝四棱，扭曲，无毛。叶互生，复叶或单叶，小叶 3 或 5，顶生小叶具柄，侧生小叶片近无柄；叶片卵形至椭圆形，先端急尖，基部楔形或圆形。聚伞花序顶生，3 ～ 25 花；苞片锥形；花萼具 5 条突起的肋，裂片锥状线形；花冠黄色，近漏斗状，裂片卵形或长圆形，先端锐尖。果球形，熟时黑色。

花果期：花期 5 ～ 9 月，果期 9 ～ 10 月。

分布：原产我国中西部地区，各地常栽培观赏。南麂主岛（大沙岙、美龄居、镇政府），栽培。

生境：坡地、山谷、林中或灌木丛。

Description: Shrubs erect or scandent, 0.4-3 m. Branchlets 4-angled, twisted, glabrous. Leaves alternate, compound or simple, leaflets 3 or 5, terminal petiolule, lateral ones subsessile; blades ovate to elliptic, apex acute, base cuneate or rounded. Cymes terminal, 3-25-flowered; bracts subulate; calyx with 5 raised ribs, lobes subulate-linear; corolla yellow, nearly funnelform, lobes ovate or oblong, apex acute. Berry ripening black, globose.

Flower and Fruit: Fl. May-Sep. , fr. Sep. -Oct.

Distribution: native in central and western regions of China, usually cultivated for ornamental. Main island (Dashaao, Meilingju, Zhenzhengfu), cultivated.

Habitat: Slopes, valleys, woods, thickets.

A366. 木樨科
Oleaceae

华素馨
Jasminum sinense
素馨属 *Jasminum*

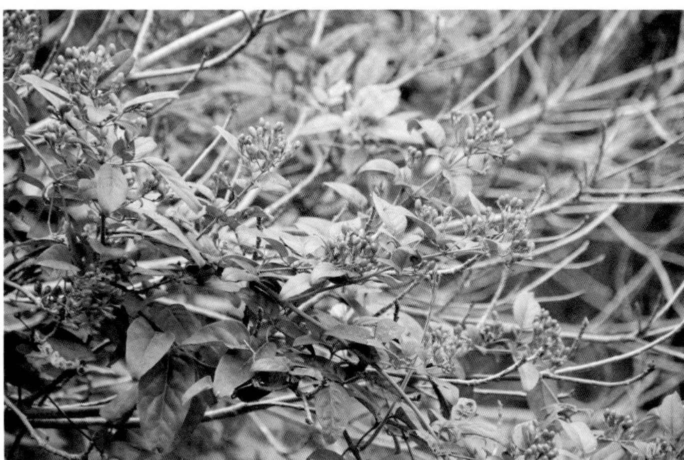

形态特征：缠绕藤本，高 1 ～ 8 m。小枝圆柱形，密被锈色长柔毛。叶对生，三出复叶；小叶纸质，卵形或卵状披针形，基部圆形，叶缘反卷，两面被锈色柔毛，羽状侧脉 3 ～ 6 对，两面明显。聚伞花序顶生或腋生，常呈圆锥状排列；花芳香，花萼被毛，裂片线形，果时稍增大；花冠白色或淡黄色，高脚碟状，花冠管细长，裂片 5，长圆形或披针形。果球形，黑色。

花果期：花期 6 ～ 10 月，果期 9 月至翌年 5 月。

分布：我国华东至西南地区。南麂主岛各处常见。

生境：山坡、灌丛或林中。

Description: Vines twining, 1-8 m. Branchlets terete, densely rusty villous.Leaves opposite, 3-foliolate; leaflet blade ovate to ovate-lanceolate, papery, base rounded, margin revolute, rusty pubescent and more densely so along veins abaxially, pinnate lateral veins 3-6 pairs, evident on both surfaces. Cymes terminal or axillary, many-flowered congested panicles; flowers fragrant, calyx pilose, lobes linear, slightly enlarged in fruit; corolla white or yellowish, salverform, corolla tube slender, lobes 5, oblong or lanceolate. Berry black, globose.

Flower and Fruit: Fl. Jun. -Oct. , fr. Sep. -May.

Distribution: East China to Southwest China. Main island of Nanji, commonly.

Habitat: Slopes, thickets, woods.

A366. 木樨科
Oleaceae

小蜡
Ligustrum sinense
女贞属 *Ligustrum*

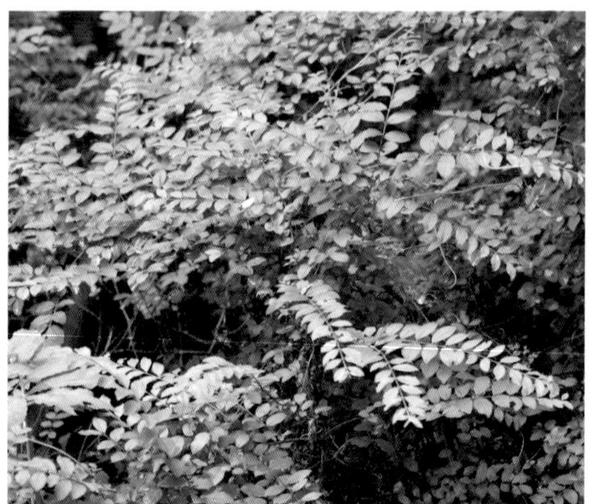

形态特征：落叶灌木或小乔木，高 2 ~ 4 m，偶见 7 m。小枝圆柱形，幼时被毛，老时近无毛。叶片纸质或薄革质，卵形、长圆状椭圆形至披针形，基部宽楔形至近圆形，常沿中脉被短柔毛，侧脉 4 ~ 8 对，上面微凹入，下面略凸起；叶柄被短柔毛。圆锥花序顶生或腋生，塔形；花萼先端呈截形或呈浅波状齿。果近球形。

花果期：花期 3 ~ 6 月，果期 9 ~ 12 月。

分布：越南。我国华东至西南地区。南麂主岛各处常见。

生境：混交林、山谷、溪边。

Description: Shrubs or small trees 2-4(7) m, deciduous. Branchlets terete, villous, pubescent, to glabrescent. Leaf blade ovate, oblong-elliptic to lanceolate, papery to somewhat leathery, base cuneate to subrounded, often pubescent along midrib, primary veins 4-8 on each side of midrib, impressed or plane adaxially, somewhat raised abaxially; petiole pubescent. Panicles terminal or axillary, turriform; calyx apex truncate or shallow wavy teeth. Fruit subglobose.

Flower and Fruit: Fl. Mar. -Jun. , fr. Sep. -Dec.

Distribution: Vietnam. East China to Southwest China. Main island of Nanji, commonly.

Habitat: Mixed forests, valleys, along streams.

A370. 车前科
Plantaginaceae

车前
Plantago asiatica
车前属 *Plantago*

形态特征：多年生草本，株高 20 ～ 60 cm。根茎短而粗，具多数须状丛生根。叶宽卵形、薄纸质，呈莲座状着基生，边缘波状，主脉 3 ～ 7 条，基部下沿收缩至细长叶柄。穗状花序排列疏松，小花白色无毛。蒴果呈椭圆形，内有 4 粒黑色种子。

花果期：花期 4 ～ 8 月，果期 6 ～ 9 月。

分布：原产东北亚，东南亚、北美洲也有分布。分布几遍我国。南麂主岛各处常见。

生境：山坡，沟壑，河岸，田野，路边，荒地，草坪。

Description: Herbs perennial , 20-60 cm tall. Stem short and thick, with many whisklike tufted roots. Leaf blade broadly ovate, thinly papery, margin undulate, main veins 3-7, base broadly cuneate to subrounded and decurrent onto petiole. Spikes loosely flowered, corolla white, glabrous. Capsule oval, with 4 black seeds.

Flower and Fruit: Fl. Apr. -Aug. , fr. Jun. -Sep.

Distribution: Northeast Asia, Southeast Asia, North America. Almost throughout China. Main island of Nanji, commonly.

Habitat: Mountain slopes, ravines, riverbanks, fields, roadsides, wastelands, lawns.

A370. 车前科
Plantaginaceae

直立婆婆纳
Veronica arvensis
婆婆纳属 *Veronica*

形态特征：一年生草本，高 5 ～ 30 cm。茎直立或上升，不分枝或铺散分枝，有两列多细胞白色长柔毛。叶常 3 ～ 5 对，下部的有短柄，中上部的无柄，卵形至卵圆形，3 ～ 5 脉，边缘具圆或钝齿，两面被硬毛。总状花序长而多花，被腺毛；苞片互生，叶状，花冠蓝紫色或蓝色。蒴果倒心形，强烈侧扁，边缘有腺毛。种子矩圆形。

花果期：花期 4 ～ 5 月。

分布：原产南欧和西南亚，世界广泛归化。常见于华东和华中地区。南麂主岛（门屿尾、三盘尾）、大檑山屿，常见。

生境：路边及荒野草地。

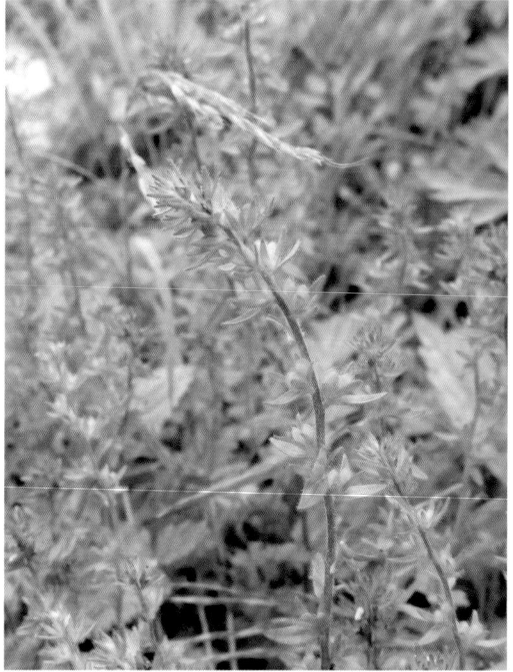

Description: Herbs annuals, 5-30 cm tall. Stems erect or ascending, simple or branched and diffuse, with white multicellular hairs along 2 lines. Leaves often 3-5 pairs, lower ones short petiolate, upper sessile; leaf blade ovate-orbicular, hirsute, margin crenate, veins 3-5. Racemes terminal, lax, many flowered, with glandular hairs; bracts alternate, leaflike, corolla blue to blue-purple, rotate. Capsule obcordate, strongly compressed, margin glandular ciliate. Seeds oblong.

Flower and Fruit: Fl. Apr. -May.

Distribution: Native to South Europe and Southwest Asia, naturalized over most of the world. Usually in East China and Central China. Main island (Menyuwei, Sanpanwei), Daleishan Island, commonly.

Habitat: Waste grassy places and along roads.

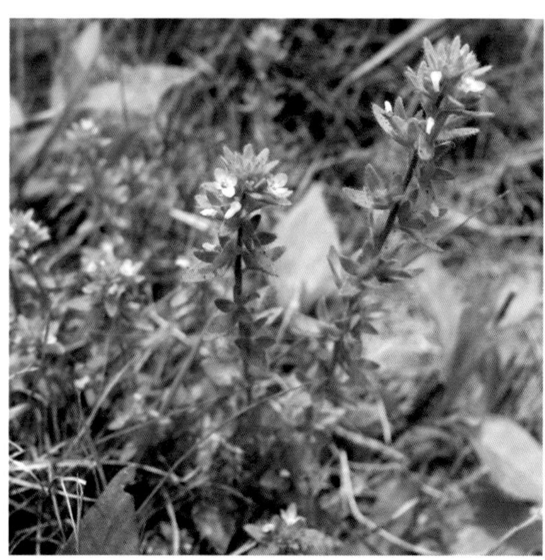

A370. 车前科
Plantaginaceae

蚊母草
Veronica peregrina
婆婆纳属 *Veronica*

形态特征： 一年生草本，高 5 ～ 25 cm。通常自基部多分枝，主茎直立，丛生状侧枝披散。叶无柄，基生叶倒披针形，茎生叶长矩圆形，全缘或中上端有三角状锯齿。松散的总状花序较长；花冠白色或浅蓝色。蒴果倒心形，明显侧扁。种子长圆形，扁平。

花果期： 花果期 5 ～ 6 月。

分布： 原产北美洲，东亚、欧洲有归化。我国南北多地有野生。南麂主岛（门屿尾、三盘尾），常见。

生境： 潮湿的荒地、路旁。

Description: Herbs annual, 5-25 cm tall. Usually branched from base, main stem erect, caespitose lateral branches spread. Leaves sessile, lower ones oblanceolate, upper narrowly oblong, margin entire or deltoid dentate above middle. Racemes elongated, lax, corolla white or pale blue. Capsule obcordate, strongly compressed. Seeds oblong, flattened.

Flower and Fruit: Fl. and fr. May-Jun.

Distribution: Native to North America, naturalized in East Asia, Europe. Most areas of the south and north China.Main island (Menyuwei, Sanpanwei), commonly.

Habitat: Moist waste lands, roadsides.

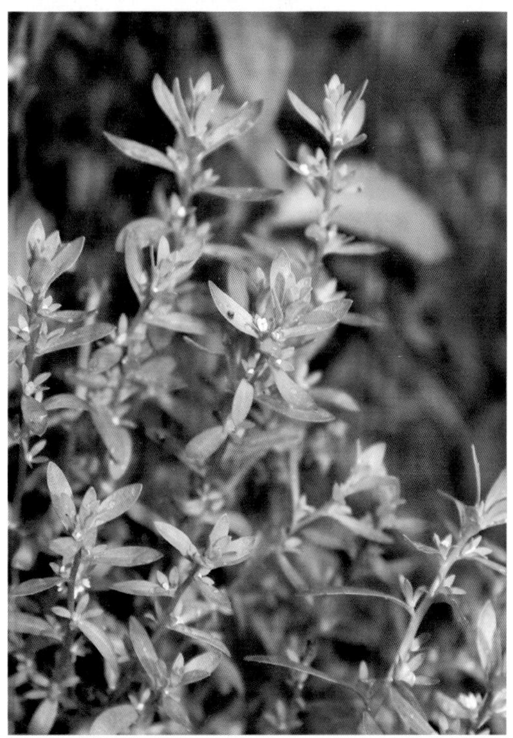

A370. 车前科
Plantaginaceae

阿拉伯婆婆纳
Veronica persica
婆婆纳属 *Veronica*

形态特征：一年生草本，偶二年生，高 10 ~ 50 cm。茎铺散，密生两列多细胞柔毛。对生叶常 3 ~ 4 对；叶片卵状披针形到近圆形，两面均匀疏生短柔毛，边缘具锯齿。总状花序顶生，长而疏散；苞片互生，叶状，有柄；花萼 4 裂，裂片卵状披针形，疏被毛，3 脉；花冠常蓝色，辐射状，喉部疏生毛，裂片卵形到圆形。蒴果倒心形，被腺毛。

花果期：花期 3 ~ 5 月。

分布：原产西南亚。我国南方地区多有归化。南麂主岛（关帝岙、门屿尾、三盘尾），常见。

生境：荒地和路边。

Description: Herbs annuals, sometimes biennials, 10-50 cm tall. Stems diffuse, densely pubescent with multicellular hairs along 2 lines. Opposite leaves in 3 or 4 paris; leaf blade ovate-lanceolate to suborbicular, evenly sparsely pubescent on both surfaces, margin obtusely crenate-serrate. Racemes terminal, lax, very long; bracts alternate, leaflike, petiolate; calyx 4-lobed; lobes ovate-lanceolate, sparsely pubescent, veins 3; corolla usually blue, rotate, throat sparsely hairy, lobes ovate to orbicular. Capsule obcordate, glandular hairy.

Flower and Fruit: Fl. Mar. -May.

Distribution: Native to Southwest Asia. Naturalized in most areas of southern China. Main island (Guandiao, Menyuwei, Sanpanwei), commonly.

Habitat: Waste fields and roadsides.

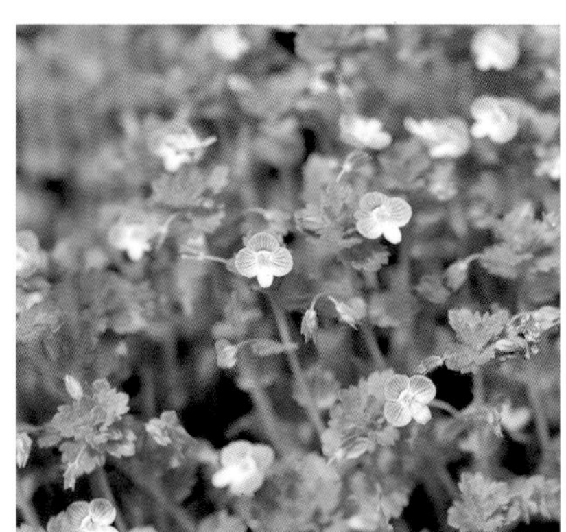

A377. 爵床科
Acanthaceae

早田氏爵床
Justicia hayatae（*Justicia procumbens* var. *ciliata*）
爵床属 *Justicia*

形态特征：草本，高 30～50 cm。茎具 4 角，有沟，沿沟两侧被短柔毛。叶片卵形，无毛，边缘全缘，先端锐尖。穗状花序大多顶生或在叶腋上部腋生；苞片倒卵状椭圆形，无毛，边缘白色，先端锐尖；小苞片披针状椭圆形。蒴果具棕黄色短柄，长圆形，无毛。种子 4，卵形，褶皱，无毛。

花果期：花期 6～8 月，果期 8～10 月。

分布：我国浙江、台湾。南麂主岛各处常见。

生境：海岸，沙滩。

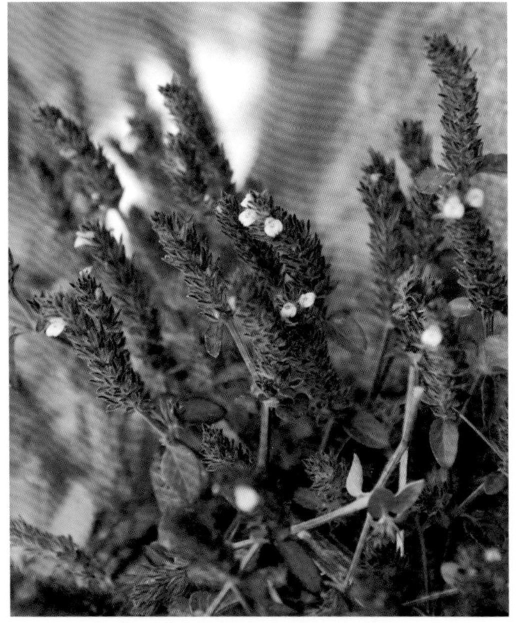

Description: Herbs, 30-50 cm tall. Stems 4-angled, sulcate, bifariously pubescent along sulcae. Leaf blade ovate, glabrous, margin entire, apex acute. Spikes mostly terminal or axillary at upper leaf axils; bracts obovate-elliptic, glabrous, margin white, apex acute; bracteoles lanceolate-elliptic. Capsule with a brownish yellow short stipe, oblong, glabrous. Seeds 4, ovate, rugose, glabrous.

Flower and Fruit: Fl. Jun. -Aug. , fr. Aug. -Oct.

Distribution: Zhejiang, Taiwan,China. Main island of Nanji, commonly.

Habitat: Seashores, sandy places.

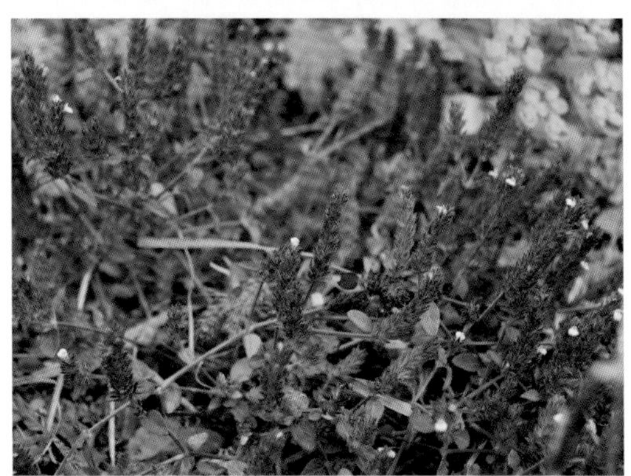

A377. 爵床科 Acanthaceae

爵床

Justicia procumbens（*Rostellularia procumbens*）

爵床属 *Justicia*

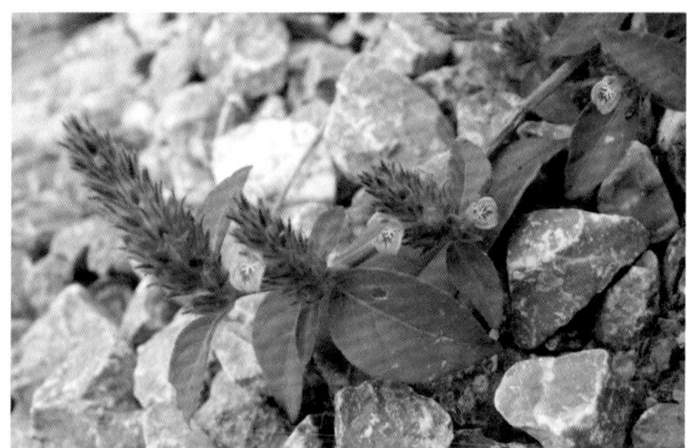

形态特征：一年生草本，高 20 ~ 50 cm。茎匍匐，四棱，具浅槽，有灰白色细柔毛。叶对生，椭圆形至椭圆状长圆形，近无毛或疏生刺毛，先端锐尖或钝，基部宽楔形或近圆形。穗状花序顶生或生上部叶腋，圆柱形；苞片卵形至披针形，有缘毛；花萼裂片 4，线形，边缘灰白色，有缘毛；花冠下唇粉色或白色带红点，3 裂。蒴果具 4 粒种子。

花果期：花期 8 ~ 10 月，果期 10 ~ 11 月。

分布：亚洲及澳大利亚。广布我国秦岭以南各地。南麂主岛各处常见。

生境：荒地、路边、草地，开阔的田边、海滨森林。

Description: Herbs annual, 20-50 cm tall. Stem procumbent, 4-angled, sulcate, pubescent. Leaves opposite, elliptic, ovate-elliptic, or elliptic-oblong, subglabrous to sparsely hispid, base broadly cuneate to subrounded, apex acute to obtuse. Spikes terminal or axillary in upper leaf axils, cylindric; bracts ovate to lanceolate, margin ciliate; calyx 4-lobed to base, lobes linear, margin yellowish white and ciliate, corolla pink or white and red-spotted on lower lip, 3-lobed. Capsule 4-seeded.

Flower and Fruit: Fl. Aug. -Oct. , fr. Oct. -Nov.

Distribution: Asia, Australia. Widely to the South of Qinling Mountains. Main island of Nanji, commonly.

Habitat: Wastelands, roadsides, lawns, open fields,

A382. 马鞭草科
Verbenaceae

马缨丹
Lantana camara
马缨丹属 *Lantana*

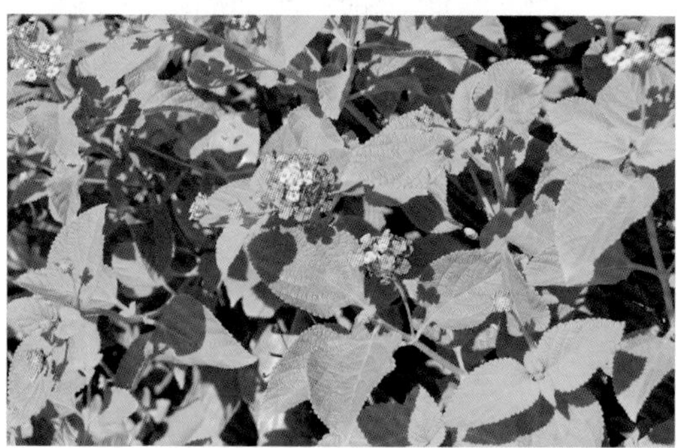

形态特征：灌木，有长弱分枝，具粗壮弯曲皮刺，被短柔毛。叶片卵形到长圆形，纸质，皱褶，很粗糙，具短硬毛，揉烂后有香味，基部圆形到近心形，边缘具圆齿。头状花序顶生，花黄色或橙色，常在开后不就变为深红色。核果深紫色，球状。

花果期：花期全年。

分布：原产热带美洲，热带亚热带地区常见归化。我国华东、华南地区有归化。南鹿主岛（大沙岙、美龄居、镇政府），栽培或归化。

生境：开阔的荒地或海边。

Description: Shrubs with long weak branches, armed with stout recurved prickles, pubescent. Leaf blade ovate to oblong, papery, wrinkled, very rough, with short stiff hairs, aromatic when crushed, base rounded to subcordate, margin crenate. Capitula terminal, flowers yellow or orange, often turning deep red soon after opening. Drupes deep purple, globose.

Flower and Fruit: Fl. year around.

Distribution: Native to tropical America, often naturalized in other tropical and subtropical regions. Naturalized in East China and South China. Main island (Dashaao, Meilingju, Zhenzhengfu), cultivated or naturalized.

Habitat: Open waste places and near coast.

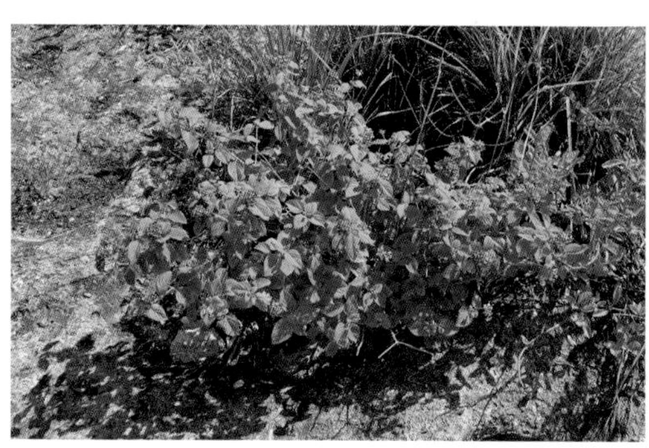

A382. 马鞭草科
Verbenaceae

马鞭草
Verbena officinalis
马鞭草属 *Verbena*

形态特征：一年生或多年生草本，直立，高 30 ~ 140 cm，被柔毛或近无毛。叶片卵形，倒卵形，或长圆形，纸质，有硬毛，背面脉上尤多，边缘有粗锯齿。穗状花序顶生或腋生，细弱；花冠蓝色到粉红色，被短柔毛。小坚果长圆形。

花果期：花果期 7 ~ 10 月。

分布：全世界温带至热带地区。分布几遍我国。南麂主岛（百亩山、国姓岙、门屿尾、三盘尾），常见。

生境：路边、山坡、溪边或林旁。

Description: Herbs annual or weakly perennial, erect, 30-140 cm tall, pubescent to subglabrous. Leaf blade ovate, obovate, or oblong, papery, hirsute especially on abaxial veins, margin coarsely dentate. Spikes terminal or axillary, long and slender. Corolla blue to pink, pubescent. Nutlets oblong.

Flower and Fruit: Fl. and fr. Jul. -Oct.

Distribution: Worldwide weed in temperate zones and tropics. Almost throughout China. Main island (Baimushan, Guoxingao, Menyuwei, Sanpanwei), commonly.

Habitat: Roadsides, hillsides, streamsides, forest margins.

A383. 唇形科
Lamiaceae

金疮小草
Ajuga decumbens
筋骨草属 ***Ajuga***

形态特征：一或二年生草本，具匍匐茎，高 10 ~ 30 cm。茎被白色长柔毛或绵状长柔毛，嫩枝部分尤多。基生叶较多，较茎生叶长；叶片先端钝到圆，基部渐狭，下延，具缘毛。轮伞花序多花，下部疏离，上部密集，花冠筒状，挺直，基部略膨大，外面被疏柔毛，内面仅冠筒被疏微柔毛。小坚果倒三棱状。

花果期：花期 3 ~ 7 月，果期 5 ~ 11 月。

分布：日本、朝鲜。我国长江以南各地。南麂主岛（关帝岙、门屿尾、三盘尾），偶见。

生境：溪边、路旁及湿润的草坡或竹林。

Description: Herbs annual or biennial, stoloniferous 10-30 cm tall. Stems white villous or lanate-villous, especially on young parts. Basal leaves numerous, longer than stem leaves; leaf blade apex obtuse to rounded, base attenuate-decurrent, ciliate. Verticillasters many flowered, basally widely spaced, apically crowded, corolla tubular, straight, basally slightly swollen, pilose, villous annulate inside. Nutlets obtriangulate.

Flower and Fruit: Fl. Mar. -Jul. , fr. May-Nov.

Distribution: Japan, Korea. To the south of Yangtze River. Main island (Guandiao, Menyuwei, Sanpanwei), occasionally.

Habitat: Streamsides, roadsides, wet grassy slopes, wet areas in bamboo forests.

A383. 唇形科
Lamiaceae

上狮紫珠
Callicarpa siongsaiensis
紫珠属 *Callicarpa*

形态特征：灌木，高约 2 m。小枝灰黄色，叶柄和花序梗疏生星状毛。叶片椭圆形或倒卵状椭圆形，顶端渐尖或短尖，基部楔形至钝圆，边缘有不明显的疏齿或近全缘，正面无毛或近无毛，背面具细小黄色腺点。聚伞花序在叶腋稍上方着生；花萼杯状，截形，无毛。果实无毛，干后暗棕色。

花果期：花期 6 ~ 8 月，果期 8 ~ 10 月。

分布：我国浙江、福建。南麂主岛各处常见。

生境：海拔 100m 以下的山坡林下或灌丛。

Description: Shrubs, 2 m tall. Branchlets gray-yellow, petioles and inflorescences sparsely stellate tomentose. Leaf blade elliptic to ovate-elliptic, apex acute to acuminate, base cuneate to obtuse, margin subentire to subserrate, adaxially glabrous or subglabrous, abaxially small yellow glandular. Cymes borne slightly above leaf axils; calyx cup-shaped, truncate, glabrous. Fruit glabrous, dark brown when dry.

Flower and Fruit: Fl. Jun. -Aug. , fr. Aug. -Oct.

Distribution: Zhejiang, Fujian. Main island of Nanji, commonly.

Habitat: Mountain slopes, forests, thickets, below 100 m.

A383. 唇形科
Lamiaceae

海州常山
Clerodendrum trichotomum
大青属 *Clerodendrum*

形态特征： 灌木或小乔木。幼枝、叶柄、花序轴等被黄褐色柔毛或近于无毛。叶片纸质，卵形、卵状椭圆形，顶端渐尖，基部宽楔形至截形，偶有心形，全缘。伞房状聚伞花序顶生或腋生，二歧分枝，疏散；苞片叶状，椭圆形，早落；花香，花冠白色或带粉红色。核果近球形，蓝紫色。

花果期： 花果期 7 ~ 11 月。

分布： 东亚、东南亚和印度。除内蒙、新疆和西藏外广布全国。南麂主岛各处常见。

生境： 山坡灌丛。

Description: Shrubs or small trees. Branchlets, petioles, peduncles yellowish brown puberulent or nearly glabrous. Leaf blade papery, ovate, ovate-elliptic, apex acuminate, base broadly cuneate to truncate, occasionally cordate, margin entire. Corymbose cymes terminal or axillary, dichotomous, lax; bracts leaflike, elliptic, deciduous; flowers fragrant, corolla white or pinkish. Drupes subglobose, blue-purple.

Flower and Fruit: Fl. and fr. Jul. -Nov.

Distribution: East Asia, Southeast Asia and India. Throughout China except Nei Mongol, Xinjiang, and Xizang. Main island of Nanji, commonly.

Habitat: Thickets on mountain slopes.

A383. 唇形科
Lamiaceae

细风轮菜
Clinopodium gracile
风轮菜属 *Clinopodium*

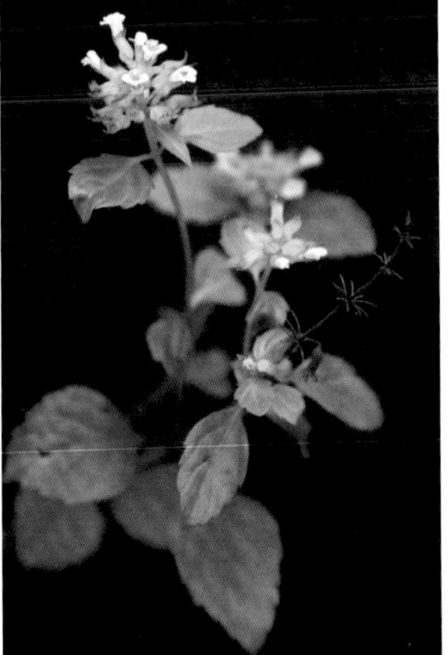

形态特征：多年生纤细草本。茎多数，四棱形，具槽，被倒向短柔毛。叶片圆卵形，边缘具疏圆齿，下面脉上疏生短毛；叶柄基部紫红色。轮伞花序顶生，短总状；花萼管状，果时增大，花冠白至紫红色，冠檐二唇形，上唇直伸，下唇3裂；花柱先端2浅裂。小坚果卵球形，褐色，光滑。

花果期：花期6～8月，果期8～10月。

分布：日本、印度和东南亚。我国长江流域以南地区。南麂各岛屿常见。

生境：溪边，开阔的草地、林缘、灌丛。

Description: Herbs slender, perennial. Stems numerous, 4-lines, grooved, retrorse pubescent. Leaf circular-ovate, margin remotely crenate, abaxially sparsely minutely hispid on veins; petiole base purple-red. Verticillasters terminal, short raceme; calyx tubular, enlarged in fruit; corolla white to purple-red, upper lip stretching, lower lip 3-lobed; style apex 2 shallow crack. Nutlets ovoid, brown, smooth.

Flower and Fruit: Fl. Jun. -Aug. , fr. Aug. -Oct.

Distribution: Japan, India and Southeast Asia. To the south of Yangtze River Basin. Throughout Nanji Islands, commonly.

Habitat: Streamsides, open grasslands, forest margins, thickets.

A383. 唇形科
Lamiaceae

风轮菜
Clinopodium chinense
风轮菜属 *Clinopodium*

形态特征：多年生草本。茎基部匍匐上部上升，多分枝，四棱形，具细条纹，密被短柔毛及腺微柔毛。叶卵圆形，边缘具整齐锯齿，下面灰白色，被疏柔毛。轮伞花序多花，密集腋生，半球状，下部疏离；苞叶线状钻形，花萼狭管状，常染紫红色；花冠紫红色，冠筒伸出；花柱先端2浅裂。小坚果倒卵形，黄褐色。

花果期：花期5～8月；果期8～10月。

分布：日本。我国华东地区。南麂各岛屿常见。

生境：山坡、林缘、草地及灌丛、林地。

Description: Herbs perennial. Base trailing and upper ascend-ing, much-branched, 4-lines, finely striate, densely pubescent, glandular puberulent. Leaf blade ovate, margin crenate-serrate, grayish white undersides, sparsely pubescence, densely on veins. Verticillasters many flowered, semiglobose; bracts linear-subulate, calyx narrowly tubular, tinged purple-red; corolla purple-red, corolla tube extends; style slightly exposed, apex 2-lobed. Nutlets obovoid, yellow-brown.

Flower and Fruit: Fl. May-Aug. , fr. Aug. -Oct.

Distribution: Japan. East China. Throughout Nanji Islands, commonly.

Habitat: Hillsides, streamsides, grassy places, thickets, forests.

A383. 唇形科
Lamiaceae

宝盖草
Lamium amplexicaule
野芝麻属 *Lamium*

形态特征：一年生或二年生草本，高 10 ~ 30 cm。茎基部多分枝，四棱形，具浅槽，常带紫蓝色。叶片圆形或肾形，先端圆，基部截形或心形，半抱茎，边缘具深锯齿。轮伞花序 6 ~ 10 花，苞片披针状钻形，具缘毛。花萼管状钟形，萼齿披针状锥形；花冠紫红或粉红色，冠檐二唇形，上唇直伸，长圆形，下唇稍长，3 裂。小坚果倒卵圆形，淡灰黄色，有白色疣突。

花果期：花期 3 ~ 5 月，果期 7 ~ 8 月。

分布：亚洲、欧洲。除东北外广布我国。南麂主岛（大山、国姓岙），偶见。

生境：路旁、林缘、沼泽草地或田间。

Description: Herbs annual or biennial, 10-30 cm tall. Stem much branched at base, 4-lines, shallow groove, tinged purple-blue. Leaf blade circular to reniform, apex rounded, base truncate or cordate, semi-clasping, margin deeply crenate. Verticillasters, 6-10 flowered, always cleistogamy flowered; bracts lanceolate-subulate, ciliate. Calyx tubular-campanulate, teeth lanceolate-subulate; corolla purple-red or reddish, crown eaves two lips, upper lip stretching, oblong, lower lip slightly longer, 3-lobed. Nutlets obovoid, grayish yellow, white tuberculate.

Flower and Furit: Fl. Mar. -May, fr. Jul. -Aug.

Distribution: Asia, Europe. Almost throughout China except Northeast China. Main island (Dashan, Guoxingao), occasionally.

Habitat: Roadsides, forest margins, marshes, sometimes weed in fields.

A383. 唇形科
Lamiaceae

滨海白绒草
Leucas chinensis
绣球防风属 *Leucas*

形态特征：灌木，高 20 ~ 30 cm，全株密被白色平伏绢毛。茎粗壮，基部分枝，直立或开展。叶片卵圆状圆形，先端钝，基部宽楔形至近心形。轮伞花序腋生，3 ~ 8 花，圆球形；苞片线形，花萼管状钟形，花冠白色，冠筒细长，喉部稍大，内面在中部以上有极斜向的稀疏毛环，下唇开张，3 裂，中裂片最大，肾形。

花果期：花期 11 ~ 12 月；果期 12 月。

分布：我国浙江、台湾、海南。南麂各岛屿常见。

生境：向阳的海滨荒地。

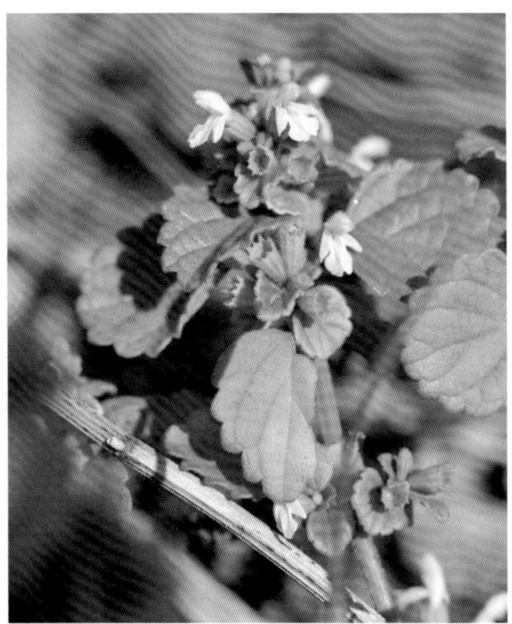

Description: Shrubs, 20-30 cm tall; densely appressed white silky-tomentose. Stems robust, branched at base; branches erect or spreading. Leaf blade ovate-orbicular, apex obtuse, base broadly cuneate to subcordate. Verticillasters axillary, 3-8-flowered, prolate; bracts linear, calyx tubular-campanulate, corolla white, tube slender, slightly enlarged at throat, obliquely villous annulate slightly above middle inside; lower lip spreading, 3-lobed, middle lobe largest, reniform.

Flower and Fruit: Fl. Nov. -Dec. , fr. Dec.

Distribution: Zhejiang, Taiwan, Hainan. Throughout Nanji Islands, commonly.

Habitat: Sunny waste areas along seashores.

A383. 唇形科
Lamiaceae

小鱼仙草
Mosla dianthera
石荠苎属 *Mosla*

形态特征：一年生草本，高 25 ~ 80 cm。茎直立，四棱，具浅槽，多分枝。叶卵状披针形，先端渐尖或急尖，基部渐狭，边缘具锐尖的疏齿。总状花序顶生，苞片针状或线状披针形，花萼钟形，果时增大；花冠淡紫色，外面被微柔毛，上唇微缺，下唇中裂片较大。小坚果灰褐色，近球形，具疏网纹。

花果期：花果期 5 ~ 11 月。

分布：东亚至东南亚。我国华东至西南地区。南麂主岛（百亩山、大山、国姓岙、门屿尾），偶见。

生境：近水的山坡、路旁。

Description: Herbs annual, 25-80 cm tall. Stems erect, 4-lines, shallow groove, much branched. Leaf blade ovate-lanceolate, apex acuminate to acute, base attenuate, margin remotely acute serrate. Racemes terminal, bracts needlelike to linear-lanceolate; calyx campanulate, corolla purplish, puberulent outside, upper lip reflexed, lower lip lobed largest. Nutlets gray-brown, subglobose, loosely netted.

Flower and Fruit: Fl. and fr. May -Nov.

Distribution: East Asia to Southeast Asia. East China to Southwest China. Main island (Baimushan, Dashan, Guoxingao,Menyuwei), occasionally.

Habitat: Hillsides, roadsides, near water.

A383. 唇形科
Lamiaceae

荔枝草
Salvia plebeia
鼠尾草属 *Salvia*

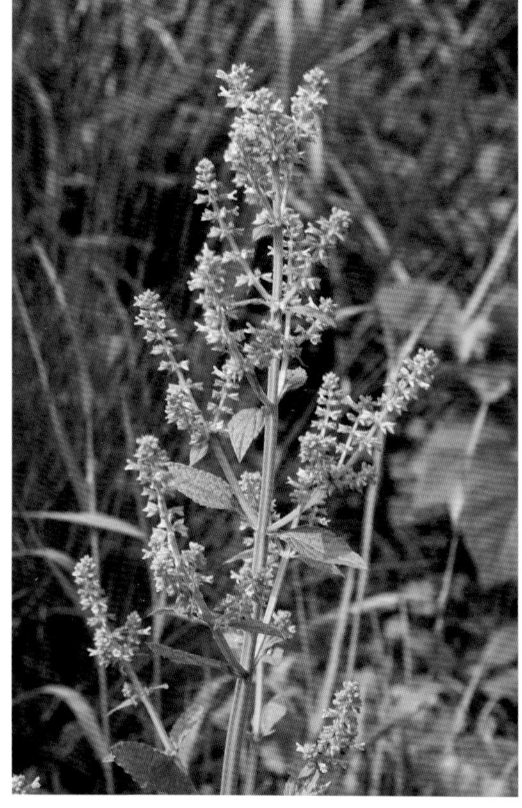

形态特征：一年生或二年生直立草本，高 15 ～ 90 cm。茎直立，多分枝，被下向灰白色疏柔毛。叶卵状椭圆形或长圆形，边缘具圆齿，背面疏生黄色腺点。轮伞花序 6 花，密集成总状或圆锥状；苞片披针形；花梗与花序轴密被疏柔毛；花萼钟形，疏被黄褐色腺点；花冠淡红、淡紫或蓝紫，稀白色。小坚果倒卵圆形，光滑。

花果期：花期 4 ～ 5 月，果期 6 ～ 7 月。

分布：东亚至东南亚。除西北地区外我国各地广布。南麂主岛（国姓岙、门屿尾、三盘尾），常见。

生境：山坡、溪边，潮湿的田野。

Description: Herbs annual or biennial, 15-90 cm. Stems erect, much branched, retrorse gray pilose. Leaf blade elliptic-ovate to oval, margin crenate, abaxially sparsely yellow glandular. Verticillasters 6-flowered, in racemes or panicles; bracts lanceolate; pedicel and inflorescences densely pilose; calyx campanulate, pilose, sparsely yellow-brown glandular; corolla reddish, purplish, purple, blue-purple, to blue, rarely white. Nutlets obovoid.

Flower and Fruit: Fl. Apr. -May, fr. Jun. -Jul.

Distribution: East Asia to Southeast Asia. Almost throughout China except Northwest China. Main island (Guoxingao, Menyuwei, Sanpanwei), commonly.

Habitat: Hillsides, streamsides, wet fields.

A383. 唇形科
Lamiaceae

韩信草（印度黄芩）
Scutellaria indica
黄芩属 *Scutellaria*

形态特征：多年生草本，全株有白色柔毛。茎直立，基部稍倾卧，四棱形，常带淡紫红色。叶心状卵形至椭圆形，基部圆形至心形，先端钝圆，边缘具圆齿，被柔毛或糙毛，背面尤密。总状花序顶生；苞片无柄，最下一对叶状，卵圆形；花萼被粗毛，花冠蓝紫色，外面疏被微柔毛。小坚果栗色或暗褐色，卵形，具瘤状突起。

花果期：花果期 2 ~ 6 月。

分布：东亚、东南亚。广布我国秦淮以南地区。南麂主岛各处常见。

生境：山坡、草地，开阔的岩石地区，路旁，疏林。

Description: Herbs perennial, with white pubescence. Stem erect, base slightly recline, 4-lines, tinged purple-red. Leaves cordate-ovate to elliptic, base rounded to cordate, apex obtuse to rounded, margin crenate, puberulent or strigose, densely so abaxially. Racemes terminal, bracts sessile, basal ones leaflike, ovate; calyx hirsute, corolla blue-purple, sparsely puberulent outside. Nutlets chestnut to dark brown, ovoid, tuberculate.

Flower and Fruit: Fl. and fr. Feb. -Jun.

Distribution: East Asia, Southeast Asia. Widely distributed in the areas to the south of Qinling and Huaihe. Main island of Nanji, commonly.

Habitat: Hillsides, grasslands, open rocky areas, roadsides, sparse forests.

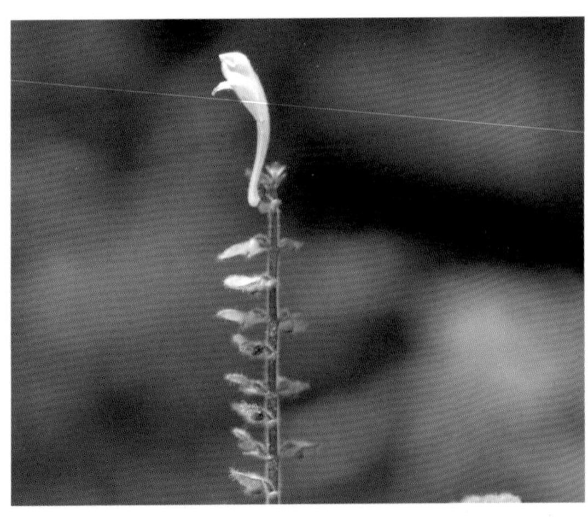

A383. 唇形科
Lamiaceae

田野水苏
Stachys arvensis
水苏属 *Stachys*

形态特征：一年生草本，高 30 ~ 50 cm。茎纤弱，四棱形，具浅槽，疏被柔毛，多分枝。叶卵圆形，先端钝，基部心形，边缘具圆齿；叶柄扁平；苞叶细小，比花萼短。轮伞花序腋生，具 2 ~ 4 花；苞片线形，被柔毛；花萼管状钟形，细小，果时花萼呈壶状增大，外面脉纹尤为显著，花冠红色。小坚果卵圆状，棕褐色。

花果期：花果期全年。

分布：欧洲、中亚、热带美洲。我国华东至西南地区。南麂主岛（门屿尾、三盘尾）、大檑山屿，常见。

生境：荒地、田野。

Description: Herbs annual, 30-50 cm. Stems slender, 4-lines, shallow groove, sparsely puberulent, much branched. Leave blade ovate, apex obtuse, base cordate, margin crenate; petiole flat; bracteoles small, shorter than calyx. Verticillasters axillary, 2(4)-flowered, bracts linear, puberulent; calyx tubular-campanulate, fine, fruiting calyx urceolate, conspicuously netted, corolla red. Nutlets brown, ovoid.

Flower and Fruit: Fl. and fr. year round.

Distribution: Europe, Central Asia, Tropical America. East China to Southwest China. Main island (Menyuwei, Sanpanwei), Daleishan Island, commonly.

Habitat: Waste areas, fields.

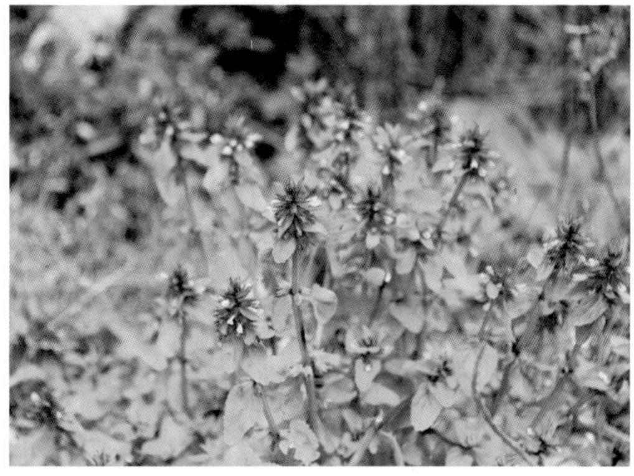

A383. 唇形科
Lamiaceae

夏枯草
Prunella vulgaris
夏枯草属 *Prunella*

形态特征：多年生草本，20 ~ 30 cm。茎上升，基部多分枝，紫红色，疏生糙毛或无毛。叶片披针形到卵形，无毛到疏生柔毛，基部截形到宽楔形下延，边缘波浪状到全缘，先端钝到圆。穗状花序无柄；苞片紫色，宽心形，沿脉疏生刚毛；花萼钟状，疏被刚毛，花冠略带紫色或白色。小坚果长圆状卵珠形，微具沟纹。

花果期：花期 4 ~ 6 月，果期 7 ~ 10 月。

分布：世界广布。我国黄河流域以南地区。南麂主岛（关帝岙、门屿尾、三盘尾），常见。

生境：开阔的荒坡、草地，湿润的溪边、林缘、灌丛。

Description: Herbs perennial, 20-30 cm. Stems ascending, base much branched, purple-red, sparsely strigose or subglabrous. Leaf blade lanceolate to ovate, glabrous to sparsely villous, base truncate to broadly cuneate-decurrent, margin undulate to entire, apex obtuse to rounded. Spikes sessile; bracts purplish, broadly cordate, veins sparsely hispid; calyx campanulate, sparsely hispid, corolla purplish or white. Nutlets oblong-ovoid, slightly 1-furrowed.

Flower and Fruit: Fl. Apr. -Jun. , fr. Jul. -Oct.

Distribution: Worldwide. To the south of Yellow River Basin. Main island (Guandiao, Menyuwei, Sanpanwei), commonly.

Habitat: Open slopes, grasslands, wet streamsides, forest margins, thickets。

A383. 唇形科
Lamiaceae

单叶蔓荆
***Vitex rotundifolia*（*Vitex trifolia* var. *simplicifolia*）**
牡荆属 *Vitex*

形态特征：落叶灌木。茎匍匐，节处常生不定根。小枝四棱形，密生细柔毛，老枝圆柱形。单叶对生，叶片倒卵形枝近圆形，先端常顿圆，基部楔形至宽楔形，全缘，下面密被灰白色短柔毛。圆锥花序顶生；花萼钟形，顶端5浅裂，花冠淡紫色或蓝紫色，两面有毛。核果近球形，熟时黑色。

花果期：花期7～9月，果期9～11月。

分布：日本、东南亚及太平洋岛屿。我国沿海各地。南麂各岛屿常见。

生境：近海的开阔沙地。

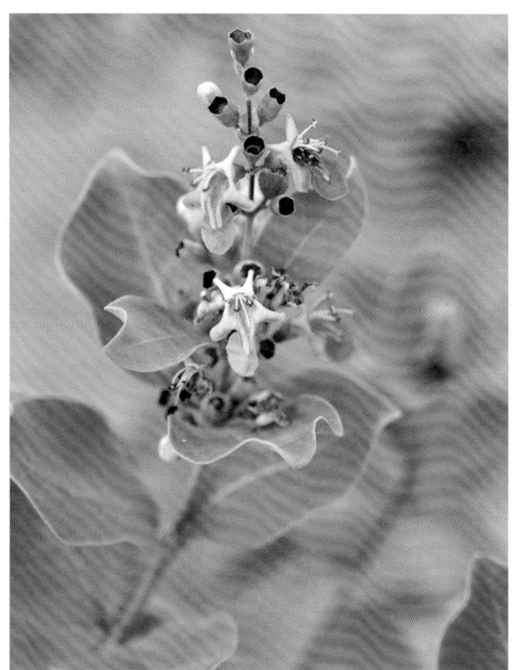

Description: Shrubs deciduous. Stems prostrate, adventitious rooting at nodes. Branchlets 4-lines, densely puberulent, old branches cylindrical. Simple and opposite leaves, blade obovate to nearly circula, apex blunt round, base cuneate to wide cuneate, margin entire, densely gray-white pubescent below. Panicle terminal; calyx campanulate, apex shallow 5-lobed, corolla mauve to lilac blue or bluish violet, both sides puberulent. Drupes subglobose, ripening balck.

Flower and Fruit: Fl. Jul. -Sep. , fr. Sep. -Nov.

Distribution: Japan; Southeast Asia, Pacific Islands. Coastal areas of China. Throughout Nanji Islands, commonly.

Habitat: Open sandy areas, usually near sea.

A384. 通泉草科 Mazaceae

通泉草
Mazus pumilus
通泉草属 *Mazus*

形态特征：一年生草本，高 3 ～ 30 cm。茎直立或斜上升，基部分支。基生叶莲座状，叶片倒卵匙形至倒卵披针形，边缘具粗锯齿或浅羽裂，基部楔形，叶柄下延。总状花序顶生，花冠白色，淡紫色，上唇 2 裂，下唇 3 裂，萼筒钟状，花萼裂片 5，卵形，急尖。蒴果球形。种子小而多数，淡黄色。

花果期：花果期 4 ～ 10 月。

分布：东亚、热带亚洲等地。遍布我国各地。南麂主岛各处常见。

生境：潮湿的草地，溪边，路旁，荒地林缘。

Description: Herbs annual, 3-30 cm tall. Stems erect or obliquely ascending, branched at base. Basal leaves rosulate, leaf blade obovate-spatulate to obovate-lanceolate, margin coarsely serrate or lobed, base cuneate, petiole decurrant. Racemes terminal, corolla white, lavender, upper lip 2-lobed, lower lip 3-lobed, calyx tubular, calyx lobes 5, ovate, acute. Capsule globose. Seeds small and numerous, light yellow.

Flower and Fruit: Fl. and fr. Apr. -Oct.

Distribution: East Asia, tropical Asia. Throughout China.Main island of Nanji, commonly.

Habitat: Wet grassland, along streams, trailsides, waste fields, edge of forests.

A387. 列当科 Orobanchaceae

野菰
Aeginetia indica
野菰属 *Aeginetia*

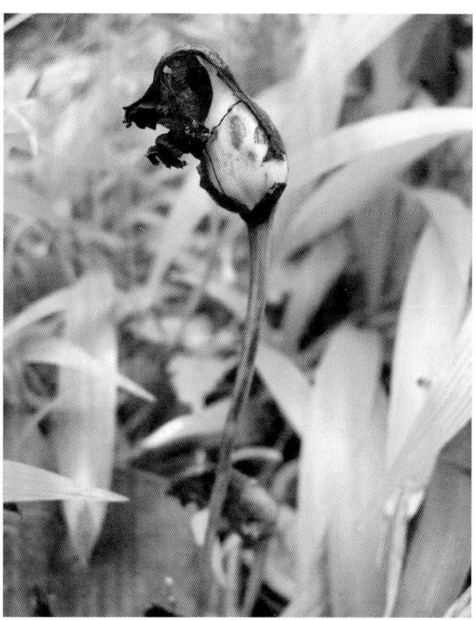

形态特征：一年生草本，高 15 ～ 40 cm。根稍肉质，具小分枝。茎不分枝或近基部分枝。叶红色，卵状披针形或披针形，无毛。花常单生，花梗直立；花冠具紫红色条纹，不明显的二唇形，管部稍弯曲；花丝紫色，无毛，花药黄色。蒴果圆锥状或长卵球状。种子椭圆形，黄色。

花果期：花期 4 ～ 8 月，果期 8 ～ 10 月。

分布：亚洲东部至南部。我国东南地区。南麂主岛（百亩山），偶见。

生境：坡地、路旁。

Description: Herbs annual, 15-40 cm. Root slightly fleshy, with small branches. Stems unbranched or branched from near base. Leaves red, ovate-lanceolate or lanceolate, glabrous. Flowers usually solitary, pedicel erect; corolla purple-red striate, indistinctly bilabiate, tube slightly curved; filaments purple, glabrous, anthers yellow. Capsule conical, or long ovoid-globose. Seeds ellipsoid, yellow.

Flower and Fruit: Fl. Apr. -Aug. , fr. Aug. -Oct.

Distribution: Eastern to southern Asia. Southeastern China.Main island (Baimushan), occasionally.

Habitat: Slopes, roadsides.

A387. 列当科
Orobanchaceae

阴行草
Siphonostegia chinensis
阴行草属 *Siphonostegia*

形态特征：一年生草本，高约 30 ～ 80 cm，密被锈色短毛。枝具棱，密被短毛。叶对生，叶片厚纸质，广卵形，密被短毛，二回羽状全裂；裂片 3 对，仅下方两枚羽裂；小羽片 1 ～ 3，线形或线状披针形，全缘。总状花序稀疏，苞片羽裂；花萼密被短毛；花冠上唇红紫色，顶端截形，下唇黄色，裂片卵形。蒴果被包于宿存的萼内，卵状长圆形，有短尖头，黑褐色。种子黑色。

花果期：花期 6 ～ 8 月。

分布：东亚。我国南北各地广布。南麂主岛（后隆），偶见。

生境：干燥山坡、草地。

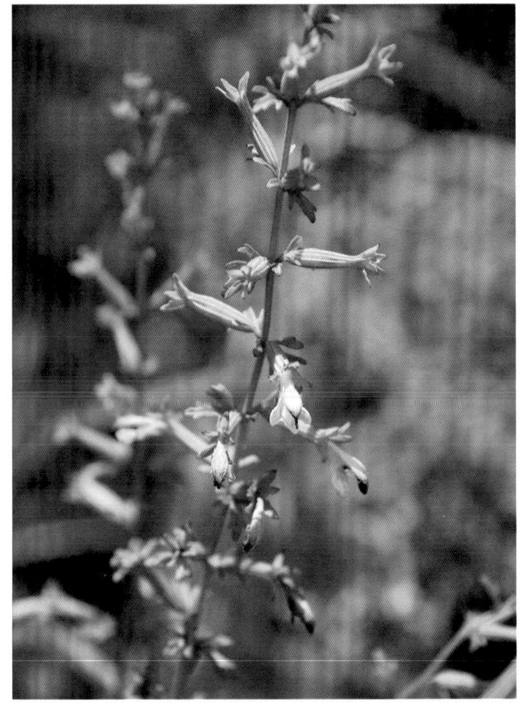

Description: Herbs annuals, 30-80 cm tall, densely tawny hairy. Branches angled, densely pubescent. Leaves opposite, leaf blade thick paper, broadly ovate, densely pubescent, 2-pinnatisect; pinnae 3 pairs, lowermost pair 2-pinnately parted; pinnules 1-3, linear to linear-lanceolate, margin entire. Racemes few flowered; bracts pinnated, calyx densely pubescent; upper lip red-purple, apex truncate, lower lip yellow, lobes ovate. Capsule enclosed in persistent calyx, ovoid-oblong, apex apiculate, dark brown. Seeds black.

Flower and Fruit: Fl. Jun. -Aug.

Distribution: East Asia. Throughout the north and south areas of China. Main island (Houlong), occasionally.

Habitat: Dry mountain slopes and grassland.

A394. 桔梗科
Campanulaceae

桔梗
Platycodon grandiflorus
桔梗属 ***Platycodon***

形态特征：多年生草本，高 20 ~ 120 cm。茎无毛，不分枝，极少上部分枝。叶轮生至互生，卵形，椭圆形或披针形，基部宽楔形，顶端急尖，背面有白粉，边缘具细锯齿。花单朵顶生，或有花序分枝而集成圆锥花序；花萼半球状或倒锥形，被白粉，裂片三角形，花冠大，蓝色或紫色。蒴果球状，或倒卵状。

花果期：花期 7 ~ 9 月，果期 8 ~ 10 月。

分布：东亚，各地广泛栽培。我国大部分地区均有分布。南麂主岛（三盘尾），偶见。

生境：阳处草丛、灌丛，林下少见。

Description: Herbs perennial, 20-120 cm tall. Stems glabrous, rarely simple, rarely branched above. Leaves whorled to alternate, leaf blade ovate, elliptic, or lanceolate, base broadly cuneate, apex acute, abaxially glaucous, margin serrate. Flowers single terminal, or several integrated false racemes, or with inflorescences branched and integrated panicles; hypanthium hemispherical, or obconic, glaucous, calyx lobes triangular, corolla large, blue or purple. Capsule globose, or obovoid.

Flower and Fruit: Fl. Jul. -Sep. , fr. Aug. -Oct.

Distribution: East Asia, widely cultivated elsewhere. Most areas of China. Main island (Sanpanwei), occasionally.

Habitat: Sunny herb communities, thickets, rarely in forests.

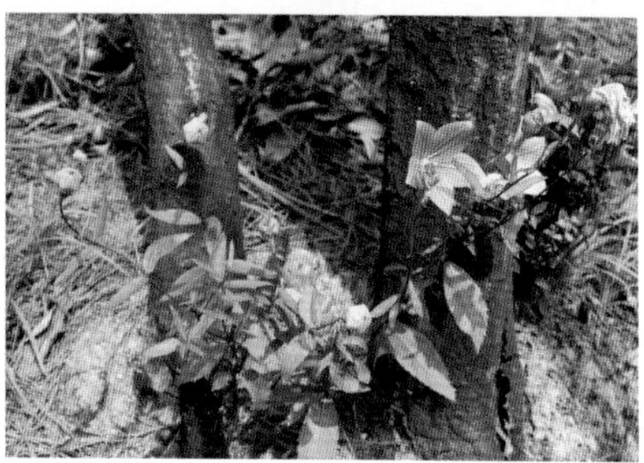

A394. 桔梗科
Campanulaceae

蓝花参
Wahlenbergia marginata
蓝花参属 *Wahlenbergia*

形态特征：多年生草本，高 20 ～ 40 cm。茎自基部多分枝，直立或匍匐状。叶互生，几无柄，常在茎下部密集；基生叶匙形，茎生叶条状披针形或椭圆形，边缘全缘、波状或具疏锯齿。花梗极长，细而伸直；花萼 5 裂，筒部倒卵状圆锥形，裂片三角状钻形，花冠钟状，蓝色。蒴果倒圆锥状或倒卵状圆锥形。

花果期：花果期 2 ～ 5 月。

分布：亚洲。我国长江流域以南等地。南麂主岛各处常见。

生境：山坡、沟边和荒地。

Description: Herbs perennial, 20-40 cm tall. Stems erect or ascending. Leaves alternate, almost sessile, dense at lower part of stem, basal leaves spatulate, stem leaves stripe-lanceolate or elliptic, margin entire, sinuate, or sparsely serrulate. Pedicels extremely long, slender and straight; calyx 5-lobed, tube obovate-conical, lobes triangular-subulate, corolla campanulate, blue. Capsule obconic or obovate conical.

Flower and Fruit: Fl. and fr. Feb. -May.

Distribution: Asia. To the south of the Yangtze River Basin.Main island of Nanji, commonly.

Habitat: Slopes, ditches, wastelands.

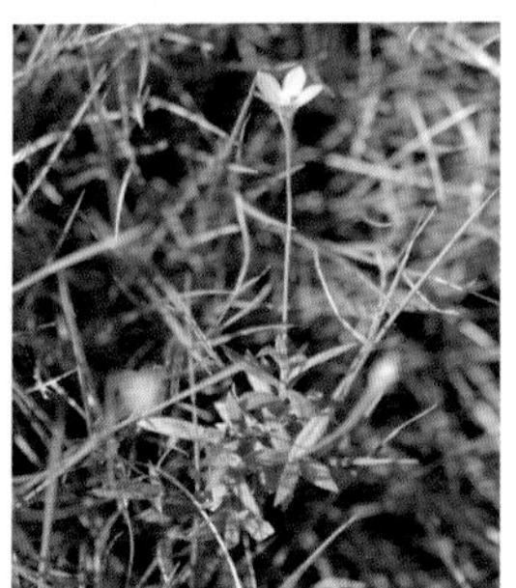

A403. 菊科
Asteraceae

藿香蓟（胜红蓟）
Ageratum conyzoides
藿香蓟属 *Ageratum*

形态特征：一年生草本，高 50 ~ 100 cm，具不明显主根。茎粗壮，不分枝或基部以上分枝。茎枝淡红色，被白色绒毛。叶对生，卵形或长圆形，基出三脉或不明显五脉。头状花序在茎顶排成通常紧密的伞房状花序，总苞钟状或半球形，2 层，长圆形或披针状长圆形；花冠外面无毛或顶端有尘状微柔毛，檐部 5 裂，淡紫色。瘦果黑褐色，5 棱。

花果期：花果期全年。

分布：原产热带美洲，广布非洲和东南亚。我国南方各地有栽培和归化。大檑山屿，常见。

生境：山谷、林下、林缘、河边、山坡草地、田边及荒地上。

Description: Herbs annual, 50-100 cm tall, with inconspicuous taproot. Stems robust, simple or branched from base or middle. Stems and branches reddish, white tomentose. Leaves opposite, sometimes upper alternate, ovate or oblong, basinerved three or inconspicuous five. Capitula in dense terminal corymbs, involucre campanulate or hemispheric, Involucral bracts 2-layered, oblong or lanceolate oblong; Corollas glabrous or apically powdery puberulent, limb purplish, 5-lobed. Achenes black, 5-angled.

Flower and Fruit: Fl. and fr. year round.

Distribution: Native to tropical America; widespread throughout Africa and Southeast Asia. Cultivated and naturalized in the south areas of China. Daleishan Island, commonly.

Habitat: Valleys, forests, forest margins on slopes, riversides, grasslands, field margins.

A403. 菊科
Asteraceae

豚草
Ambrosia artemisiifolia
豚草属 *Ambrosia*

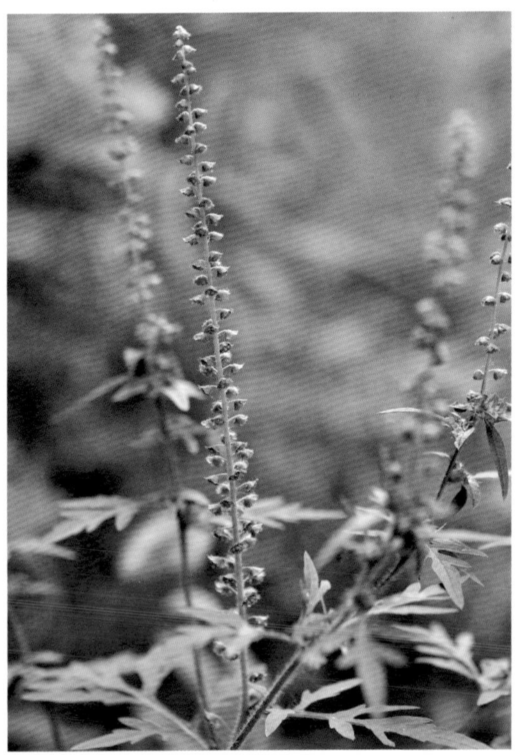

形态特征：一年生草本，高 20 ～ 150 cm。茎直立，疏生糙毛。叶对生，三角形至披针形，1 ～ 2 回羽状分裂，被糙伏毛，两面具腺点。雄头状花序半球形或卵形，具短梗，下垂，在枝端密集成总状花序；总苞宽半球形或碟形，花冠淡黄色，管部短，上部钟状，裂片宽。瘦果倒卵形，无毛，藏于坚硬的总苞中。

花果期：花期 7 ～ 10 月，果期 9 ～ 10 月。

分布：原产中北美洲，欧亚地区广布。我国广布。南麂主岛各处常见。

生境：路旁或空旷草丛中。

Description: Herbs annual, 20-150 cm tall. Stem erect, abaxially sparsely strigillose. Leaves opposite, blade deltate to lanceolate, pinnately 1-or 2-lobed, strigillose, both surfaces gland-dotted. Male capitula hemispherical or ovate, shortly pedunculate, pendulous, densely clustered into racemes at branch ends; involucre broadly hemispherical or saucer-shaped, corolla pale yellow, with short tube, upper campanulate, with broad lobes. Achene obovate, glabrous, hidden in rigid involucre.

Flower and Fruit: Fl. Jul. -Oct. , fr. Sep. -Oct.

Distribution: Native to Central and North America; introduced and widely distributed in Asia and Europe. Widely distributed in China. Main island of Nanji, commonly.

Habitat: Roadside or open grass.

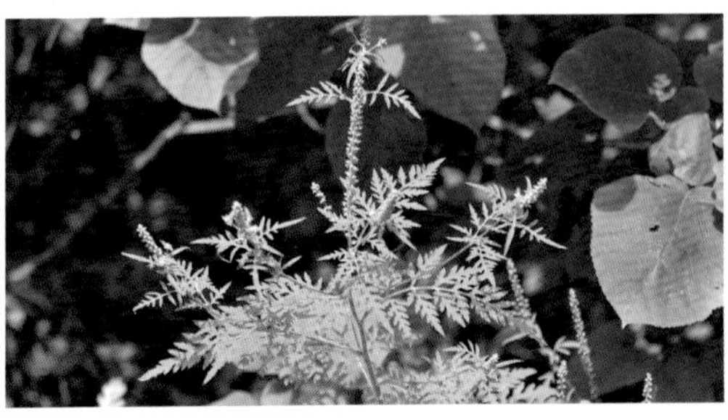

A403. 菊科
Asteraceae

茵陈蒿
Artemisia capillaris
蒿属 *Artemisia*

形态特征： 二年生或多年生草本，高 30 ~ 80 cm；根茎垂直，木质。茎单生或少数，纤细，直立，浅紫色或红褐色。基生叶具丝状毛，具短柄，中部茎生叶几无柄；叶片长圆状卵形，1 ~ 2 回羽状全裂，裂片丝状。复合花序圆锥状，头状花多数，下垂；总苞卵形，3 ~ 4 层，松散覆瓦状，无毛，小花 8 ~ 12，黄色，缘花 3 ~ 5，雌性，盘花 5 ~ 7，雄性。瘦果褐色，长圆状卵形。

花果期： 花果期 7 ~ 10 月。

分布： 东亚、东南亚。我国东部湿润地区。南麂各岛屿常见。

生境： 湿润的山坡、梯田、路边或河岸。

Description: Herbs biennial or perennial, 30-80 cm tall; rootstock vertical, woody. Stems usually 1 to few, slender, erect, pale purplish or reddish brown. Basal leaves silky hairy, shortly petiolate, middle stem leaves almost sessile; leaf blade oblong-ovate, 1-or 2-pinnatisect, segments filiform. Synflorescence a narrow to wide panicle, capitula many, nodding; involucre ovoid, phyllaries 3-or 4-seriate, laxly imbricate, glabrous; florets 8-12, yellow, marginal florets 3-5, female, disk florets 5-7, male. Achenes brown, oblong-ovate.

Flower and Fruit: Fl. and fr. Jul. -Oct.

Distribution: East Asia, Southeast Asia. Humid areas of the eastern China. Throughout Nanji Islands, commonly.

Habitat: Humid slopes, hills, terraces, roadsides, riverbanks.

A403. 菊科
Asteraceae

牡蒿
Artemisia japonica
蒿属 *Artemisia*

形态特征：多年生草本，高 30 ～ 120 cm。基生叶与茎下部叶长匙形，3 ～ 5 深裂，基部楔形，常有线形假托叶，先端圆顿，具不规则牙齿；上部叶小，上端具 3 浅裂或不分裂。花序狭圆锥状，头状花多数，总苞卵球形或近球形；小花 12 ～ 20，黄色，缘花 3 ～ 8，雌性，盘花 5 ～ 10，雄性。瘦果倒卵形，暗褐色。

花果期：花期 8 ～ 10 月，果期 12 月至翌年 1 月。

分布：东亚至中亚。分布几遍我国。南麂各岛屿常见。

生境：林缘、荒地、灌丛、丘陵山坡、路旁。

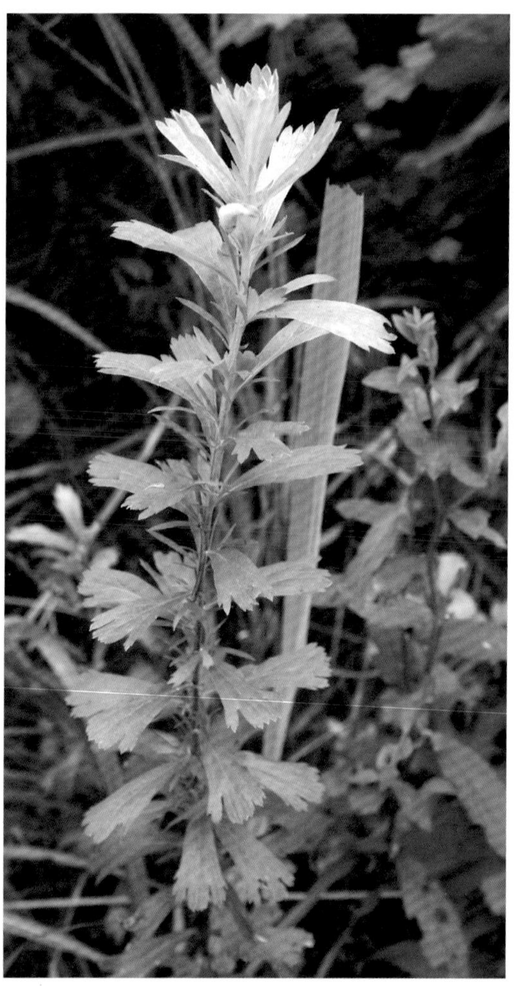

Description: Herbs perennial, 30-120 cm tall. Basal and lower stem leaves blade spatulate, 3-5 deep partite, basal cuneate, false stipules often linear, apex rounded, toothed; uppermost leaves small,3-cleft or entire. Synflorescence a ± narrow panicle, capitula many, involucre ovoid or subglobose; florets 12-20, yellow, marginal female florets 3-8, disk florets 5-10, male. Achenes dark brown, obovoid.

Flower and Fruit: Fl. Aug. -Oct. , fr. Dec. -Jan.

Distribution: East Asia to Central Asia. Throughout China.Throughout Nanji Islands, commonly.

Habitat: Forest margins, waste areas, shrublands, hills, slopes, roadsides.

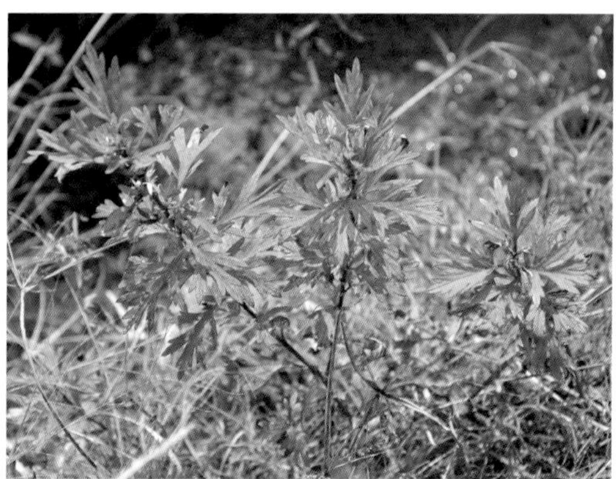

A403. 菊科
Asteraceae

矮蒿
Artemisia lancea
蒿属 *Artemisia*

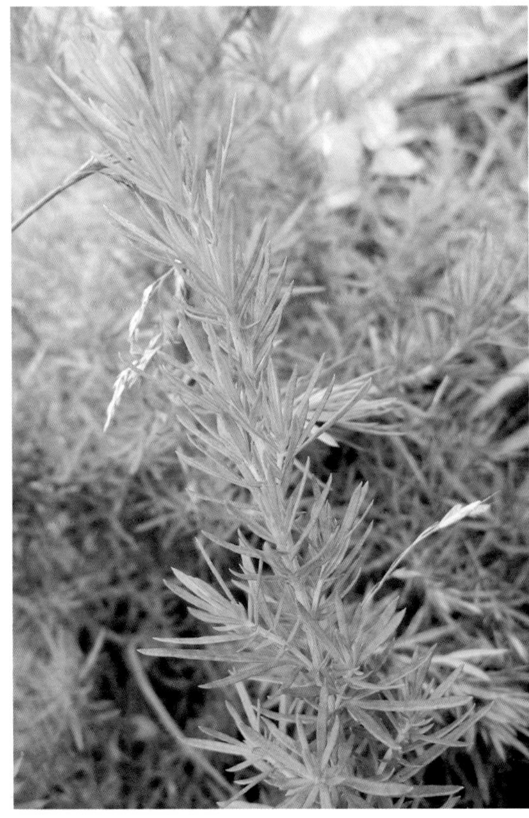

形态特征：多年生草本，高 80 ～ 150 cm。茎多数，成丛。基生叶与中下部茎生叶卵圆形，二回羽状全裂，小裂片线状披针形或线形；上部茎生叶与叶状苞 3 ～ 5 裂或全缘，裂片披针形或线状披针形。花序圆锥状，头状花多数，无梗，缘花 1 ～ 3，雌性，盘花 2 ～ 5，两性，花冠狭管状，紫红色。瘦果小，长椭圆形。

花果期：花果期 8 ～ 10 月。

分布：东亚，印度。分布于我国各地。南麂主岛（打铁礁、门屿尾），偶见。

生境：林缘、路旁、荒坡、干燥的田野荒地。

Description: Herbs perennial, 80-150 cm tall. Stems numerous, often into tufts. Basal and middle lower cauline leaves ovoid, 2-pinnatisect, lobules linear-lanceolate or linear; uppermost leaves and leaflike bracts 3-5-lobed or entire, lobes lanceolate or linear-lanceolate. Synflorescence panicle, Capitula many, involucre ovoid or ellipsoid-ovoid, sessile, marginal female florets 1-3, disk florets 2-5, bisexual; corolla narrowly tubular, purplish red. Achenes small, oblong.

Flower and Fruit: Fl. and fr. Aug. -Oct.

Distribution: East Asia, India. Throughout China. Main island (Datiejiao, Menyuwei), occasionally.

Habitat: Forest margins, roadsides, slopes, dry fields, waste areas.

A403. 菊科
Asteraceae

野艾蒿
Artemisia lavandulifolia
蒿属 *Artemisia*

形态特征：多年生草本或灌木，高 50 ~ 120 cm。茎粗壮、直立，被灰色毛。茎下部叶 1 ~ 2 回羽状分裂，具长柄；中部叶卵形，一回羽状深裂，裂片条状披针形，被毛，上面密生白腺点；上部叶渐小，条形，全缘。圆锥花序狭窄，叶状，具多数红褐色头状花序，总苞片椭圆形或长圆形，密被蛛丝毛；缘花 4 ~ 9，雌性，盘花 10 ~ 20，两性，紫色。瘦果倒卵形，无毛。

花果期：花果期 7 ~ 10 月。

分布：东亚。除新疆、青海、西藏外广布我国。南麂各岛屿常见。

生境：路边、林缘、河岸或湖边灌丛。

Description: Herbs perennial, or shrubs, 50-120 cm tall. Stems robust, erect, gray arachnoid pubescent. Lower cauline leaves 1 to 2-pinnatisect, long petiolate; middle leaves ovate, 1-pinnatisect, lobs linear-lanceolate, pubescent, adaxially white gland-dotted; uppermost leaves small, linear, entire. Panicle narrow, leafy, capitula many, reddish brown, involucre ellipsoid or oblong, phyllaries densely arachnoid tomentose; marginal female florets 4-9, disk florets 10-20, bisexual, purple. Achenes obovoid, glabrous.

Flower and Fruit: Fl. and fr. Jul. -Oct.

Distribution: East Asia. Almost throughout China except Xinjiang, Qinghai and Xizang. Throughout Nanji Islands, commonly.

Habitat: Roadsides, forest margins, slopes, riverbanks or lakesides, brushlands.

A403. 菊科
Asteraceae

猪毛蒿
Artemisia scoparia
蒿属 *Artemisia*

形态特征：多年生或一、二年生草本，高 40 ~ 90 cm。茎下部多分枝，枝叶被灰色或淡黄色绢毛，后脱落，植株有浓烈香气。下部茎生叶 2 ~ 3 回羽状全裂，裂片 3 ~ 4 对，小裂片 1 ~ 2 对；中部茎生叶 1 ~ 2 回羽状全裂，裂片 2 ~ 3 对，小裂片丝状，常弯曲；最上部叶和叶状的苞片 3 ~ 5 全裂。复合花序宽圆锥状，头状花多数；总苞近球形，缘花 5 ~ 7，雌性，盘花 4 ~ 10，雄性。瘦果倒卵球形或长圆形。

花果期：花果期 7 ~ 10 月。

分布：欧亚大陆温带与亚热带地区。分布于我国各地。南麂各岛屿常见。

生境：山坡、林缘、路旁。

Description: Herbs perennial, biennial, or annual, 40-90 cm tall. Much branched from lower on stem, branches and leaves gray or yellowish sericeous-pubescent, later glabrescent, strongly aromatic. Lower cauline leaves 2-or 3-pinnatisect, segments 3 or 4 pairs, lobules 1 or 2 pairs; middle cauline leaves 1-or 2-pinnatisect, segments 2 or 3 pairs, lobules filiform, usually curved; uppermost leaves and leaflike bracts 3-5-sect. Synflorescence a broad panicle, capitula many; involucre subglobose, marginal female florets 5-7, disk florets 4-10, male. Achenes obovoid or oblong.

Flower and Fruit: Fl. and fr. Jul.-Oct.

Distribution: Temperate and subtropical regions of Europe and Asia. Throughout China. Throughout Nanji Islands, commonly.

Habitat: Slopes, forest margins, roadsides.

A403. 菊科
Asteraceae

普陀狗娃花
Aster arenarius
紫菀属 *Aster*

形态特征：二年生或多年生草本，高 15 ～ 70 cm，主根木质化。茎平卧或斜升，自基部分枝。叶匙形，基部渐狭成长柄，全缘或有时疏生粗齿，有缘毛，质厚，下部茎生叶在花期枯萎。头状花序单生枝端，基部稍膨大，有苞片状小叶；舌状花 1 层，雌性，条状矩圆形，淡蓝色或淡白色；管状花两性，黄色，裂片 5。瘦果倒卵形，浅黄褐色，扁，被绢状柔毛。

花果期：花果期 7 ～ 10 月。

分布：日本。我国东部沿海岛屿。南麂各岛屿常见。

生境：海岸沙地、岩石荒坡。

Description: Herbs, biennial or perennial, 15-70 cm tall; taproot woody. Stems procumbent or ascending, branched from base. Leaf blade spatulate, base cuneate to attenuate, margin entire or sometimes remotely serrate, scabrous-ciliate, thick; lower and basal cauline leaves withered by anthesis. Capitula solitary at ends of branches, peduncle apex dilated, bracts leaflike; ray florets 1-layered, female, lamina linear-oblong, bluish or whitish; disk florets bisexual, yellow, lobes 5. Achenes obovoid, yellowish brown, compressed, strigose.

Flower and Fruit: Fl. and fr. Jul. -Oct.

Distribution: Japan. Offshore islands in East China. Throughout Nanji Islands, commonly.

Habitat: Sandy seashore, rocky barren slope.

A403. 菊科
Asteraceae

白舌紫菀
Aster baccharoides
紫菀属 *Aster*

形态特征：多年生草本或亚灌木，高 15 ～ 150 cm，基部木质。茎直立，密被毛，具柄腺。下部叶花后枯萎，中上部叶柄具窄翅；叶片狭卵形到披针形，上面被短糙毛；侧脉 3 ～ 4 对。头状花序顶生或腋生，呈圆锥伞房状；总苞倒锥状，4 ～ 7 层，背面被短密毛，有缘毛；舌状花白色。瘦果狭长圆形，稍扁，密被短毛。

花果期：花期 7 ～ 10 月，果期 8 ～ 11 月。

分布：我国华东至华南地区。南麂主岛（百亩山、国姓岙），偶见。

生境：山坡、草原、灌丛、沙质地带、海崖和路旁。

Description: Low shrubs or perennial herbs, 15-150 cm tall, basally woody. Stems erect, finely striate, densely strigose, minutely stipitate glandular. Lowest leaves withered by anthesis, lower to upper leaves shortly winged petiolate; blade narrowly ovate to lanceolate, short strigose adaxially; lateral veins 3-4. Capitula numerous, in terminal or axillary corymbiform to paniculiform synflorescences; involucres campanulate, phyllaries 4-7-seriate, abaxially moderately to densely strigose, margin villous-ciliate; ray florets white. Achene narrowly oblong, slightly flattened, densely short hairy.

Flower and Fruit: Fl. Jul. -Oct. , fr. Aug. -Nov.

Distribution: East China to South China. Main island (Baimushan, Guoxingao), occasionally.

Habitat: Slopes, grasslands, shrublands, sandy areas, sea bluffs, roadsides.

A403. 菊科
Asteraceae

马兰
***Aster indicus*（*Kalimeris indica*）**
紫菀属 *Aster*

形态特征：多年生草本，高 30 ～ 70 cm。茎直立，被微柔毛，从上部或下部分枝。基部叶开花枯萎；茎生叶长翅具叶柄，倒卵形或倒披针形，边缘有锯齿或羽状浅裂，先端钝或锐尖；上部叶无柄，小，稍薄，基部突然变细，边全缘，具短硬毛。头状花序 8 ～ 12 个呈伞房状；总苞半球形。瘦果棕色，倒卵球形长圆形，压扁。

花果期：花果期 5 ～ 11 月。

分布：亚洲南部及东部。广布我国中东部地区。南麂主岛（百亩山、国姓岙、门屿尾、三盘尾），常见。

生境：林缘、草地、河岸、田边、路旁、受保护的栖息地。

Description: Herbs perennial, 30-70 cm tall. Stems erect, puberulent above, branched from upper or lower part. Basal leaves withered by anthesis; cauline leaves long winged petiolate, obovate or oblanceolate, margin serrate or pinnately lobed, apex obtuse or acute; upper leaves sessile, small, slightly thin, base abruptly attenuate, margin entire and hispidulous. Capitula 8-12 in corymbiform synflorescences; involucres hemispheric. Achenes brown, obovoid-oblong, compressed.

Flower and Fruit: Fl. and fr. May-Nov.

Distribution: Eastern and southern Asia. Widely distributed in central and eastern China. Main island (Baimushan, Guoxingao, Menyuwei, Sanpanwei), commonly.

Habitat: Forest margins, grasslands, riverbanks, field margins, roadsides, protected shaded habitats.

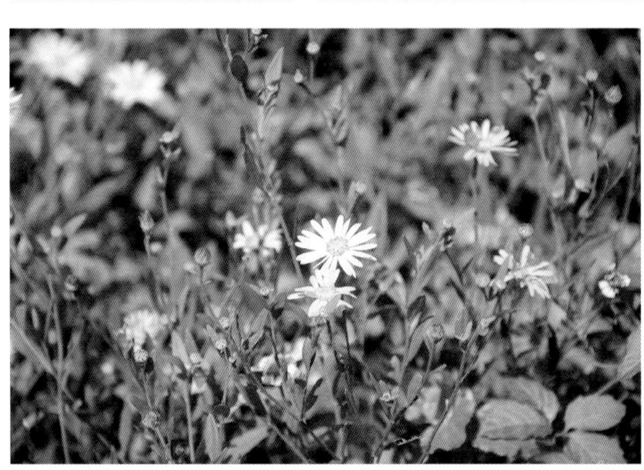

A403. 菊科
Asteraceae

琴叶紫菀
Aster panduratus
紫菀属 *Aster*

形态特征：多多年生草本，高 14 ~ 100 cm。茎直立，单生或丛生，具紫色条纹，多少密被长柔毛和细小的长柄腺体。叶茎生，质厚，密被毛，中下部叶匙状长圆形，基部扩大成心形或耳状，半抱茎；上部叶渐小，卵状长圆形；全部叶稍厚质，两面被长贴毛和密短毛。头状花序 3 ~ 40，在枝端成疏松伞房花序，很少单生。瘦果倒卵形，稍扁，疏生具柄腺体，有 2 肋。

花果期：花期 2 ~ 9 月，果期 6 ~ 10 月。

分布：我国长江流域以南地区。南麂主岛(门屿尾)，偶见。

生境：灌木丛，山坡上的草地，运河边，路旁，田缘。

Description: Herbs perennial, 14-100 cm tall. Stems erect, simple or branched upward, purplish striate, ± densely villous, ± densely minutely long-stipitate glandular.. Leaves cauline, slightly thick, densely villous; lower and middle leavesoblong-spatulate, base expanded cordate or orbicular, hemiculus; upper leaves decuminate, ovate-oblong. Capitula 3-40, in terminal, laxly corymbiform synflorescences, rarely solitary. Achenes obovoid, slightly compressed, sparsely strigillose, minutely stipitate glandular, 2-ribbed.

Flower and Fruit: Fl. Feb. -Sep. , fr. Jun. -Oct.

Distribution: To the south of Yangtze River Basin. Main island (Menyuwei), occasionally.

Habitat: Thickets, grasslands on slopes, canal sides, roadsides, field margins.

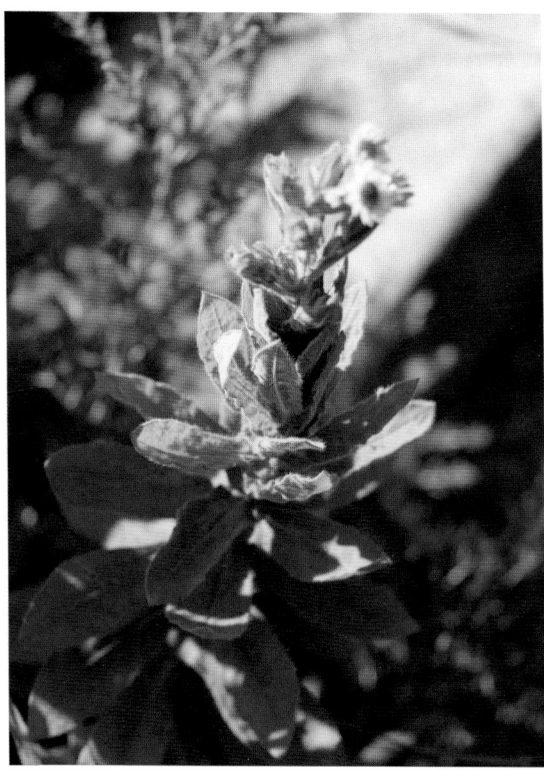

A403. 菊科
Asteraceae

陀螺紫菀
Aster turbinatus
紫菀属 *Aster*

形态特征：多年生草本，高 60 ～ 100 cm。茎直立，粗壮，单生，被粗毛。基部叶花期枯落，卵圆形或卵圆披针形，有疏齿，基部渐狭成具宽翅柄；叶厚纸质，两面被短糙毛；中脉在下面凸起。头状花序单生或 2 ～ 3 个簇生上部叶腋，总苞钟状，苞片 5 ～ 6 列，不等长，紫色，具缘毛。瘦果狭倒卵形，4 ～ 5 肋，具糙伏毛。

花果期：花期 8 ～ 10 月，果期 10 ～ 11 月。

分布：特产于我国华东地区。南麂主岛各处常见。

生境：开阔的森林、灌丛、草地、山坡、河岸与林阴地。

Description: Herbs perennial, 60-100 cm tall. Stems erect, thick, simple, villous. Basal leaves withered by anthesis, ovate to ovate-lanceolate, sparsely toothed, base narrowly winged petiolate; leaves thick papery, short hispid on both sides; midrib bulges below. Capitula solitary or 2-3 fascicled upper leaf axils, involucres campanulate, phyllaries 5 or 6 seriate, unequal, purplish, ciliate. Achenes narrowly obovoid, 4-or 5-ribbed, strigillose.

Flower and Fruit: Fl. Aug. -Oct. , fr. Oct. -Nov.

Distribution: Endemic in East China. Main island of Nanji, commonly.

Habitat: Open forests, thickets, grasslands, hillsides, stream banks, shaded places.

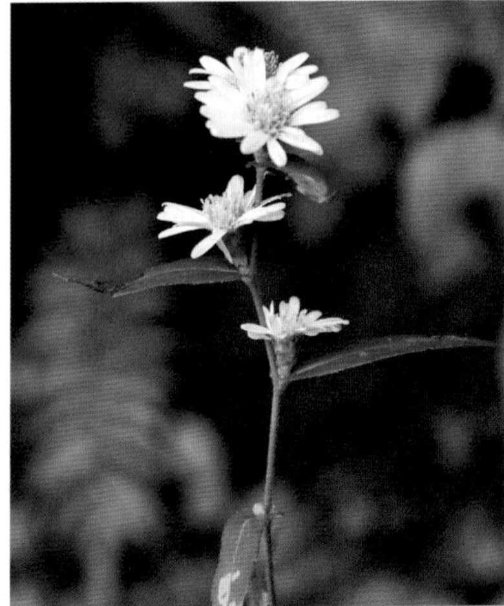

A403. 菊科
Asteraceae

金盏银盘
Bidens biternata
鬼针草属 *Bidens*

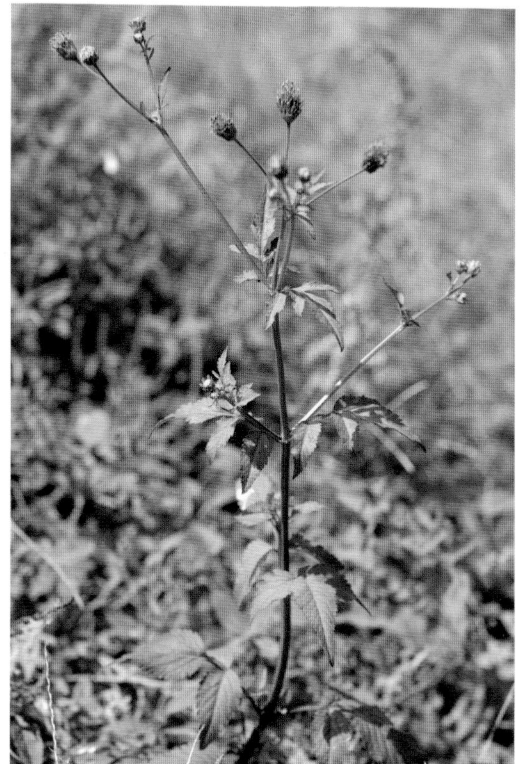

形态特征：一年生草本，高 30 ~ 150 cm。茎直立，稍具四棱，无毛或稍被短柔毛。一回羽状复叶，顶生小叶卵形至长圆状卵形或卵状披针形。头状花序，总苞基部有短柔毛，草质，条形，先端锐尖，外面背面密被短柔毛，内层苞片长椭圆形；舌状花 0 ~ 5 朵，淡黄色；盘花筒状。瘦果条形，黑色，具四棱，压扁，具短糙伏毛，顶端芒刺 3 ~ 4 枚，具倒刺毛。

花果期：花果期 9 ~ 11 月。

分布：亚洲、非洲和大洋洲。我国南北各地广布。南麂主岛各处常见。

生境：路边、荒地。

Description: Herbs annual, 30-150 cm tall. Stems erect, subtetragonal, loosely crisp pilose. Simple pinnate leaf, terminal leaflets ovate to oblong-ovate or ovate-lanceolate. Capitula, involucre base pubescent, grassy, striate, apex acute, outside abaxially densely pubescent, inner bracts long elliptic; ray florets 0 to 5, lamina yellow; disc florets tubular. Achenes linear, black, 4-angled, compressed, shortly strigose, pappus awns 3-4, barbellate.

Flower and Fruit: Fl. and fr. Sep. -Nov.

Distribution: Asia, Africa, Oceania. Throughout the north and south areas of China. Main island of Nanji, commonly.

Habitat: Roadsides, waste fields.

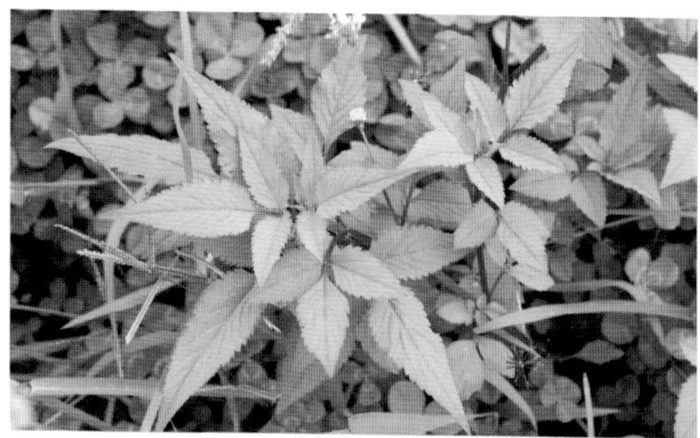

A403. 菊科
Asteraceae

大狼杷草
Bidens frondosa
鬼针草属 *Bidens*

形态特征：一年生草本，高 20 ～ 90 cm。茎直立，分枝，常带紫色。叶对生，具柄，一回羽状复叶，披针形，先端渐尖，边缘有粗锯齿，背面被稀疏短柔毛。头状花序单生茎端和枝端，总苞钟状或半球形，披针形或匙状倒披针形，叶状，边缘有缘毛，内层苞片长圆形，无舌状花或舌状花不发育，极不明显，筒状花两性。瘦果扁平，狭楔形，顶端芒刺 2 枚，有倒刺毛。

花果期：花期 8 ～ 9 月。

分布：原产北美洲，世界性入侵杂草。我国华东地区。南麂主岛各处常见。

生境：潮湿的树林、草地、灌木丛、田野、路边、铁路、溪边、池塘、沼泽、沟渠中。

Description: Herbs annual, 20-90 cm tall. Stem erect, branched, often purplish. Leaves opposite, stipitate, simple pinnate leaves lanceolated, apex acuminate, margin coarsely serrate, abaxially usually sparsely pubescent. Capitula solitary terminal, involucre campanulate or hemispherical, lanceolate or spatulate oblanceolate, leaflike, margin ciliate, inner bracts oblong, ligulate or ligulate flowers undeveloped, inconspicuous, tubular flowers bisexual. Achene flat, cuneate, pappus awns 2, bar bellate.

Flower and Fruit: Fl. Aug. -Sep.

Distribution: Native to North America, cosmopolitan invasive weed. East China. Main island of Nanji, commonly.

Habitat: Moist woods, meadows, thickets, fields, roadsides, railroads, borders of streams, ponds, sloughs, swamps, ditches.

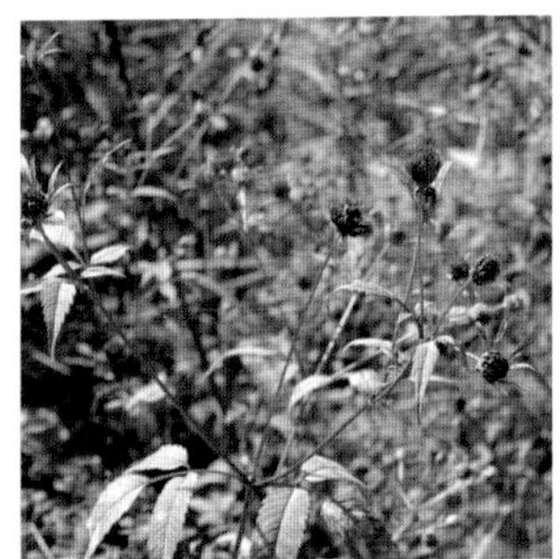

A403. 菊科
Asteraceae

鬼针草（三叶鬼针草）
Bidens pilosa
鬼针草属 *Bidens*

形态特征：一年生草本，高 30 ～ 180 cm。叶片卵形至披针形，不裂或 1 回羽状，裂片 3 ～ 7，两面被疏毛或无毛，基部截形至楔形，边缘有锯齿。头状花序单生或成疏松伞房花序，辐射状或盘状，副萼苞片 7 ～ 9，贴伏，匙形至线形，具缘毛；总苞陀螺状到钟状，苞片 8 或 9，披针形至倒披针形；舌状花无或 5-8，舌片白色至浅粉色，管状花 20 ～ 40，花冠淡黄色。瘦果黑色，条形，略扁，顶端芒刺 3 ～ 4 枚，具倒刺毛。

花果期：花期全年。

分布：热带亚热带地区。广布我国南北各地。南麂各岛屿常见。

生境：路边、田野、村旁。

Description: Herbs annuals, 30-180 cm tall. Stems erect, glabrous or very sparsely pubescent in upper part. Leaf blade simple or pinnate with 3-7 lobes, ovate to lanceolate, both surfaces pilosulose or glabrate, bases truncate to cuneate, margin serrat. Capitula solitary or in lax corymbs, radiate or discoid, calycular bracts 7-9, appressed, spatulate to linear, margins ciliate; involucres turbinate to campanulate, phyllaries 8 or 9, lanceolate to oblanceolate; ray florets absent or 5-8, lamina whitish to pinkish, disk florets 20-40, corollas yellowish. Achenes blackish, linear,± flat, pappus awns 3-4, barbellate.

Flower and Fruit: Fl. year-round.

Distribution: Tropical and subtropical regions. Widely distributed in the north and south of China. Throughout Nanji Islands, commonly.

Habitat: Roadsides, fields, villages.

A403. 菊科
Asteraceae

野菊
Chrysanthemum indicum
菊属 *Chrysanthemum*

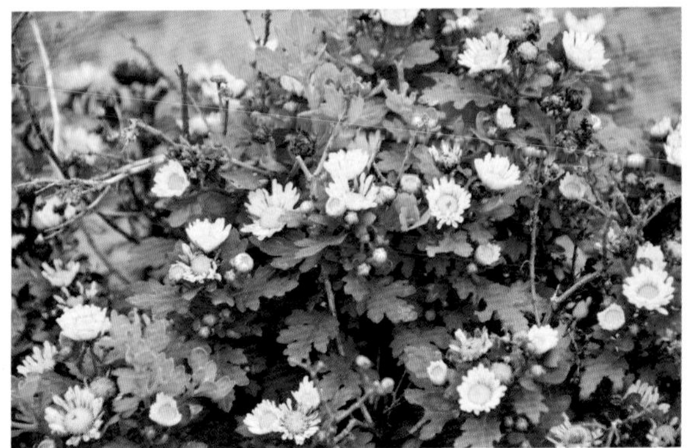

形态特征：多年生草本，高 25 ~ 90 cm。茎直立或铺散，分枝，被疏毛。基生叶和下部叶花期脱落；中部茎叶长卵形或卵状椭圆形，一回羽状分裂或分裂不明显但有浅锯齿。头状花序多或少，在枝顶成疏松的、平顶的聚伞花序；总苞片卵形或卵状三角形，白色或褐色，边缘宽膜质；缘花舌状，黄色，雌性；盘花管状，两性。

花果期：花果期 6 ~ 11 月。

分布：东亚、南亚。除西北外，广布我国南北各地。南麂各岛屿常见。

生境：山坡草地、灌丛、河边水湿地、田边及路旁、滨海盐渍地。

Description: Herbs perennial, 25-90 cm tall. Stems erect or diffuse, branched, sparsely pilose. Basal leaf and lower leaves withered at anthesis; middle stem leaves long ovate, or elliptic-ovate, pinnatifid, pinnatilobed, or inconspicuously divided, edges lightly serrated. Capitula many or few, formed lax terminal flat-topped cyme; phyllaries ovate or ovate-triangular, scarious margin broad, white or brown; ray floret lamina yellow, female; disc flower tubular, bisexual.

Flower and Fruit: Fl. and fr. Jun. -Nov.

Distribution: East Asia, South Asia. Widely distributed in China except Northwest China. Throughout Nanji Islands, commonly.

Habitat: Grasslands on mountain slopes, thickets, wet places by rivers, fields, roadsides, saline places by seashores, under shrubs.

A403. 菊科
Asteraceae

蓟
Cirsium japonicum
蓟属 *Cirsium*

形态特征：多年生草本，高 30 ～ 50 cm。茎灰白色，直立，被多细胞长节毛。基生叶花期宿存，卵形或倒卵形，羽状深裂，基部渐窄成翼柄，边缘有刺状锯齿；中下部茎生叶渐小，与基生叶同形，无柄，两面绿色，基部半抱茎。头状花序直立，顶生，红色或紫色；总苞钟状，覆瓦状排列。瘦果扁，冠毛浅褐色。

花果期：花果期 6 ～ 9 月。

分布：东亚。广泛分布于我国各地。南麂主岛（门屿尾、三盘尾）、大檑山屿，常见。

生境：山坡林中、林缘、灌丛中、草地、荒地、田间、路旁或溪旁。

Description: Herbs perennial, 30-50 cm tall. Stems grayish white, erect, with sparse to dense long multicellular hairs mixed with dense felt under capitula. Basal leaves present at anthesis, leaf blade ovate, obovate, pinnately divided, stem base with winged petiole, wing spiny or with spiny teeth; lower and middle cauline leaves similar but sessile, gradually smaller upward, both sides green, semiamplexicaul. Capitula erect, terminal, red or purple; involucre campanulate, phyllaries imbricate. Achene flat, pappus bristles pale brown.

Flower and Fruit: Fl. and fr. Jun. -Sep.

Distribution: East Asia. Widely distributed in China. Main island (Menyuwei, Sanpanwei), commonly.

Habitat: Forests, forest margins, thickets, grasslands, wastelands, farmlands, roadsides, streamsides.

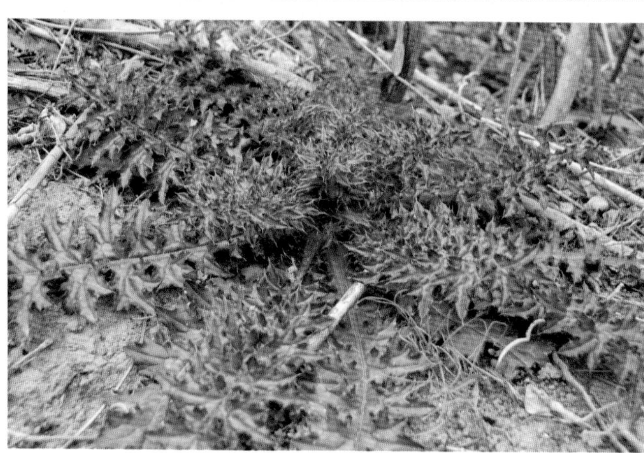

A403. 菊科
Asteraceae

野茼蒿（革命菜）
Crassocephalum crepidioides
野茼蒿属 *Crassocephalum*

形态特征：一年生草本。茎直立，高 30 ~ 80 cm。茎有纵棱，无毛。叶椭圆形或长圆状椭圆形，边缘有不规则锯齿或重锯齿，或有时基部羽状裂，两面无毛。头状花序数个在茎端排成伞房状；总苞圆柱形，基部截形，总苞片线状披针形，顶端有簇状毛；小花管状，两性，花冠红褐色或橙红色，5 齿裂。瘦果狭圆柱形，赤红色，有肋，被毛。

花果期：花果期 7 ~ 11 月。

分布：原产非洲，泛热带地区广泛分布。我国长江中下游各地。南麂主岛各处常见。

生境：斜坡、路旁、溪边或灌丛。

Description: Herbs annual, 30-80 cm tall. Stems striate, glabrous. Leaf blade elliptic or oblong-elliptic, margin irregularly serrate or double-serrate, sometimes pinnately lobed at base, both surfaces glabrous or subglabrous. Capitula several to numerous in terminal corymbiform cymes; involucres cylindric, basally truncate, phyllaries linear-lanceolate, apically puberulent; florets tubular, bisexual, corolla red-brownish or orange, lobes 5. Achenes brownish, narrowly oblong, ribbed, hairy.

Flower and Fruit: Fl. and fr. Jul. -Nov.

Distribution: Native to Africa; widely distributed in pantropics. The middle and lower reaches of Yangtze River Basin. Main island of Nanji, commonly.

Habitat: Slopes, roadsides, streamsides, thickets.

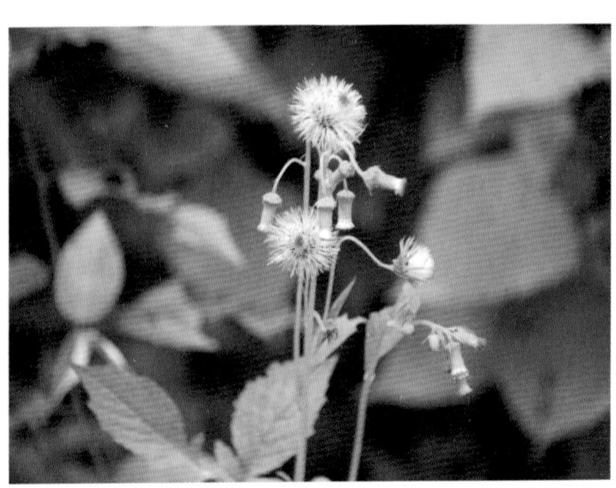

A403. 菊科
Asteraceae

假还阳参
Crepidiastrum lanceolatum
假还阳参属 *Crepidiastrum*

形态特征：多年生草本，高 10 ~ 20 cm，莲座状，具木质根茎。基生叶匙形，不裂或羽状裂，侧裂片卵形至披针形，无毛，基部收窄，先端圆钝。复合花序伞房状，具数枚头状花序；头状花序具 8 ~ 12 小花，花序梗柔弱；总苞狭圆柱状，总苞片无毛；舌状小花黄色。瘦果棕色，近纺锤形。

花果期：花果期 9 ~ 11 月。

分布：日本。我国江苏、浙江沿海岛屿。南麂各岛屿常见。

生境：沿海地区的多石山坡。

Description: Herbs perennial, 10-20 cm tall, rosulate, with a woody caudex or rootstock. Rosette leaves spatulate, undivided or pinnatifid to pinnatisect with ovate to lanceolate lateral lobes, glabrous, base cuneately attenuate, apex usually rounded. Synflorescence corymbiform, with few to several capitula; capitula with 8-12 florets. peduncle slender; involucre narrowly cylindric, phyllaries glabrous; ray florets yellow. Achene brownish, subfusiform.

Flower and Fruit: Fl. and fr. Sep. -Nov.

Distribution: Japan. Coastal islands of Jiangsu and Zhejiang. Throughout Nanji Islands, commonly.

Habitat: Rocky situations on hillsides in coastal areas.

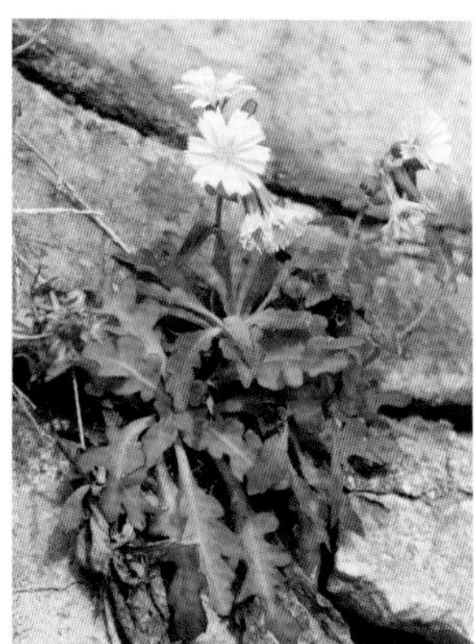

A403. 菊科
Asteraceae

芙蓉菊
Crossostephium chinense
芙蓉菊属 *Crossostephium*

形态特征：亚灌木，高 10 ~ 40 cm。枝、叶密被灰色柔毛。叶互生，聚生枝顶，窄匙形，质厚，全缘或 3 ~ 4 裂。头状花序盘状，在枝端叶腋排成总状；总苞半球形，外层和中层总苞片椭圆形，草质，内层长圆形，边缘宽膜质；小花管状，密生腺点，边花雌性，顶端 2 ~ 3 齿裂；盘花两性，花冠管状，顶端 5 裂。瘦果长圆形，基部窄，5 棱，被腺点。

花果期：花果期全年。

分布：日本。我国东南沿海地区。南麂各岛屿常见。

生境：海滨岩缝。

Description: Subshrubs, 10-40 cm tall. Branches and leaves densely gray-white pubescent. Leaves alternate, aggregated at apex of branches, blade narrowly spatulate, thick, margin entire or sometimes apex 3-or 4-lobed. Capitula disciform, in a frondose raceme along branches; involucres hemispheric, outer and mid phyllaries elliptic, herbaceous, inner oblong, margin broadly scarious; florets tubular, gland-dotted outside, marginal florets female, apex 2-or 3-denticulate; disk florets bisexual, 5-lobed. Achene oblong, 5-tibed, gland-dotted.

Flower and Fruit: Fl. and fr. year-round.

Distribution: Japan. Coast areas of southeast China. Throughout Nanji Islands, commonly.

Habitat: Littoral rock crevices.

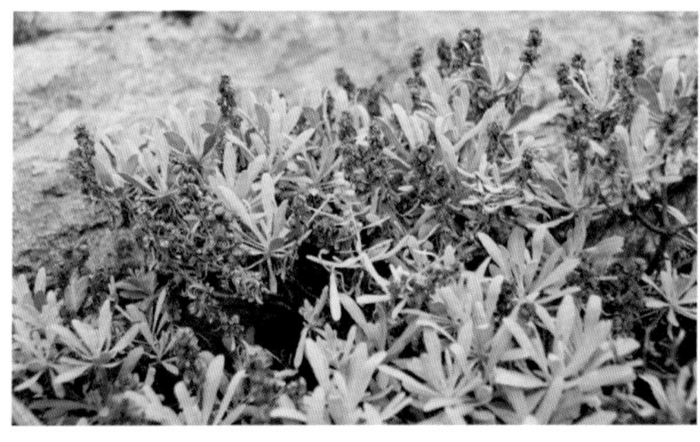

A403. 菊科
Asteraceae

鳢肠（墨旱莲）
Eclipta prostrata
鳢肠属 *Eclipta*

形态特征：一年生草本，高 50 cm。茎直立，上升或匍匐，贴生糙毛，基部分枝。叶披针形，纸质，边缘有细锯齿，两面密被糙毛。头状花序 1 ~ 2 腋生或顶生，花序梗细；总苞球状钟形，果期变大，总苞片 5 ~ 6 个排成 2 层，绿色，草质，长圆形，外层较长；舌状花 2 层，舌片二裂或全缘；管状花多数，花冠 4 裂。瘦果边缘具白肋。

花果期：花期 6 ~ 9 月。

分布：原产美洲，现世界广布。我国南北各地广泛分布。南麂各岛屿常见。

生境：河边、田野、废弃池塘或路旁。

Description: Herbs annual, 50 cm tall. Stems erect, ascending or prostrate, strigose-pilose, branched at base. Leaves lanceolate, papery, margin serrulate, densely strigose-pubescent on both surfaces. Capitula 1 to 2, axillary or terminal, peduncle slender; involucre globose-campanulate, enlarging in fruit, phyllaries 5 or 6, 2-seriate, green, grassy, oblong, outer longer; ray florets 2-seriate, lamina bifid or entire; disk florets many, corolla 4-lobed. Achenes margin ribbed.

Flower and Fruit: Fl. Jun. -Sep.

Distribution: Native to America; widely introduced worldwide. Widely distributed in China. Throughout Nanji Islands, commonly.

Habitat: Riversides, fields, abandoned ponds, roadsides.

A403. 菊科
Asteraceae

一点红
Emilia sonchifolia
一点红属 *Emilia*

形态特征：一年生草本，高 10 ~ 40 cm。茎直立或斜升，常基部分枝。下部叶密集，大头羽状，背面变紫，两面被卷毛；中部叶疏生，较小，卵状披针形或长圆状披针形，无柄，基部箭状抱茎，全缘或有细齿；上部叶少，线形。头状花序花前下垂，花后直立，疏伞房状；总苞圆柱形，长圆状线形或线形，黄绿色；管状花粉红或紫色。瘦果圆柱形，肋间被毛。

花果期：花果期 7 ~ 11 月。

分布：泛热带。我国长江流域以南地区。南麂各岛屿常见。

生境：草坡、路边、田埂、沙地。

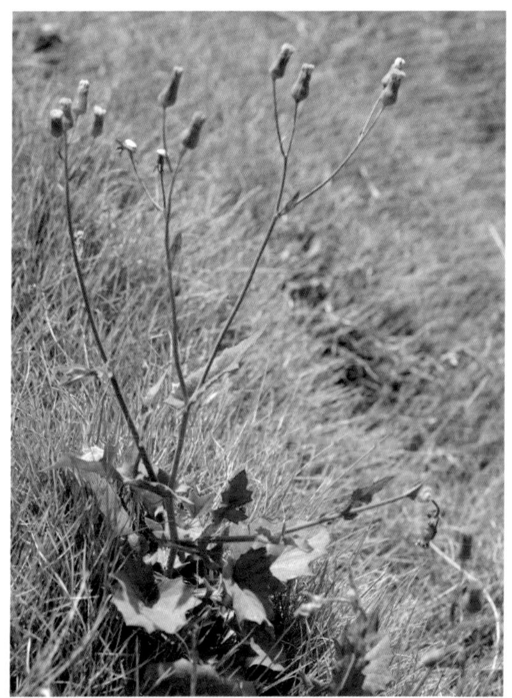

Description: Herbs annual, 10-40 cm tall. Stems erect or ascending, usually branching from base. Lower leaves crowded, lyrate-pinnatilobed, abaxially often purple, both surfaces crisped-hairy; median stem leaves lax, sessile, smaller, ovate-lanceolate or oblong-lanceolate, basally hastately semiamplexicaul, margin entire or irregularly denticulate; upper leaves few, linear. Capitula pendulous before anthesis, erect later, in terminal lax corymbs; involucres cylindric, oblong-linear or linear, yellow-green; disk florets pink or purplish. Achenes cylindric, puberulent between ribs.

Flower and Fruit: Fl. and fr. Jul. -Nov.

Distribution: Pantropical. To the south of the Yangtze River Basin. Throughout Nanji Islands, commonly.

Habitat: Weedy slopes, roadsides, field margins, sandy places.

A403. 菊科
Asteraceae

一年蓬
Erigeron annuus
飞蓬属 *Erigeron*

形态特征：一年生草本，高 30 ～ 100 cm。茎粗壮，直立，上部多分枝，疏被刚毛。基生叶花期枯萎，长圆形或宽卵形，叶柄具翅；中上部叶较小，长圆状披针形，顶部叶线形。头状花序呈疏圆锥花序状；总苞半球形，2 ～ 3 层，偶见 4 层，背面被粗毛和小腺点，披针形；舌状花舌片平展，白色，线形；管状花黄色。瘦果披针形，扁平，被疏贴柔毛。

花果期：花期 6 ～ 9 月。

分布：原产北美洲，现世界广布。我国各地有归化，分布广泛。南麂各岛屿常见。

生境：荒野、路边。

Description: Herbs annual, 30-100 cm tall. Stems erect, branched in upper part, sparsely hispid. Basal leaves withered at anthesis, blade elliptic or broadly ovate, winged petiolate; mid and upper leaves smaller, blade oblong-lanceolate, uppermost linear. Capitula in loose paniculiform; involucre hemispheric, phyllaries 2-or 3(or 4)-seriate, abaxially sparsely hirsute, minutely glandular, lanceolate; ray florets lamina white, linear, flat; disk florets yellow, lobes glabrous. Achenes lanceolate, flattened, sparsely strigillose.

Flower and Fruit: Fl. Jun. -Sep.

Distribution: Native to North America; widely introduced worldwide. Widely spread and naturalized in China.Throughout Nanji Islands, commonly.

Habitat: Wastelands, roadsides.

A403. 菊科
Asteraceae

小蓬草（加拿大蓬、小飞蓬）
Erigeron canadensis（*Conyza canadensis*）
飞蓬属 *Erigeron*

形态特征：一年生草本，高 50 ～ 100 cm。茎直立，被疏长硬毛，上部多分枝。叶密集，基生叶花期常枯萎，下部叶倒披针形，两面或背面疏生柔毛；中上部叶小，线状披针形，全缘或 1 ～ 2 锯齿。头状花序多数，成顶生的大型圆锥花序；花序梗细；舌状花多数，舌状，白色，舌片线形；管状花淡黄色，花冠裂片 4 ～ 5。瘦果线状披针形，扁压，疏被微毛。

花果期：花期 5 ～ 9 月。

分布：原产北美洲，现世界广布。我国南北各地有归化。南麂各岛屿常见。

生境：常见杂草，开阔地、田边、路旁或干涸的溪边。

Description: Herbs annual, 50-100 cm tall. Stems erect, sparsely hirsute, branched above. Leaves densely, basal withered at anthesis, lower blade oblanceolate, surfaces or only adaxial sparsely pilose; mid and upper, blade linear-lanceolate, smaller, margin entire or rarely 1-or 2-toothed. Capitula numerous, in terminal, large paniculiform synflorescences; peduncles slender; ray florets white; disk florets yellowish, lobes 4 or 5. Achenes linear-lanceoloid, compressed, sparsely strigillose.

Flower and Fruit: Fl. May-Sep.

Distribution: Native to North America; widely introduced worldwide. Naturalized in the north and south areas of China. Throughout Nanji Islands, commonly.

Habitat: Common weed of open places, field margins, roadsides, dry streamsides.

A403. 菊科
Asteraceae

苏门白酒草
Erigeron sumatrensis
飞蓬属 *Erigeron*

形态特征：一年生或二年生草本，高 80 ~ 150 cm。茎直立，粗壮，中部以上分枝，密被灰白色糙伏毛。叶密集，基生叶花期凋落，下部叶倒披针形，被糙短毛，中上部叶少，狭披针形或线形。头状花序多数，成大而长的圆锥花序；总苞钟状或坛状，3 层，灰绿色，线形；舌状花多数，舌片淡黄色或淡紫色，极短细，丝状；管状花 6 ~ 11，淡黄色。瘦果线状披针形，扁压，被微毛。

花果期：花期 5 ~ 10 月。

分布：原产南美洲，现热带和亚热带地区广泛分布。我国东南至西南地区。南麂各岛屿常见。

生境：山坡草地、旷野、路旁、溪边，是一种常见杂草。

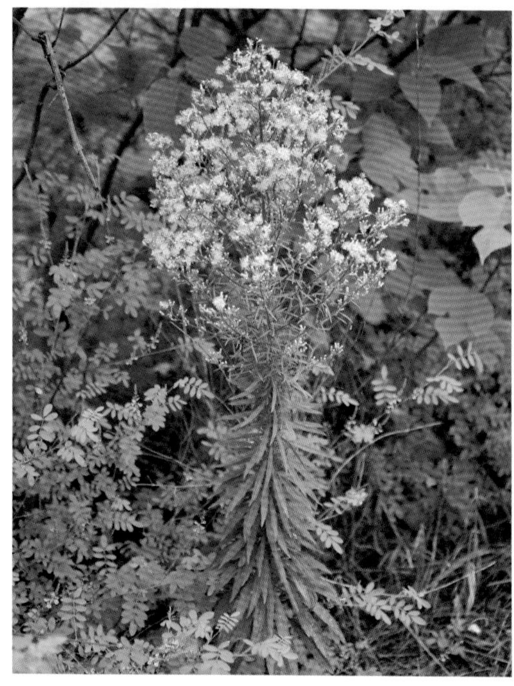

Description: Herbs annual or biennial, 80-150 cm tall. Stems erect, thick, branched above middle, densely gray-white strigose. Leaves densely, basal withered at anthesis, lower cauline blade oblanceolate, densely strigose, mid and upper reduced, blade narrowly lanceolate to linear. Capitula numerous, in large and long paniculiform synflorescences; involucre campanulate to urceolate, phyllaries 3-seriate, gray-green, linear; ray florets numerous, lamina yellowish or purplish, short, filiform; disk florets 6-11, yellowish. Achenes linear-lanceoloid, compressed, strigillose.

Flower and Fruit: Fl. May-Oct.

Distribution: Native to South America, widely distributed as a weed in tropical and subtropical regions. Southeast to southwest China. Throughout Nanji Islands, commonly.

Habitat: Common weed of grasslands on slopes, open places, roadsides, streamsides.

A403. 菊科
Asteraceae

多须公（华泽兰）
Eupatorium chinense
泽兰属 *Eupatorium*

形态特征： 多年生草本，或小灌木、半灌木状，高 70 ~ 100 cm。茎直立，多分枝，被污白色柔毛。叶对生，中部叶单生或三裂，卵形或宽卵形，羽状脉，叶两面被白色柔毛及黄色腺点。头状花序在枝顶成大型疏散复伞房花序；总苞钟状，上部及边缘白色，背面有黄色腺点；花白、粉或红色，被黄色腺点。瘦果淡黑褐色，椭圆状，被黄色腺点。

花果期： 花果期 7 ~ 11 月。

分布： 东亚至印度。我国东南至西南地区。南麂主岛（百亩山、国姓岙、门屿尾），常见。

生境： 山坡林缘、草地或灌丛。

Description: Herbs perennial, or small shrubs or subshrubs, 70-100 cm tall. Stems erect, well branched, sordid-white puberulent. Leaves opposite, median cauline leaves simple or 3-lobed, ovate or broadly ovate, pinnately veined, both surfaces white puberulent and yellow glandular. Capitula terminal, of large laxly compound corymbs; involucre campanulate, apically and marginally white, membranous, abaxially sparse yellow glands; corolla white, pink, red, with sparse yellow glands. Achenes pale black-brown, elliptic, yellow glandular.

Flower and Fruit: Fl. and fr. Jul. -Nov.

Distribution: East Asia to India. Southeast to southwest China. Main island (Baimushan, Guoxingao, Menyuwei, Sanpanwei), commonly.

Habitat: Forest margins, thickets or grasslands on slopes.

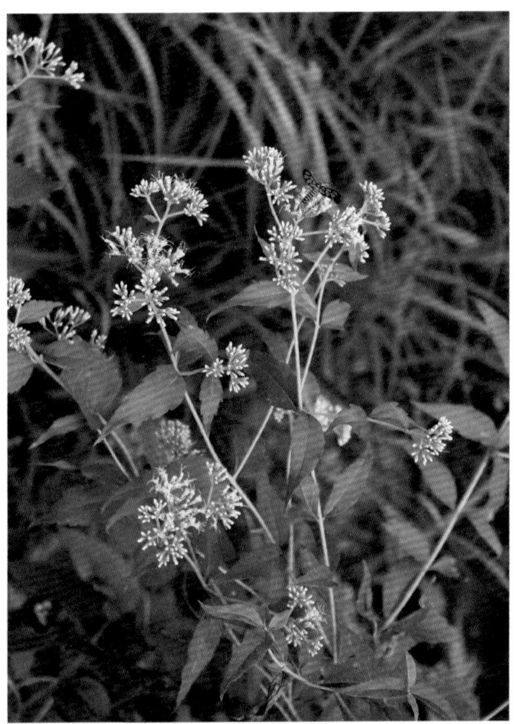

A403. 菊科
Asteraceae

白头婆（泽兰）
Eupatorium japonicum
泽兰属 *Eupatorium*

形态特征：多年生草本，高 50 ~ 200 cm。茎直立，紫红色，不分枝或上部伞房花序状分枝，被白色波状毛。叶对生，中部茎叶椭圆形或披针形，两面被柔毛和黄色腺点，边缘有粗锯齿。头状花序顶生，呈紧密伞房状；总苞钟状，总苞片覆瓦状，3 层，绿色或带紫红色；小花 5，白色、红紫色或粉红色，密被黄色腺点。瘦果淡黑褐色，椭圆状，5 棱，被黄色腺点，无毛。

花果期：花果期 6 ~ 11 月。

分布：东亚。除新疆未见记录外，遍及我国各省。南麂主岛（百亩山、国姓岙、门屿尾），常见。

生境：山坡草地、开阔林地、灌丛，湿地、河岸。

Description: Herbs perennial, 50-200 cm tall. Stems erect, purplish red, usually simple or corymbose synflorescence branched in upper part, white crisped-puberulent. Leaves opposite, median stem leaves elliptic or lanceolate, both surfaces scabrid, puberulent and yellow glandular, margin coarsely or doubly serrate. Capitula apically densely corymbose; involucre campanulate, phyllaries imbricate, 3-seriate, green or purple-tinged; florets 5, white, red-purple, or pink, with dense yellow glands. Achenes black-brown, elliptic, 5-angled, with many yellow glands, glabrous.

Flower and Fruit: Fl. and fr. Aug. -Nov.

Distribution: East Asia. Throughout China except Xinjiang. Main island (Baimushan, Guoxingao, Menyuwei, Sanpanwei), commonly.

Habitat: Grasslands on slopes, open forests, thickets, wet places, riverbanks.

A403. 菊科
Asteraceae

大吴风草
Farfugium japonicum
大吴风草属 *Farfugium*

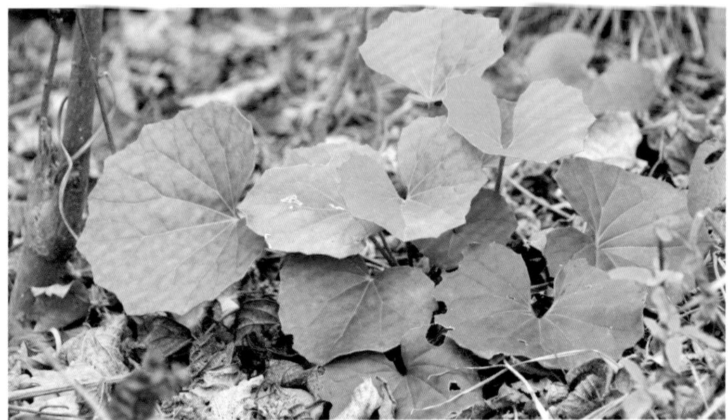

形态特征：多年生草本，高 30 ～ 70 cm。根茎粗壮，花葶幼时被密的淡黄色柔毛。叶全部基生，莲座状，有长柄，幼时被毛；叶片肾形，先端圆形，全缘或有小齿至掌状浅裂，基部弯缺宽，叶质厚，近革质，两面幼时被灰色柔毛，上面绿色，下面淡绿色。头状花序辐射状，排列成伞房状花序。瘦果圆柱形，有纵肋，被毛。

花果期：花果期 8 月至翌年 3 月。

分布：日本。我国华东至华南地区。南麂各岛屿常见。

生境：低海拔地区，林下、草坡、山谷，常栽培。

Description: Herbs perennial, 30-70 cm tall. Rhizomes stout, scape initially densely shortly pale yellow pilose. Leaves all basal, rosulate, long petiolate, initially densely pilose; leaf blade reniform, apex rounded, margin entire or dentate to palmatilobed, base cordate, leaf thickness, subleathery, initially gray puberulent, adaxially green, abaxially pale green. Capitula radiate, arranged into corymbose. Achenes terete, ribbed, hairy.

Flower and Fruit: Fl. and fr. Aug. -Mar.

Distribution: Japan. East China to South China. Throughout Nanji Islands, commonly.

Habitat: Forests, grassy slopes, valleys, sometimes cultivated in gardens; low elevations.

A403. 菊科
Asteraceae

牛膝菊
Galinsoga parviflora
牛膝菊属 *Galinsoga*

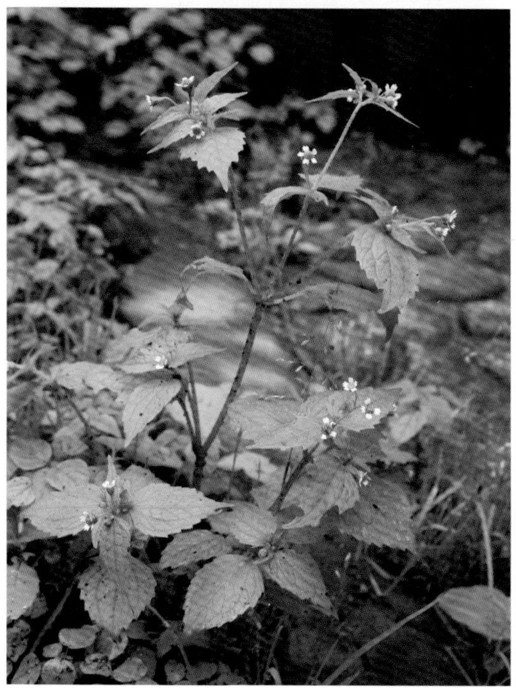

形态特征：一年生草本，高 10 ～ 80 cm。茎纤细，不分枝或自基部分枝。叶对生，卵形或长椭圆状卵形，基部圆形、宽楔形，基出三脉或不明显五出脉。头状花序半球形，有长花梗；总苞半球形或宽钟状，宿存；舌状花 5，舌片苍白色或粉色，顶端 3 齿裂；管状花 15 ～ 50，黄色。瘦果无冠毛。

花果期：花果期 7 ～ 10 月。

分布：原产南美洲。广泛分布于我国。南麂主岛各处常见。

生境：田野、溪边或疏林下。

Description: Herbs annual, 10-80 cm tall. Stem slender, unbranched or branched from base. Leaves opposite, ovate or long elliptic ovate, base round or broad cuneate, trinerved or unconspicuous quinquenerved. Capitula hemispherical, long pedicle, involucres hemispherical or campanulate; phyllaries persistent; ray florets 5, corollas usually dull white or pink, 3-lobed; disk florets 15-50, yellow. Achenes pappus absent.

Flower and Fruit:Fl. and fr. Jul. -Oct.

Distribution: Native to South America. Widely distributed in China. Main island of Nanji, commonly.

Habitat: Fields, streamsides, sparse forests.

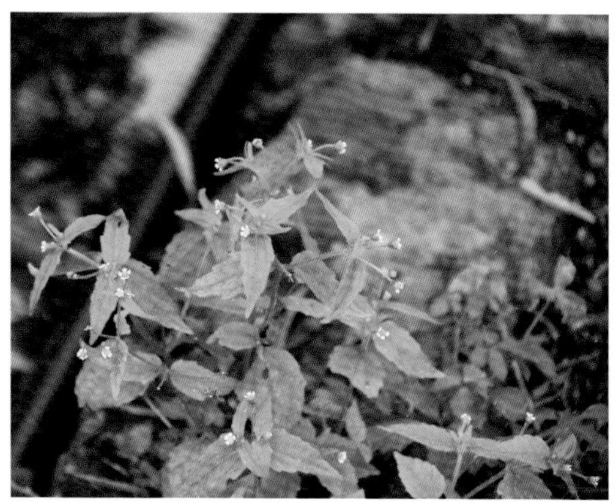

A403. 菊科
Asteraceae

匙叶合冠鼠曲草（匙叶鼠麴草）
Gamochaeta pensylvanica（*Gnaphalium pensylvanicum*）
合冠鼠曲草属 *Gamochaeta*

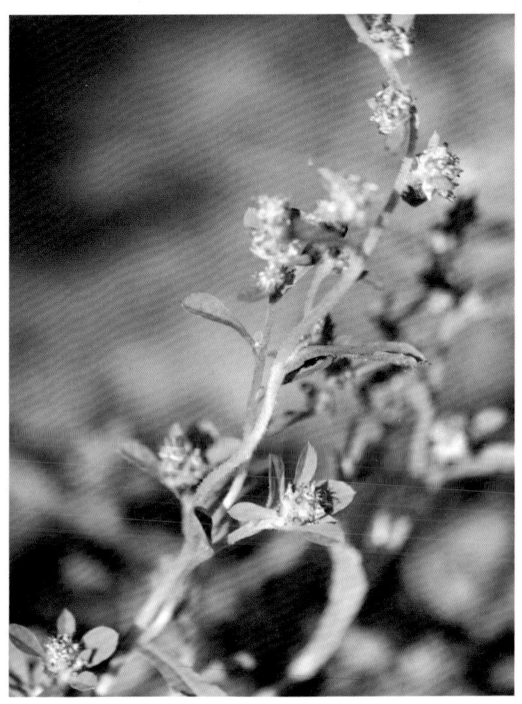

形态特征：一年生草本，高 20 ～ 40 cm。茎直立或基部斜倾分枝或不分枝，被白色绵毛。下部叶无柄，倒披针形或匙形，基部长渐狭，下延，顶端钝、圆，全缘或微波状，上面被疏毛，下面密被灰白色绵毛，侧脉 2 ～ 3 对，细弱；中部叶倒卵状长圆形或匙状长圆形。头状花序数个成束腋生，排列成穗状圆锥花序；总苞卵形。瘦果褐色，椭圆形，有小腺体。

花果期：花果期 12 月至翌年 5 月。

分布：世界广布。我国华东、华南、西南等地。南麂主岛各处常见。

生境：荒地、路边。

Description:Herbs annual, 20-40 cm tall. Stems erect, simple or more often branching from base, grayish tomentose. Lower leaves sessile, oblanceolate to spatulate, base length attenuate, decurrent, apex rounded to obtuse, margin entire or slightly wavy, adaxially loosely arachnoid, abaxially densely grey lanate, lateral veins 2-3 pairs, thin; middle leaves obovate-oblong or spatulate oblong. Capitula numerous in axillary clusters, forming spicate panicles; involucre ovate. Achenes brown, elliptic, minutely glandular.

Flower and Fruit:Fl. and fr. Dec.-May.

Distribution:Worldwide. East China, South China, Southwest China. Main island of Nanji, commonly.

Habitat:Waste fields, roadsides.

A403. 菊科
Asteraceae

细叶鼠曲草（细叶鼠麴草）
Gnaphalium japonicum
鼠曲草属 *Gnaphalium*

形态特征：多年生草本，高 8 ～ 25 cm。茎直立，密被白色绵毛。叶多基生，茎生叶少；基生叶莲座状，线状倒披针形，上面绿色，疏被棉毛，中部茎生叶线形，上部叶近花序，披针形。头状花序 10 个以上，成密集头状伞房花序；总苞近钟形，苞片红褐色，3 层，顶端钝；缘花雌性，花冠丝状，短于花柱，盘花顶端带粉色。瘦果纺锤状圆柱形，冠毛粗糙，白色。

花果期：花期 5 ～ 10 月。

分布：日本、朝鲜、大洋洲。我国长江流域以南各地。南麂主岛各处常见。

生境：草地、荒地。

Description: Herbs perennial, 8-25 cm tall. Stem erect, white lanate. Leaves chiefly radical, few cauline; radical leaves rosulate, linear-oblanceolate, adaxially green, thinly lanate; median cauline leaves linear, uppermost leaves subtending synflorescence, lanceolate. Capitula 10 to numerous, in a dense headlike corymb; involucre campanulate, phyllaries reddish brown, imbricate, 3-seriate, obtuse; marginal florets female; corolla filiform, shorter than style; disk florets apex pinkish. Achenes oblong, pappus of capillary bristles, white.

Flower and Fruit: Fl. May -Oct.

Distribution: Japan, Korea, Oceania. To the south of the Yangtze River Basin. Main island of Nanji, commonly.

Habitat: Grasslands, waste fields.

A403. 菊科
Asteraceae

白子菜（白背三七草）
Gynura divaricata
菊三七属 *Gynura*

形态特征：多年生草本，高 30 ～ 60 cm。茎直立，基部多少斜升，木质，干时具棱，不分枝或上部有花序枝，稍带紫色。叶质厚，通常集中于下部；叶柄有短柔毛，基部有卵形或半月形具齿的耳。叶片卵形，椭圆形或倒披针形，顶端钝或急尖，两面被短柔毛；小花橙黄色，有香气，略伸出总苞。瘦果圆柱形，褐色。

花果期：花期 8 ～ 10 月。

分布：越南北部。我国华南、西南等地，各地常栽培，偶有逸生。南麂主岛（关帝岙、三盘尾），偶见。

生境：山坡草地、荒坡和田边潮湿处。

Description: Herbs perennial, 30-60 cm tall. Stems erect, ± ascending from base, woody, striate when dry, simple or with synflorescence branched in upper part, purplish. Leaves thick, usually crowded in lower part; petiole shortly pubescent, with ovate or half-moon-shaped, dentate auricle at base; blade adaxially green, ovate, apex obtuse or acute. Floret orange-yellow, fragrant, slightly exceeding involucres. Achenes brown, cylindric.

Flower and Fruit: Fl. Aug. -Oct.

Distribution: North Vietnam. South China and Southwest China, usually cultivated, occasionally naturalized. Main island (Guandiao, Sanpanwei), occasionally.

Habitat: Grassy and weedy slopes, wet places by fields, seaside rocks.

A403. 菊科
Asteraceae

旋覆花
Inula japonica
旋覆花属 *Inula*

形态特征：多年生草本，高 15 ~ 100 cm。茎直立。中部叶长圆形、长圆状披针形或披针形，基部常有圆形半抱茎小耳，无柄，有小尖头状疏齿或全缘，上面有疏毛或近无毛，下面有疏伏毛和腺点，中脉和侧脉有较密长毛；上部叶线状披针形。头状花序成疏散伞房状，花序梗细长；舌状花黄色，舌片线形；盘花管状花多数，顶端裂片三角状披针形。瘦果圆柱形，被疏毛。

花果期：花果期 7 ~ 10 月。

分布：东亚。广泛分布于我国南北各地。南麂主岛（大沙岙、门屿尾、三盘尾），常栽培。

生境：山坡、草地、河岸溪边、田野、林下。

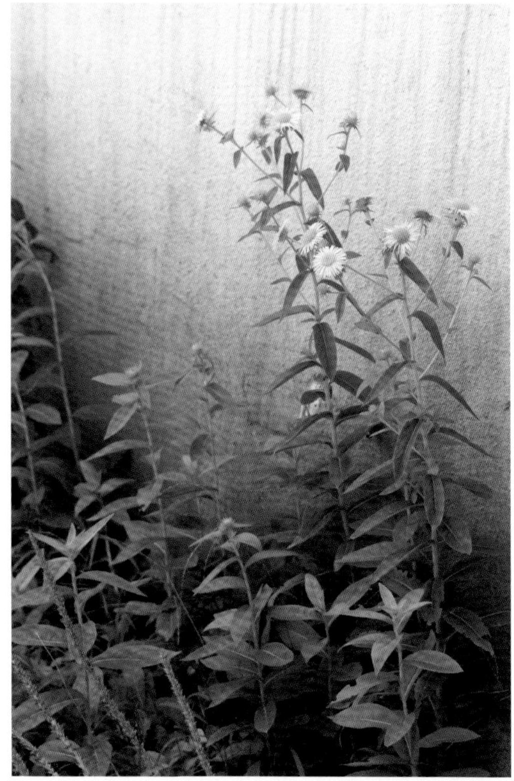

Description: Herbs perennial, 15-100 cm tall. Stems erect. Median leaves oblong, oblong lanceolate, or lanceolate, base often rounded semiamplexicaul auricle, sessile, sparsely toothed or entire, sparsely hairy or subglabrous above, sparsely hairy and glandular below, midvein and lateral veins densely hairy; upper leaves linear lanceolate. Capitula of laxly corymbs, peduncle slender; ray floret yellow, ligule linear; disc flower tubular flowers many, apical lobes triangular lanceolate. Achenes cylindric, pilose.

Flower and Fruit: Fl. and fr. Jul. -Oct.

Distribution: East Asia. Widely distributed in the north and south of China. Main island (Dashaao, Menyuwei, Sanpanwei), usually cultivated.

Habitat: Montane slopes, grasslands, riverbanks, fields, broad-leaved forests, streamsides

A403. 菊科
Asteraceae

小苦荬
Ixeridium dentatum
小苦荬属 *Ixeridium*

形态特征：多年生草本，高 20 ～ 50 cm。根壮茎短缩，生多数等粗的细根。茎直立，单生，分枝，无毛，叶稀疏。基生叶密集，不裂成羽状深裂；茎叶披针形，不分裂。头状花序多数，在茎枝顶端排成伞房状花序；头状花具小花 5 ～ 7，总苞圆柱状，背面无毛；小花黄色，少白色。瘦果纺锤形，褐色。

花果期：花果期 4 ～ 8 月。

分布：俄罗斯远东地区、日本、朝鲜。分布于我国东部湿润地区。南麂主岛各处常见。

生境：山坡林下、潮湿处或田边。

Description: Herbs perennial, 20-50 cm tall. Rhizomes shortly oblique, with fibrous roots. Stems solitary or few, slender, erect, branched, glabrous, sparsely leafy. Basal leaves crowded, undivided or pinnatipartite; stem leaves lanceolate, undivided. Synflorescence corymbiform, with some to many capitula; capitula with 5-7 florets; involucre cylindric, phyllaries abaxially glabrous; florets yellow or rarely white. Achene brown, fusiform.

Flower and Fruit: Fl. and fr. Apr. -Aug.

Distribution: East Russian, Japan, Korea. Humid area of eastern China. Main island of Nanji, commonly.

Habitat: Forests on mountain slopes, moist places, fields.

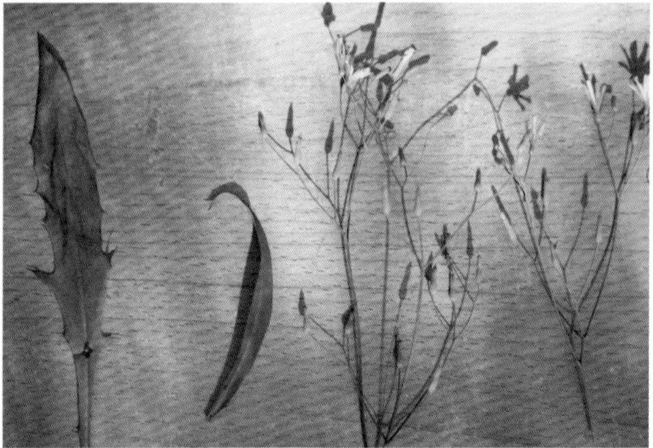

A403. 菊科
Asteraceae

台湾翅果菊
***Lactuca formosana*（*Pterocypsela formosana*）**
莴苣属 *Lactuca*

形态特征：一年生或多年生草本，高 0.5 ~ 1.5 m。茎单生，直立，上部疏散分枝。叶两面粗糙，下面沿脉有小刺毛；中下部茎叶椭圆形或倒披针形，基部羽状抱茎；顶裂片长披针形或三角形；侧裂片 2 ~ 5 对，椭圆形或宽镰刀状，边缘有锯齿。头状花序多数，排成稀疏伞房状；舌状小花黄色。瘦果卵状椭圆形，压扁。

花果期：花果期 5 ~ 10 月。

分布：我国南北各地广布。南麂主岛（门屿尾、三盘尾），偶见。

生境：山坡草地、灌丛、田间路旁。

Description: Herbs annual or perennial, 0.5-1.5 m tall. Stem solitary, erect, loosely branched apically. Leaves hirsute, main rib echinulate; lower and middle stem leaves elliptic, or oblanceolate, with narrow petiole-like amplexicaul basal portion; apical lobes long lanceolate or triangular, lateral lobes 2-5 pairs, elliptic or broadly falcate, faintly to strongly dentate on margin. Synflorescence loosely corymbose, with many capitula; ray florets yellow. Achene ellipsoid, compressed.

Flower and Fruit: Fl. and fr. May-Oct.

Distribution: Throughout the north and south areas of China. Main island (Menyuwei, Sanpanwei), occasionally.

Habitat: Grasslands, thickets on mountain slopes, fields, along trails.

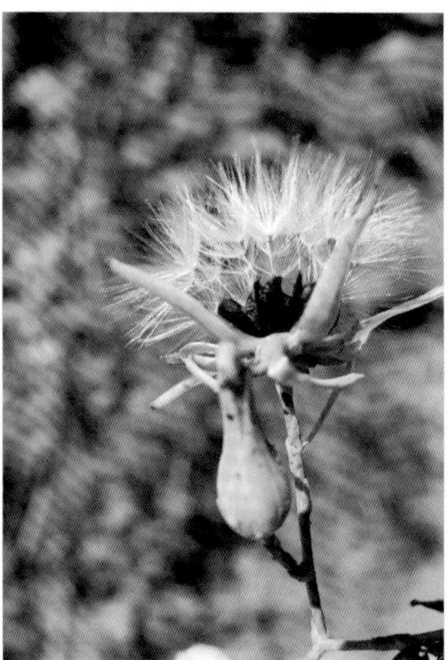

A403. 菊科
Asteraceae

翅果菊（多裂翅果菊）
Lactuca indica（*Pterocypsela indica*）
莴苣属 *Lactuca*

形态特征：一年生或多年生草本，高 0.4 ～ 2 m。茎单生，直立，顶部分枝，无毛。茎生叶线形，无柄，两面无毛，边缘全缘或中部以下两侧有小尖头或疏锯齿，中下部叶缘具稀疏点状或三角状锯齿。头状花序果期卵球形，多数，排成圆锥花序或总状圆锥花序；舌状小花淡黄色。瘦果椭圆形，黑色，压扁，边缘有宽翅。

花果期：花果期 4 ～ 11 月。

分布：原产亚洲，世界范围内广泛引种。分布我国各地。南麂主岛（门屿尾、三盘尾），偶见。

生境：山谷、林缘、灌丛、沟壑、草地、田野或荒地。

Description: Herbs annual or perennial, 0.4-2 m tall. Stem solitary, stout, erect, branched apically, glabrous. Cauline leaves linear-oblong, glabrous, margin entire, base or below middle margin with small cusp or sparse serration or tine on both sides; middle and lower cauline leaves margin sparsely pointed or triangular serrate. Synflorescence paniculiform to racemiform-paniculiform, with numerous capitula, ovoid in fruit, arranged along the tips of stem branches; ray florets, pale yellow. Achene, oval, black, compressed, broadly winged.

Flower and Fruit: Fl. and fr. Apr. -Nov.

Distribution: Native in Asia, widely introduced in the world. Throughout China. Main island (Menyuwei, Sanpanwei), occasionally.

Habitat: Mountain valleys, forest margins, thickets, ravines, grasslands, fields, wastelands

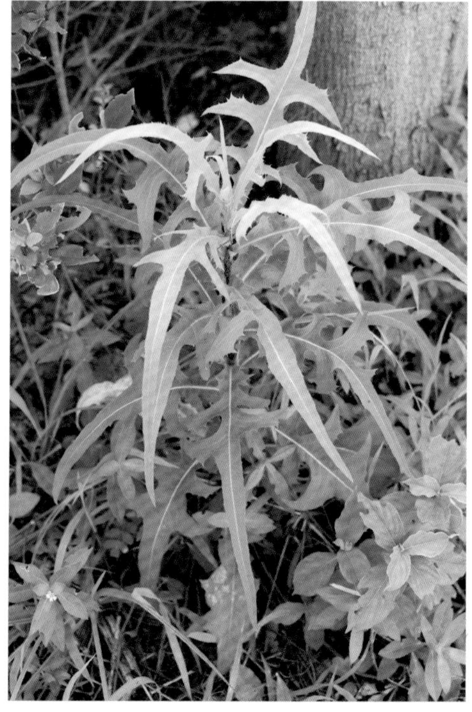

A403. 菊科
Asteraceae

卤地菊
Melanthera prostrata（*Wedelia prostrata*）
卤地菊属 *Melanthera*

形态特征： 一年生草本。茎匍匐，长 20 ～ 90 cm，分枝。叶无柄或有短柄，叶片披针形或长圆状披针形，边缘有不规则锯齿，两面密被短糙毛。头状花序常单生茎顶；总苞近球形，绿色，总苞片 1 层，卵形至卵状长圆形，被糙毛；舌状花黄色，先端 3 裂；盘花管状，5 裂，疏被短毛。瘦果倒卵状三棱形，顶端被短毛。

花果期： 花期 6 ～ 10 月。

分布： 日本、朝鲜、泰国、越南。我国华东、华南的沿海岛屿。南麂各岛屿常见。

生境： 海滨干燥沙土地。

Description: Herbs annual. Stems creeping, 20-90 cm long, branched. Leaves sessile or short-stalked, leaf blade lanceolate or oblong-lanceolate, margin irregularly dentate, both surfaces densely short hispid. Capitula usually solitary, terminal; involucre hemispheric, green, phyllaries 1-seriate, ovate to ovate oblong, coarsely strigose; ray corolla yellow, apex 3-lobed; disk corolla tubular, 5-lobed, sparsely hairy. Achenes obovate triangulate, apically strigillose.

Flower and Fruit: Fl. Jun. -Oct.

Distribution: Japan, Korea, Thailand, Vietnam. Coastal islands in East and South China. Throughout Nanji Islands, commonly.

Habitat: Littoral sand dunes, sandy seashores.

A403. 菊科
Asteraceae

鼠曲草（鼠麴草）
Pseudognaphalium affine（*Gnaphalium affine*）
拟鼠麴草属 *Pseudognaphalium*

形态特征：二年生草本，高 10 ～ 40 cm。茎直立，密被白色绒毛。叶匙形，两面被白色绒毛，基部渐狭，无柄，边全缘，顶端圆，具刺尖头。头状花序多数，在枝端密集成伞房花序；总苞球形，总苞片 3 齿裂，浅黄色，外层短，阔卵形，内层长圆形，先端钝；中央小花 5 ～ 10。瘦果卵形，扁平，有乳头状突起。

花果期：花果期 9 ～ 10 月。

分布：东亚、东南亚。除东北外，广布我国大部分地区。南麂主岛各处常见。

生境：旱地或湿润草地。

Description: Herbs biennial, 10-40 cm tall. Stems erect, densely white lanate tomentose. Leaves spatulate, white lanate on both surfaces, base decurrent, narrowed, sessile, margin entire, apex rounded, mucronulate. Capitula numerous, densely aggregated in terminal corymbs; involucre globose-campanulate, phyllaries 3-seriate, pale yellow, outer ones shorter, broadly ovate, inner ones oblong, apex obtuse; central florets 5-10. Achenes oblong, compressed, papillose.

Flower and Fruit: Fl. and fr. Sep. -Oct.

Distribution: East Asia, Southeast Asia. Widely distributed in most areas of China except Northeast China. Main island of Nanji, commonly.

Habitat: Dry lands or moist grasslands.

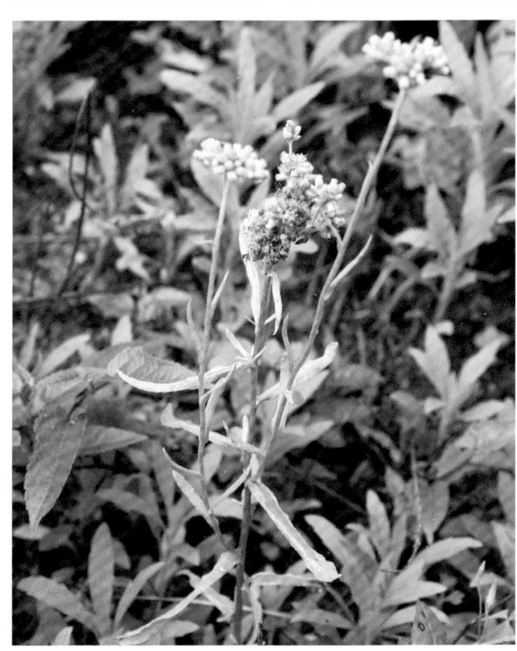

A403. 菊科
Asteraceae

毛梗豨莶
Sigesbeckia glabrescens
稀莶属 *Siegesbeckia*

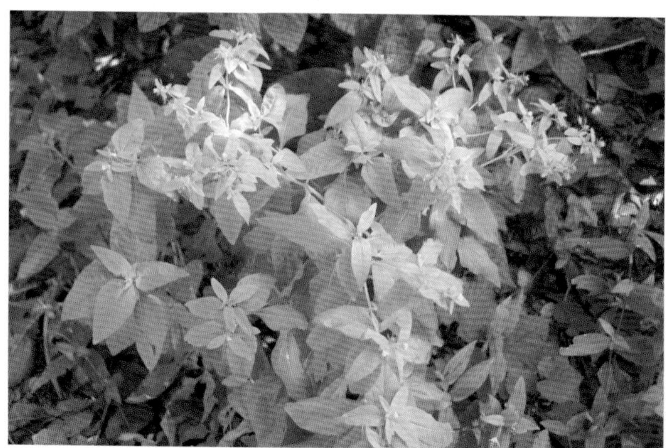

形态特征：一年生草本，高 35 ~ 100 cm。茎直立，较细弱，通常上部分枝，被平伏柔毛。基部叶花期枯萎；中部叶叶柄具翅，叶片卵状三角形，边缘有不规则锯齿；上部叶无柄，长圆形，顶部叶线形。头状花序辐射状，花序梗密被短柔毛；总苞钟状，苞片 2 层，外层匙形，密被腺毛，内层倒卵状长圆形；盘花花冠钟状。瘦果倒卵形，4 棱。

花果期：花期 4 ~ 9 月，果期 6 ~ 9 月。

分布：日本、朝鲜。我国东部湿润地区。南麂主岛各处常见。

生境：路边、旷野、灌丛。

Description: Herbs annual, 35-100 cm tall. Stems erect, slender, upper branched, shortly appressed pilose. Basal leaves withered when blooming, median cauline leaves with winged petiole, blade ovate-deltate, irregularly toothed; upper leaves sessile, oblong; uppermost leaves linear. Capitula radiate, peduncle densely shortly pubescent; involucre campanulate, 2-seriate, outer phyllaries spatulate, densely glandular pilose, inner phyllaries obovate-oblong; disk corolla campanulate. Achenes obovate, 4-ribed.

Flower and Fruit: Fl. Apr. -Sep. , fr. Jun. -Sep.

Distribution: Japan, Korea. Moist areas of the eastern China. Main island of Nanji, commonly.

Habitat: Roadsides, fields, thickets.

A403. 菊科
Asteraceae

裸柱菊
Soliva anthemifolia
裸柱菊属 *Soliva*

形态特征：一年生草本。茎极短，平卧。叶互生，基生莲座状，2～3回羽状分裂，裂片线形，被长柔毛或近无毛。头状花序近球形，无梗，生于茎基部；总苞片2层，长圆形或披针形；边缘雌花多数，无花冠，花柱宿存；中央两性花少数，花冠管状，黄色，顶端3齿裂，不结实。瘦果倒披针形，扁平，有厚翅。

花果期：花果期全年。

分布：原产于南美洲。我国华东地区有归化。南麂主岛（三盘尾），偶见。

生境：荒地、田野。

Description: Herbs annual. Stem very short, prostrate. Leaves alternate, in basal rosettes, leaf blade 2-or 3-pinnatifid; ultimate lobes linear, sparsely villous or subglabrous. Capitula subglobose, sessile, born at stem base; involucres 2-seriate, oblong or lanceolate; ray florets numerous, female, corolla absent, styles persistent; disk florets few, corolla tubular, yellow, apex 3-lobed, fruitless. Achenes oblanceolate, dorsiventrally flattened, with thick corky lateral wings.

Flower and Fruit: Fl. and fr. year-round.

Distribution: Native to South America. Naturalized in East China. Main island (Sanpanwei), occasionally.

Habitat: Waste lands and fields.

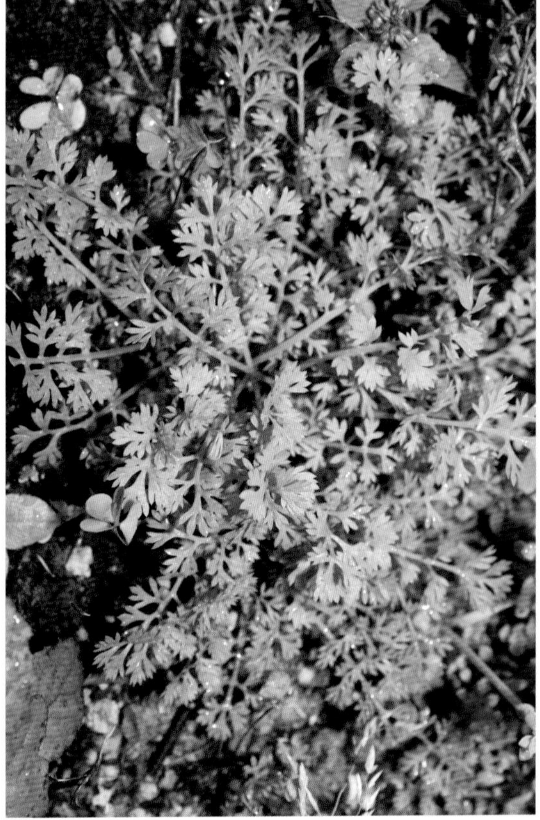

A403. 菊科
Asteraceae

花叶滇苦菜（续断菊）
Sonchus asper
苦苣菜属 *Sonchus*

形态特征：一年生草本，高 20 ~ 50 cm。茎直立，单生或簇生。基生叶和下部叶变化极大，倒卵形、匙形或椭圆形，不裂或稍不规则羽状全裂；中上部叶匙形至披针形，不裂，基部扩大呈圆形抱茎，紧贴。头状花序数个，成稠密伞房状；总苞宽钟状，背面光滑，很少有腺毛；舌状小花黄色。瘦果阔倒柱状，强烈压扁，多少具翅。

花果期：花果期 5 ~ 10 月。

分布：全球广布。我国各地有归化。南麂主岛各处常见。

生境：山坡、林缘、水边、田边、荒地。

Description: Herbs annual, 20-50 cm tall. Stem erect, solitary or fascicled. Basal and lower stem leaves extremely variable, obovate, spatulate, or elliptic, undivided or ± irregularly pinnatisect; middle and upper stem leaves spatulate to lanceolate, base auriculately clasping with conspicuous rounded and appressed auricles. Synflorescence densely corymbiform, with few to some capitula; involucre campanulate, phyllaries abaxially glabrous, rarely glandular hairy; corolla yellow. Achene broadly obcolumnar, strongly compressed, ± winged.

Flower and Fruit: Fl. and fl. May-Oct.

Distribution: Widely distributed almost worldwide. Naturalized in almost all over China. Main island of Nanji, commonly.

Habitat: Mountain slopes, forest margins, by water, field margins, ruderal areas.

A403. 菊科
Asteraceae

苦苣菜
Sonchus oleraceus
苦苣菜属 *Sonchus*

形态特征：一年生或二年生草本，高 40 ~ 150 cm。基生叶和下部茎生叶基部渐狭成叶柄状，多小于中部叶；中上部叶叶形多变，椭圆形、倒披针形或披针形，全缘或不规则羽状全裂，叶耳明显平展，边缘具刺状粗锯齿。头状花序少数，成短伞房状或总状；总苞钟形，无毛或具少量腺毛；花冠黄色。瘦果倒柱状，明显压扁，无翅，成熟时有明显皱纹。

花果期：花果期 5 ~ 12 月。

分布：全球广布。我国各地有归化。南麂主岛各处常见。

生境：山坡或山谷林缘、林下或平地田间、空旷处或近水处。

Description: Herbs annual or biennial, 40-150 cm tall. Basal and lower stem leaves with basal portion petiole-like and attenuate, mostly smaller than middle stem leaves; middle and upper stem leaves extremely variable, elliptic, oblanceolate, or lanceolate, entire or irregularly pinnatisect, base auriculately clasping with auricles usually acutely prostrate, margin coarsely spinulosely dentate. Synflorescence shortly corymbiform or racemiform, with few to several capitula on branchlets; involucre campanulate, phyllaries glabrous or with few glandular hairs; corolla yellow. Achene obcolumnar, distinctly compressed, distinctly rugose when fully mature.

Flower and Fruit: Fl. and fr. May-Dec.

Distribution: Widely distributed almost worldwide. Naturalized in almost all over China. Main island of Nanji, commonly.

Habitat: Mountain slopes, forests, forest margins, fields, near water, open land, ruderal areas.

A403. 菊科
Asteraceae

蟛蜞菊
Sphagneticola calendulacea（*Wedelia chinensis*）
蟛蜞菊属 *Sphagneticola*

形态特征：多年生草本。茎匍匐，上部近直立，基部各节生出不定根。叶对生，无柄，叶片线状长圆形至披针形，纸质，两面疏被贴生的短糙毛，边缘疏生具短尖细锯齿，先端锐尖。头状花序单生于枝顶；总苞半球形，总苞片5，1层；舌状花1层，黄色，花冠2～3齿，管状花花冠5裂。瘦果倒卵形，顶部有糙毛。

花果期：花期3～9月。

分布：南亚至东亚。我国东部和南部各地及沿海岛屿。南麂各岛屿常见。

生境：田边、低洼湿地、草地，沿海地区常见。

Description: Herbs perennial. Stem prostrate, upper erect, basal nodes with adventitious roots. Leaves opposite, sessile, blade linear-oblong to lanceolate, papery, appressed pilose on both surfaces, margin sparsely mucronulate-serrulate, apex acute. Capitula solitary on erect branches; involucre hemispheric, phyllaries 5, 1-seriate; ray florets 1-layer, yellow, corolla 2-3-dentate; disk florets corolla 5-lobed. Achenes obovoid, coarsely hairy at tip.

Flower and Fruit: Fl. Mar. -Sep.

Distribution: South Asia to East Asia. Eastern and southern areass and coastal islands of China. Throughout Nanji Islands, commonly.

Habitat: Paddy ridges, in grassy fields and moist lowland depressions, common in littoral areas.

A403. 菊科
Asteraceae

钻叶紫菀（钻形紫菀）
Symphyotrichum subulatum（*Aster subulatus*）
联毛紫菀属 *Symphyotrichum*

形态特征：一年生草本，高 16 ～ 150 cm。茎直立，有时略带紫色，无毛，无腺体。基生叶披针形至卵形，常在花期脱落；茎叶线状披针形，全缘或有细锯齿，先端锐尖。头状花序多数，成圆锥状；总苞圆柱形，总苞片 3 ～ 5 层，披针形至线状披针形，不等长；舌状花多数 1 层，蓝紫色；管状花黄色，裂片直立，三角形。瘦果披针形，疏生糙伏毛。

花果期：花果期 8 ～ 10 月。

分布：原产北美洲。我国秦淮以南大部分温暖湿润地区有归化。南麂主岛（百亩山、后隆、门屿尾），常见。

生境：易受干扰地区，路边、草地、灌溉渠、田边。

Description: Herbs annual, 16-150 cm tall. Stems erect, sometimes purplish, glabrous, eglandular. Basal leaves lanceolate to ovate, usually fallen at anthesis; cauline leaves linear-lanceolate, margin serrulate to entire, apex acute. Capitula numerous, in paniculiform synflorescences; involucre cylindric, phyllaries 3-5-seriate, lanceolate to linear-lanceolate, strongly unequal; ray florets numerous, 1-seriate, lamina purplish blue; disk florets yellow, lobes erect, triangular. Achenes lanceoloid, sparsely strigillose.

Flower and Fruit: Fl. and fr. Aug. -Oct.

Distribution: Native to North America. Naturalized in most warm and humid areas to the south of Qinling and Huaihe.Main island (Baimushan, Houlong, Menyuwei), commonly.

Habitat: Disturbed areas, roadsides, grassy fields, irrigation ditches, field margins.

A403. 菊科
Asteraceae

苍耳
Xanthium strumarium（*Xanthium sibiricum*）
苍耳属 *Xanthium*

形态特征：一年生草本。茎直立，高 30 ～ 60 cm。茎下部圆柱形，上部有纵沟，叶片三角状卵形或心形，近全缘，边缘有不规则的粗锯齿，上面绿色，下面苍白色，被糙伏毛。雄头状花序球形，总苞片长圆状披针形，花冠钟形；雌头状花序椭圆形，外层总苞片小，披针形，喙坚硬，锥形。瘦果倒卵形。

花果期：花期 7 ～ 8 月，果期 9 ～ 10 月。

分布：泛热带杂草，广布于新旧大陆。分布于我国各地。南麂主岛各处常见。

生境：潮湿或季节性潮湿的碱性土壤、荒地、农田边缘。

Description: Herbs annual. Stem erect, 30-60 cm tall. Stem lower terete, upper with longitudinal groove. Leaf blade triangular-ovate or cordate-shaped, subentire, margin irregularly coarsely serrate, green upper, pale below, strigose. Male capitula spherical, involucre oblong-lanceolate, corolla bell-shaped; female capitula elliptic, outer involucre small, lanceolate, beak hard, conical. Achene obovate.

Flower and Fruit: Fl. Jul. -Aug. , fr. Sep. -Oct.

Distribution: Pantropical weed, widely distributed in both Old and New Worlds. Throughout China. Main island of Nanji, commonly.

Habitat: Damp or seasonally wet often alkaline soils, wastelands, margins of agriculture.

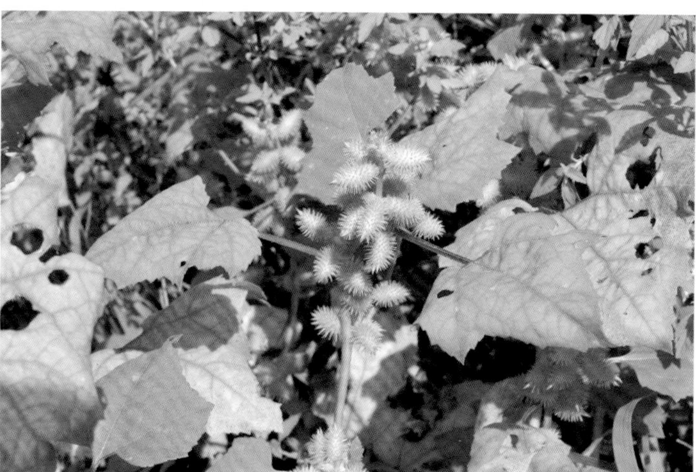

A403. 菊科
Asteraceae

黄鹌菜
Youngia japonica
黄鹌菜属 *Youngia*

形态特征：一年生草本，高 20 ~ 60 cm。茎直立。叶基生，倒披针形，提琴状羽裂；裂片有深波状齿，叶柄微具翅。头状花序有柄，排成伞房状、圆锥状和聚伞状，总苞圆筒形，外层总苞片远小于内层；舌状花花冠黄色。瘦果纺锤状，稍扁，冠毛白色。

花果期：花果期 4 ~ 10 月。

分布：亚洲温带和热带地区。除东北、西北外广布我国。南麂主岛各处常见。

生境：山坡、路边、林缘和荒野。

Description: Herbs annual, 20-60 cm tall. Stem erect. Leaves basal, oblanceolate, fiddle-like pinnatifid; lobes with deep undulate teeth, petiole slightly winged. Inflorescences stalked, arranged in corymbose, paniculate and cymose, involucre cylindrical, outer involucre much smaller than inner; ray florats, corolla yellow. Achenes fusiform, slightly flat, pappus white.

Flower and Fruit: Fl. and fl. Apr. -Oct.

Distribution: Temperate and tropical Asia. Throughout China except Northeast and Northwest China. Main island of Nanji, commonly.

Habitat: Hillsides, roadsides, forest edges and wilderness.

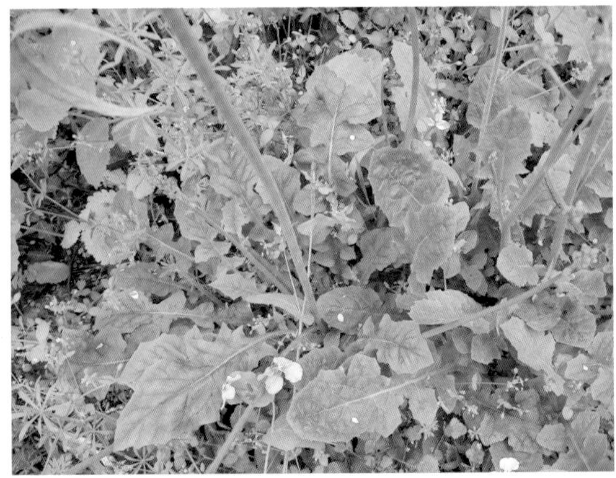

A408. 五福花科
Adoxaceae

日本珊瑚树（法国冬青）
Viburnum odoratissimum* var. *awabuki
荚蒾属 *Viburnum*

形态特征：常绿灌木或小乔木。叶对生，叶柄带红色，叶片有光泽，椭圆状倒卵形，厚革质，顶端钝或急狭而钝头，基部宽楔形，边缘具不规则的波状浅钝锯齿。花序圆锥状，常生于具 1 对叶的幼枝顶，花序轴无毛；花冠钟状，白色无毛，裂片反折，边缘全缘。果红色，熟时变黑，无毛；果核倒卵形。

花果期：花期 5 ～ 6 月，果熟期 9 ～ 10 月。

分布：日本和菲律宾。我国浙江、台湾等地，长江下游各地常见栽培。南麂主岛（美龄居、镇政府），栽培。

生境：林下。

Description: Shrubs or small trees, evergreen. Leaves always opposite, petiole reddish, leaf blade lustrous, elliptic-obovate, thickly leathery, apex obtuse or sharp acute, base broadly cuneate, margin irregularly serrate. Inflorescence paniculate, pyramidal, at apices of short lateral branchlets with 1-jugate leaves, axes glabrous; corolla campanulate, white, glabrous, lobes reflexed, margin entire. Fruit initially turning red, maturing nigrescent, glabrous; pyrenes obovoid.

Flower and Fruit: Fl. May-Jun. , fr. Sep. -Oct.

Distribution: Japan, Philippines. Zhejiang, Taiwan,China, commonly cultivated in the lower reaches of the Yangtze River. Main island (Meilingju, Zhenzhengfu), cultivated.

Habitat: Forests.

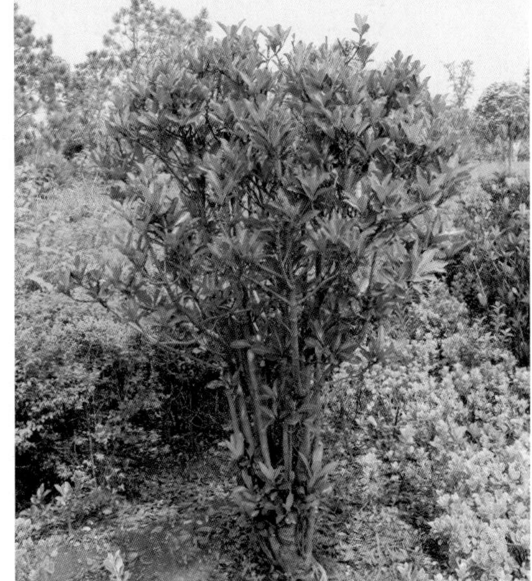

A409. 忍冬科
Caprifoliaceae

忍冬（金银花）
Lonicera japonica
忍冬属 ***Lonicera***

形态特征：半常绿藤本。小枝中空。小枝、叶柄和花梗密被黄褐色硬毛。叶纸质，卵形、或长圆形至披针形，两面被毛，基部圆或近心形。花芳香，成对腋生枝顶，花冠二唇形，白色，后变黄，上唇不规则4裂，下唇反曲；雄蕊和花柱均高出花冠。浆果球形，熟时蓝黑色，有光泽。种子卵圆形或椭圆形，褐色。

花果期：花期4～6月，果期10～11月。

分布：日本、韩国，东南亚广泛栽培。我国各地均有分布，多为栽培。南麂主岛各处常见。

生境：灌丛、疏林中、山坡、乱石堆或路旁。

Description: Climbers, semievergreen. Branches becoming hollow. Branches, petioles, and peduncles with dense, yellow-brown spreading stiff hairs. Leaf blade ovate or oblong to lanceolate, hairy on both sides, base rounded to subcordate. Flowers fragrant, paired and axillary toward apices of branchlets; corolla bilabiate, white, becoming yellow, upper lip irregularly 4-lobed, lower lip recurved; stamens and style subequaling to exceeding corolla. Berries black when mature, glossy, globose. Seeds brown, ovoid or ellipsoid.

Flower and Fruit: Fl. Apr. -Jun. , fr. Oct. -Nov.

Distribution: Japan, Korea; widely cultivated in Southeast Asia. Throughout China, usually cultivated. Main island of Nanji, commonly.

Habitat: Scrub, sparse forests, mountain slopes, stony places, roadsides.

A409. 忍冬科
Caprifoliaceae

攀倒甑
Patrinia villosa
败酱属 *Patrinia*

形态特征：多年生或二年生草本，高 50 ~ 120 cm。基生叶莲座状，具长柄，叶片宽卵形至长圆状披针形，先端渐尖，边缘具粗钝齿，基部楔形下延，不分裂或大头羽状深裂，裂片 1 ~ 2 对；茎生叶对生，上部叶较窄，常不分裂。聚伞花序顶生，分枝成伞房圆锥状；花冠钟形，白色，5 深裂。瘦果倒卵形，基部宿存翅状苞片。

花果期：花期 8 ~ 10 月，果期 10 ~ 12 月。

分布：日本。我国东北至西南地区。南麂主岛（百亩山、国姓岙、后隆、门屿尾），常见。

生境：林缘、灌丛、草地或路边。

Description: Herbs perennial or biennial, 50-120 cm tall. Basal leaves rosulate, long petiolate, blade broadly ovate to oblong-lanceolate, apex acuminate, margin coarsely obtuse, base cuneate decurrent, undivided or pinnately lobed, often 1-2 pairs of lobes; cauline leaves opposite, upper leaves narrower, often undivided. Cymes terminal, branching into corymbose panicles; corolla campanulate, white, deeply 5-lobed. Achenes obovoid, base persistent winged bracts.

Flower and Fruit: Fl. Aug. -Oct. , fr. Oct. -Dec.

Distribution: Japan. Northeast to Southwest China. Main island (Baimushan, Guoxingao, Houlong, Menyuwei), commonly.

Habitat: Forest margins, thickets, grassy areas, roadsides.

A413. 海桐科
Pittosporaceae

海桐
Pittosporum tobira
海桐属 *Pittosporum*

形态特征：常绿灌木或小乔木，高达 6 m。嫩枝被褐色柔毛，有皮孔。叶聚生于枝顶，革质，倒卵形或倒卵状披针形，全缘。伞形花序或伞房状伞形花序顶生或近顶生，密被黄褐色柔毛；花瓣离生，初白色，后变黄。蒴果圆球形，有棱或呈三角形，3 裂。种子多数多角形，红色。

花果期：花期 3 ~ 5 月，果期 5 ~ 10 月。

分布：原产我国台湾北部、日本南部和朝鲜南部。我国长江以南滨海各省有栽培或归化。南麂各岛屿常见。

生境：林地、石灰岩地区、斜坡、沙滩、路边。

Description: Shrubs or small trees to 6 m tall. Young branchlets brown puberulent, lenticellate. Leaves clustered at branchlet apex, leathery, obovate or obovate-lanceolate, margin entire. Inflorescences terminal or near so, umbellate or corymbose, densely tawny puberulent; petals free, white at first, becoming yellow later, flowers fragrant. Capsule globose, angular, dehiscing by 3 valves. Seeds numerous, red, angular.

Flower and Fruit: Fl. Mar. -May, fr. May-Oct.

Distribution: Native to North Taiwan,China, South Japan and South Korea. Cultivated for ornament and possibly naturalized in coastal provinces to the south of Yangtze River. Throughout Nanji Islands, commonly.

Habitat: Forests, limestone areas, slopes, sandy seashores, roadsides.

A414. 五加科
Araliaceae

常春藤（中华常春藤）
Hedera nepalensis* var. *sinensis
常春藤属 ***Hedera***

形态特征：攀缘灌木。幼枝具铁锈色的鳞片。叶二形，不育枝上全缘或3裂，三角状卵形；能育枝上椭圆状卵形或椭圆状披针形；叶片无毛或背面具稀疏鳞片，叶脉两面明显，叶片基部宽楔形，边缘全缘，先端渐尖。伞形花序顶生或成小总状花序，具铁锈色鳞片；花萼边缘近全缘。果球形，成熟时红色或黄色。

花果期：花期9～11月，果期3～5月。

分布：老挝，越南。我国南北各地广布。南麂各岛屿常见。

生境：林中，路旁或岩石坡，通常攀援在树上或岩石上。

Description: Shrubs scandent. Young branches with ferruginous scales. Leaves dimorphic, entire or 3-lobed on sterile branches, usually triangular-ovate; elliptic-ovate or elliptic-lanceolate on fertile branches; blade glabrous or with sparse scales abaxially, venation distinct on both surfaces, base broadly cuneate, margin entire, apex acuminate. Inflorescence a terminal umbel or a small raceme, with ferruginous scales; calyx rim subentire. Fruit globose, red or yellow at maturity.

Flower and Fruit: Fl. Sep. -Nov. , fr. Mar. -May.

Distribution: Laos, Vietnam. Throughout the north and south areas of China. Throughout Nanji Islands, commonly.

Habitat: Forests, roadsides, rocky slopes, usually climbing on trees or rocks.

A414. 五加科 Araliaceae

鹅掌柴（鸭脚木）
Schefflera heptaphylla（*Schefflera octophylla*）
鹅掌柴属 *Schefflera*

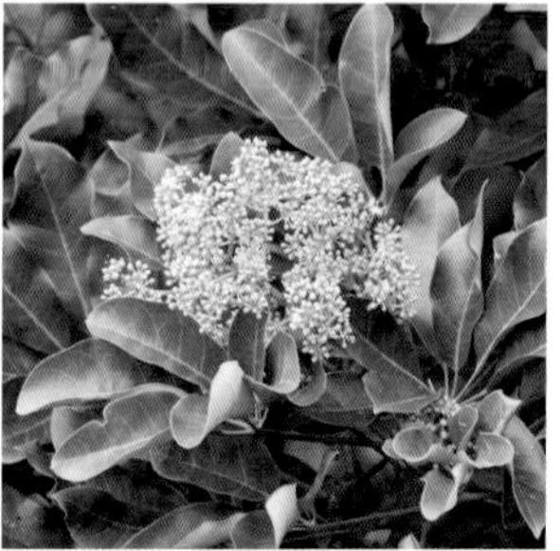

形态特征：常绿乔木或灌木。幼枝、幼叶、花序轴和花梗密被星状毛，后脱落。掌状复叶有小叶 6 ～ 9，偶见 11，椭圆形至倒卵状椭圆形，纸质至革质。伞形花序呈顶生圆锥状，小花 10 ～ 15，白色，花萼全缘或有 5 ～ 6 小齿。果球形，黑色，干时有不明显的棱，花柱宿存。

花果期：花期 9 ～ 12 月；果期 12 月至翌年 2 月。

分布：印度、日本、泰国、越南。我国华东和华南地区。南麂各岛屿常见。

生境：常绿阔叶林或向阳山坡。

Description: Evergreen trees or shrubs. Branchlets, leaflets, rachis and pedicels densely stellate pubescent, glabrescent. Palmately compound leaves with leaflets 6-9(11), elliptic to obovate-elliptic, papery to leathery. Inflorescence a terminal panicle of umbels, with 10-15-flowered, flowers white, calyx pubescent at first, entire or 5-or 6-toothed. Fruit globose, black, inconspicuously angled when dry; styles persistent.

Flower and Fruit: Fl. Sep. -Dec. , fr. Dec. -Feb.

Distribution: India, Japan, Thailand, Vietnam. East China, South China. Throughout Nanji Islands, commonly.

Habitat: Evergreen broad-leaved forests or the sunny mountain slopes.

A416. 伞形科
Apiaceae

积雪草（老鸦碗）
Centella asiatica
积雪草属 *Centella*

形态特征：多年生草本。茎匍匐，细长，节上生根。叶片圆形或肾形，基部阔心形，边缘有钝锯齿，两面无毛或背面沿脉疏生柔毛；掌状脉 5 ～ 7，隆起。伞形花序 2 ～ 4，聚生叶腋；苞片 2，卵形，果期宿存；小花 3 ～ 4，头状，无梗或近无柄，花瓣卵形，紫红色或乳白色。果实球形。

花果期：花果期 4 ～ 10 月。

分布：热带亚热带地区广布。我国东部至西部大部分地区。南麂主岛各处常见。

生境：阴湿的草地、水沟边。

Description: Herbs perennial. Stems creeping, slender, rooting at nodes. Leaf blade orbicular or reniform, base broadly cordate, coarsely toothed, both surfaces glabrous or abaxially sparsely pubescent on the veins; palmate veins 5-7, prominent. Umbels 2-4, clustered axillary; bracts 2, ovate, persistent in fruit; 3-4-flowered, capitate, sessile or subsessile, petals white or rose-tinged. Fruit globose.

Flower and Fruit: Fl. and fr. Apr. -Oct.

Distribution: Widespread throughout tropical and subtropical areas. Most areas across east to west of China. Main island of Nanji, commonly.

Habitat: Shady, wet, grassy places, river margins.

A416. 伞形科
Apiaceae

芫荽（香菜）
Coriandrum sativum
芫荽属 *Coriandrum*

形态特征：一年生或二年生草本，高 60 cm，有强烈气味。基部和下部叶片 1 ～ 2 回羽状全裂，叶柄基部短鞘状，叶片卵形，羽片宽卵形或扇形，末回裂片宽；中上部茎生叶 2 ～ 3 回羽状全裂，末回裂片线形至丝状，全缘。伞形花序顶生或与叶对生，伞辐 2 ～ 8；小苞片 2 ～ 5，线形，全缘；小伞形花序 3 ～ 9 花。

花果期：花果期 4 ～ 11 月。

分布：原产地中海地区，世界各地广泛种植。我国各地普遍栽培。南麂主岛（东方岙、国姓岙、后隆、门屿尾），栽培。

生境：田野、路旁、荒地。

Description: Herbs annul or biennial, up to 60 cm tall, strongly aromatic. Basal and lower leaves pinnate to 2-pinnatisect; petiole shortly sheathing at base; blade ovate, pinnae broadly ovate or flabelliform, ultimate segments broad; mid and upper cauline leaves ternate-2-3-pinnatisect, ultimate segments linear to filiform, entire. Umbels terminal or opposite to leaves, rays 2-8; bracteoles 2-5, linear, entire; umbellules 3-9-flowered.

Flower and Fruit: Fl. and fr. Apr. -Nov.

Distribution: Native to the Mediterranean region, cultivated worldwide. Almost throughout China. Main island (Dongfangao, Guoxingao, Houlong, Menyuwei), cultivated.

Habitat: Fields, roadsides, wastelands.

A416. 伞形科
Apiaceae

细叶旱芹
***Cyclospermum leptophyllum*（*Apium leptophyllum*）**
细叶旱芹属 *Cyclospermum*

形态特征：一年生草本，高 25 ~ 45 cm。茎多分枝，无毛。叶长圆形至长圆状卵形，3 ~ 4 回羽状多裂，末回裂片线形至丝状；茎生叶三出式羽状多裂，裂片线形。伞形花序顶生或腋生，伞辐 2 ~ 3；小伞形花序 5 ~ 23 花，花柄不等长，中心小花常无柄；花瓣白色或绿白色，卵圆形，顶端内折。果球形。

花果期：花期 4 ~ 5 月，果期 6 ~ 7 月。

分布：原产南美洲，热带和温带地区广泛归化。分布于我国华东、华南地区。南麂主岛（百亩山、关帝岙、门屿尾、三盘尾），常见。

生境：溪沟、荒地，杂草地区。

Description: Herbs annul, 25-45 cm, stems many-branched, glabrous. Leaf blade oblong to oblong-ovate, 3-4 pinnatifid, ultimate segments linear to filiform; cauline leaves ternate-pinnately decompound, segments linear. Umbels terminal or axillary, rays 2-3; umbellules 5-23-flowered, pedicels unequal, central flower often almost sessile; petals white or greenish-white, oval, apex inflexed. Fruit globose.

Flower and Fruit: Fl. Apr. -May, fr. Jun. -Jul.

Distribution: Native to South America; widely naturalized as a weed in tropical and temperate regions. East and South China. Main island (Baimushan, Guandiao, Menyuwei, Sanpanwei), commonly.

Habitat: Streamsides, wastelands, ruderal areas.

A416. 伞形科
Apiaceae

野胡萝卜
Daucus carota
胡萝卜属 *Daucus*

形态特征：二年生草本，高 120 cm。叶长圆形，2 ~ 3 回羽状全裂，末回裂片线形至披针形，无毛或叶缘及脉上有糙毛，顶端尖锐，有小尖头。复伞形花序，花序梗具反曲糙毛；苞片叶状羽裂，裂片线形；伞辐不等长，小苞片 5 ~ 7，线形，全缘或 2 ~ 3 浅裂，具缘毛；花瓣白色，有时黄色或带粉色。果实圆卵形，棱上有白色刺毛。

花果期：花期 5 ~ 7 月。

分布：欧洲、东南亚。我国华东至西南地区。南麂主岛（百亩山、国姓岙、门屿尾、三盘尾），常见。

生境：山坡，杂草地区。

Description: Herbs biennial, up to 120 cm. Leaves oblong, 2-3-pinnatisect, ultimate segments linear to lanceolate, glabrous to hispid especially on the veins and margins, acute, mucronate. Compound umbels, peduncles retrorsely hispid; bracts foliaceous, pinnate, rarely entire, lobes linear, margin scarious; rays unequal, bracteoles 5-7, linear, entire or 2-3-lobed, ciliate; petals white, sometimes yellow or pinkish. Fruit ovate, with white bristles on ribs.

Flower and Fruit: Fl. May-Jul.

Distribution: Europe and Southeast Asia. East to Southwest China. Main island (Baimushan, Guoxingao, Menyuwei, Sanpanwei), commonly.

Habitat: Mountain slopes, ruderal areas.

A416. 伞形科
Apiaceae

天胡荽
Hydrocotyle sibthorpioides
天胡荽属 *Hydrocotyle*

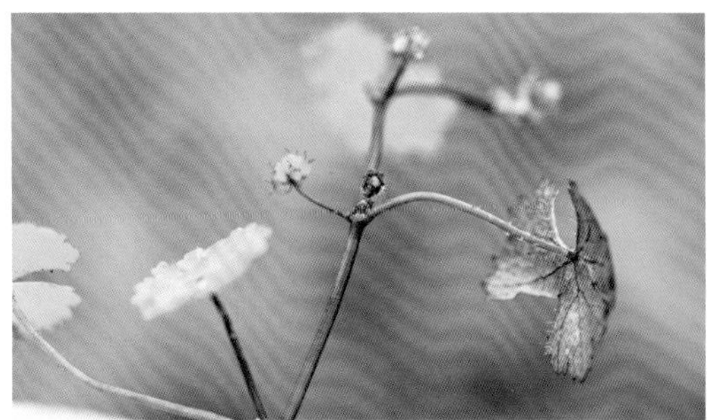

形态特征：多年生草本，有气味。茎细长而匍匐，平铺分枝。叶片肾圆形，膜质，基部心形，不裂或 5 ～ 7 浅裂，裂片圆形。伞形花序单生，5 ～ 8 花；花序梗纤细，苞片卵形至卵状披针形，膜质，有黄色透明腺点；花瓣卵形，绿白色，有腺点。果实阔球形，幼时草黄色，熟时有紫色斑点。

花果期：花果期 4 ～ 9 月。

分布：东南亚、热带非洲。我国长江流域以南地区。南麂主岛（百亩山、国姓岙、门屿尾、三盘尾），常见。

生境：湿润的草地、河沟边、林下。

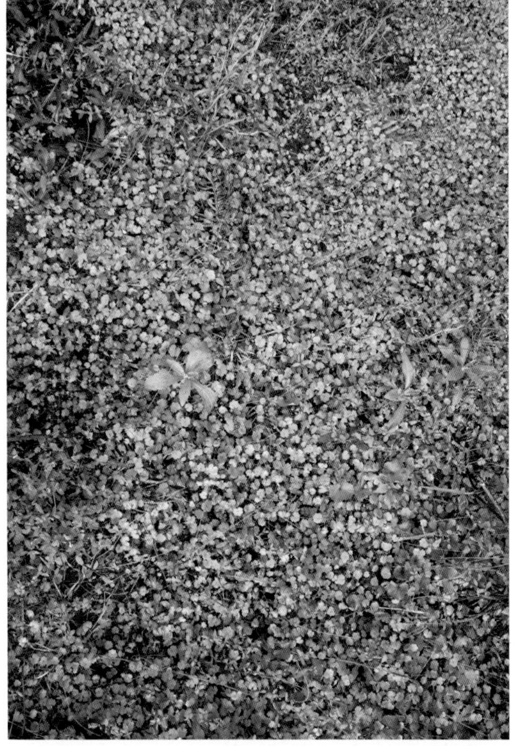

Description: Herbs perennial, strongly aromatic. Stem weak, slender, creeping, diffusely branched. Leaf blade reniform-rounded, membranous, base cordate, shallowly 5-7-lobed or nearly entire, crenate. Umbel solitary at the nodes, each umbel 5-8-flowered; peduncle filiform, bracts ovate to ovate-lanceolate, membranous, with bright yellow glands; petals greenish white, with yellow glands. Fruit broadly globose, greenish yellow when young, covered with purplish stains when mature.

Flower and Fruit: Fl. and fr. Apr. -Sep.

Distribution: Southeast Asia, tropical Africa. To the south of Yangtze River Basin. Main island (Baimushan, Guoxingao, Menyuwei, Sanpanwei), commonly.

Habitat: Forests, wet grassy places, stream banks.

A416. 伞形科
Apiaceae

滨海前胡
Peucedanum japonicum
前胡属 *Peucedanum*

形态特征：多年生草本。茎常曲折，多分枝，无毛。叶片宽卵状三角形，薄革质，1～2 回三出；小叶卵圆形，3 深裂，有白霜，中裂片倒卵状楔形，侧裂片斜卵形。伞形花序，苞片 2～3 或无，卵状披针形，被短毛；伞辐 15～30，不等长，被微柔毛；伞形花 20 余朵，花瓣紫色或白色，背部具短硬毛。果实长卵形至椭圆形，具短硬毛。

花果期：花期 6～7 月，果期 8～9 月。

分布：日本、朝鲜、菲律宾。我国华东滨海地区。南麂各岛屿常见。

生境：滨海滩地或近海山地。

Description: Herbs perennial. Stem often flexuous, much branched, essentially glabrous. Leaf blade broadly ovate-triangular, thinly coriaceous, 1-2-ternate; leaflets ovate-orbicular, 3-parted, glaucous, central segments obovate-cuneate, lateral segments oblique-ovate. Umbels, bracts 2-3 or absent, ovate-lanceolate, pubescent; rays 15-30, unequal, puberulous; umbellules ca. 20-flowered, petals purple or white, abaxially hispidulous. Fruit oblong-ovate or ellipsoid, hirsute.

Flower and Fruit: Fl. Jun-Jul, fr. Aug-Sep.

Distribution: Japan, Korea, Philippines. Coastal areas of East China. Throughout Nanji Islands, commonly.

Habitat: Coastal beaches, seashores.

A416. 伞形科
Apiaceae

小窃衣（破子草）
Torilis japonica
窃衣属 *Torilis*

形态特征：一年或多年生草本，高 20 ～ 120 cm。茎有纵条纹及刺毛。叶片三角状卵形至卵状披针形，1 ～ 2 回羽状分裂，羽片卵状披针形。复伞形花序顶生或腋生，总梗倒生刺毛；苞片少，线形，伞辐 4 ～ 12，开展，被刚毛；小苞片 5 ～ 8，线形或钻形；小伞形花序 4 ～ 12 花。果实卵圆形，熟时黑紫色。

花果期：花期 5 ～ 8 月，果期 5 ～ 10 月。

分布：亚欧广布杂草。除黑龙江、内蒙古和新疆外遍布我国。南麂主岛各处常见。

生境：山谷混交林、草地，特别是易受干扰的地区。

Description: Herbs annual or perennial, 20-120 cm. Stem with longitudinal stripe and bristles. Leaf blade triangular-ovate to ovate-lanceolate, 1-2-pinnatifid, pinnae ovate-lanceolate. Compound umbels terminal or axillary, peduncles retrorse hispid; bracts few, linear; rays 4-12, spreading, bristly; bracteoles 5-8, linear or subulate;umbellules 4-12-flowered. Fruit globose-ovoid, blackish purple when mature.

Flower and Fruit: Fl. May-Aug. , fr. May-Oct.

Distribution: Widespread as a ruderal in Asia and Europe. Throughout China except Heilongjiang, Nei Mongol, and Xinjiang. Main island of Nanji, commonly.

Habitat: Mixed forests in valleys, grassy places, especially in disturbed areas.

A416. 伞形科
Apiaceae

窃衣
Torilis scabra
窃衣属 *Torilis*

形态特征：一年生或多年生草本，高 90 cm。基部和下部茎生叶具叶柄；叶卵形，羽片披针形到狭卵形。复伞形花序顶生或腋生，苞片通常无；伞辐 2 ~ 4，粗壮，具脊，密被贴伏毛；小苞片 2 ~ 6，钻形；小伞形花序 2 ~ 6 花，花梗被粗毛。果长圆形，常深绿色，偶带深紫色。

花果期：花果期 4 ~ 11 月。

分布：日本、朝鲜。我国华东至西南和西北地区。南麂主岛各处常见。

生境：山坡或山谷混交林、路旁，特别是易受干扰的地区。

Description: Herbs annal or perennial to 90 cm tall. Basal and lower cauline leaves petiolate; blade ovate in outline; pinnae lanceolate to narrowly ovate. Compound umbels terminal or axillary, bracts usually absent; rays 2-4, stout and ridged, densely appressed-strigose; bracteoles 2-6; subulate, umbellules 2-6-flowered, pedicels hirsute. Fruit oblong, usually dark green, occasionally tinged dark purple.

Flower and Fruit: Fl. and fr. Apr. -Nov.

Distribution: Japan, Korea. East China to Southwest and Northwest China. Main island of Nanji, commonly.

Habitat: Mixed forests on mountain slopes or in valleys, roadsides, especially in disturbed areas.

中文名索引 INDEX OF CHINESE NAMES

T

W

X

Y

Z

学名索引 INDEX OF SCIENTIFIC NAMES

D

T

附录 APPENDIX

岛屿景观俯瞰 Aerial view of islands landscape

门屿 Menyu

海龙山 Hailongshan

门屿尾 Menyuwei

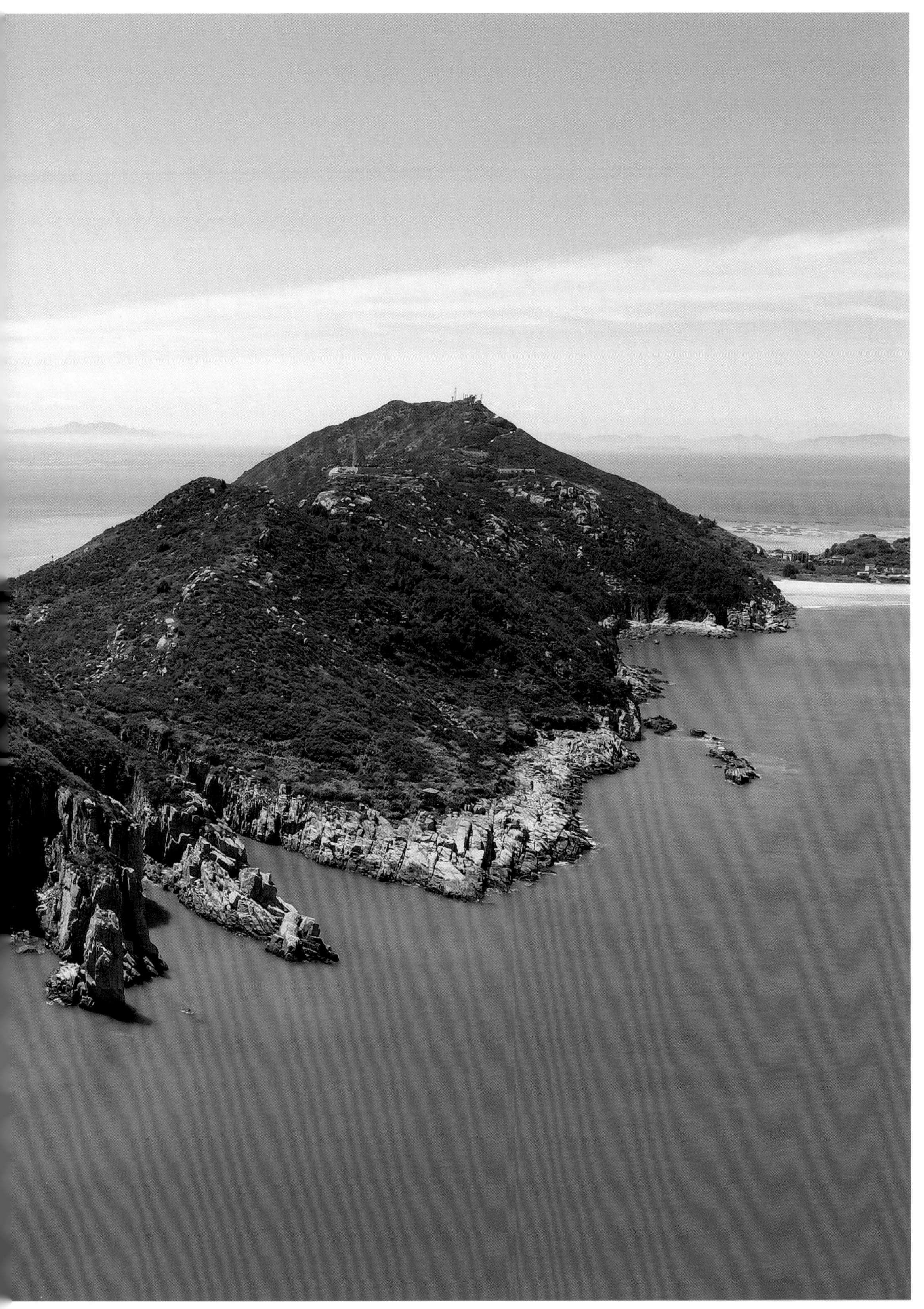

主要植被群落景观 Main landscape of plant community

水仙群落 Community of *Narcissus tazetta* var. *chinensis*

黑松林 Forest of *Pinus thunbergii*

木麻黄林 Forest of *Casuarina equisetifolia*

台湾相思林 Forest of *Acacia confusa*

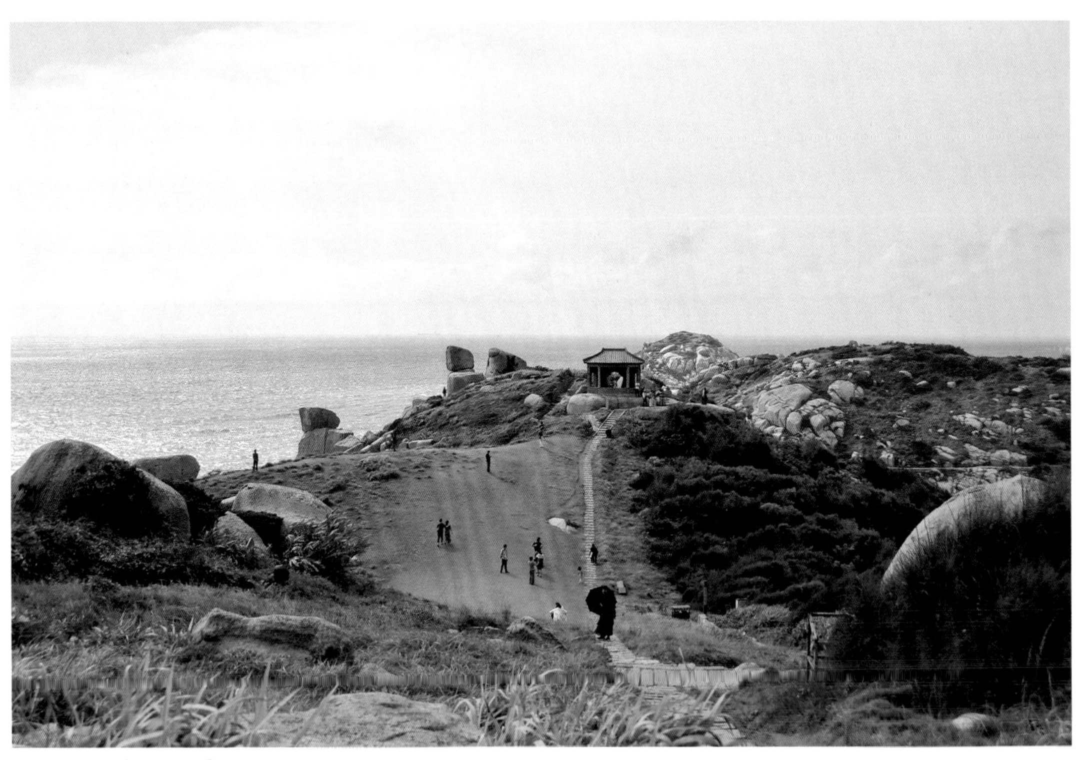

三盘尾天然草坪（夏季）Natural meadow of Sanpanwei (Summer)

三盘尾天然草坪（冬季）Natural meadow of Sanpanwei (Winter)

道路沿途植物景观 Plant landscape along the road

池杉 *Taxodium distichum* var. *imbricatum*

台湾相思 *Acacia confuse* 与木麻黄 *Casuarina equisetifolia*

桑 *Morus alba*

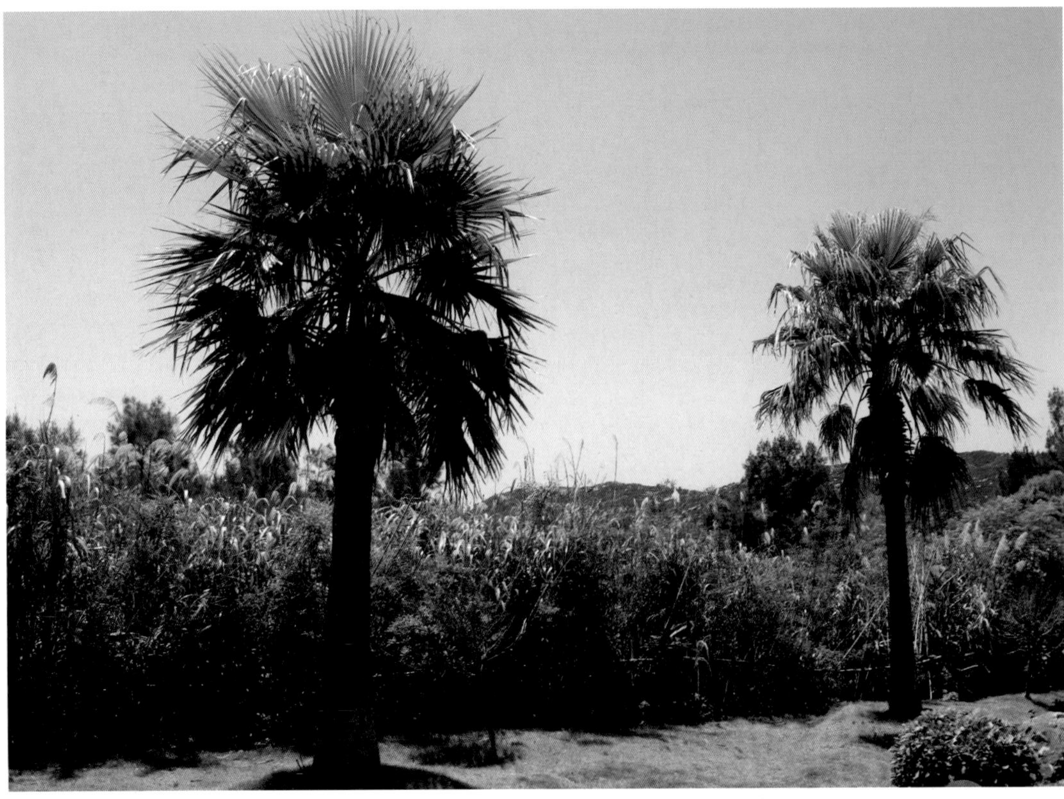

大丝葵 *Washingtonia robusta*